LANDSCAPE EVOLUTION, NEOTECTONICS AND QUATERNARY ENVIRONMENTAL CHANGE IN SOUTHERN CAMEROON

Palaeoecology of Africa

International Year book of Landscape Evolution
and Palaeoenvironments

Volume 31

Editor in Chief

J. Runge, Frankfurt, Germany

Editorial board

G. Botha, Pietermaritzburg, South Africa
E. Cornellissen, Tervuren, Belgium
F. Gasse, Aix-en-Provence, France
P. Giresse, Perpignan, France
K. Heine, Regensburg, Germany
S. Kröpelin, Köln, Germany
T. Huffmann, Johannesburg, South Africa
E. Latrubesse, Austin, Texas, USA
J. Maley, Montpellier, France
J.-P. Mund, Eberswalde, Germany
D. Olago, Nairobi, Kenya
F. Runge, Altendiez, Germany
L. Scott, Bloemfontein, South Africa
I. Stengel, Windhoek, Namibia
F.A. Street-Perrott, Oxford, UK

Landscape Evolution, Neotectonics and Quaternary Environmental Change in Southern Cameroon

Editor

Jürgen Runge
*Centre for Interdisciplinary Research on Africa (CIRA/ZIAF),
Johann Wolfgang Goethe University, Frankfurt am Main, Germany*

CRC Press
Taylor & Francis Group
Boca Raton   London   New York   Leiden

CRC Press is an imprint of the
Taylor & Francis Group, an **informa** business

A BALKEMA BOOK

Front cover: The Dehane waterfalls (*Chutes de Dehane*) close to the South Cameroonian coast: the Nyong River traverses Precambrian rocks by two steps of approximately 1 m in height, rectangular to flow direction. Pepples and gravel fragments below the waterfalls are fixed in a poriferous matrix. Photograph by Mark Sangen, February 2005.

CRC Press/Balkema is an imprint of the Taylor & Francis Group, an informa business

© 2012 Taylor & Francis Group, London, UK

Typeset by V Publishing Solutions Pvt Ltd., Chennai, India
Printed and bound in Great Britain by CPI Group (UK) Ltd, Croydon, CR0 4YY

Published by: CRC Press/Balkema
             P.O. Box 447, 2300 AK Leiden, The Netherlands
             e-mail: Pub.NL@taylorandfrancis.com
             www.crcpress.com – www.taylorandfrancis.com

Library of Congress Cataloging-in-Publication Data

Landscape evolution, neotectonics and quaternary environmental change in southern Cameroon/ editor, Jürgen Runge.—1st ed.

       p. cm.—(Palaeoecology of Africa : international year book of landscape evolution and palaeoenvironments, ISSN 0168-6208; v. 31)

Includes bibliographical references and index.
   ISBN 978-0-415-67735-6 (hardback : alk. paper)
   1. Paleoecology—Cameroon. 2. Paleoecology—Quaternary.
3. Rain forest ecology—Cameroon. 4. Savanna ecology—Cameroon. I. Runge, Jürgen, 1962- II. Series: Palaeoecology of Africa; v. 31.

QE720.2.C17L36 2012

560'.45096711—dc23

2012011032

ISBN: 978-0-415-67735-6 (Hbk)
ISBN: 978-0-203-12020-0 (eBook)

Contents

Foreword

This year's book for the series 'Palaeoecology of Africa', volume (31), brings together four contributions on long- and short-term landscape evolution in Southern Cameroon, and represents the outcome of the German-Cameroonian earth scientist cooperation during the period 2003–2009. In fact, this cooperation is an ongoing and effective one. In 2003, the German Research Foundation (DFG, Deutsche Forschungsgemeinschaft) agreed to establish an interdisciplinary research group on "Ecological and Cultural Change in West and Central Africa" composed of archaeologists, palaeobotanists and physical geographers from the Universities of Tübingen and Frankfurt. Aside from sharing common scientific research interests, it was and it remains crucial for Germany's research politics to have a strong focus on intercultural cooperation and academic capacity building of local partners. In the framework of the "Rain Forest Savanna Contact" subproject (in German "ReSaKo") a close and fruitful relationship and cooperation developed in particular between the University of Yaoundé I and the University of Frankfurt. An outcome of this subproject were two PhD theses, one by Joachim Eisenberg on 'neotectonics and landscape evolution' in Southern Cameroon, the other, by Mark Sangen, on the 'Quaternary environment' as evidenced by alluvia in river valleys in the rain forest and along transitional areas into savannas. In 2009 and 2010, respectively these theses were formally published in German in partial fulfillment of PhD requirements. Thereafter, it occurred to me that these valuable studies should be made available to the wider international scientific community. For this reason, I proposed in 2011 to re-publish the two PhD works in condensed English versions. Within the framework of the 'Palaeoecology of Africa' (PoA) annual series, it was possible to publish these extensive papers, which normally would have had no place in scientific journals. Underlining and reinforcing the excellent cooperation we have enjoyed with our colleagues from Cameroon, I invited Boniface Kankeu to contribute an introductory chapter on the Palaeozoic evolution of the study area. Finally, Mesmin Tchindjang, who is strongly focused on applied questions and development problems ('environmental risks') of Cameroon, was asked to contribute a longer paper looking at present and future challenges of the wider Cameroonian study area. By bringing these issues together in this book, we have managed to achieve one of our goals: to form a "linkage" between basic scientific research and problem-oriented applications.

Within the University of Frankfurt Physical Geography working group, all manuscripts have been reviewed and corrected several times. The English was consistently checked and corrected by Dr. Dawn Frame (C.N.R.S. Montpellier). Formatting of the papers for PoA layout and style was done by Mrs. Blanka Kunz. Apart from the cartographic art work done by the principal authors themselves, Ursula Olbrich revised numerous figures and helped by carrying out additional cartographic work for the book. The Taylor & Francis team in Leiden (NL) and especially Janjaap Blom and

Richard Gundel, helped streamline the editing process. To these workers and to other colleagues, who indirectly contributed and helped to make this book a reality, I am greatly obliged.

Jürgen Runge
Frankfurt and Yaoundé
February 2012

Contributors

Samuel Aimé Abossolo

Department of Geography, University of Yaoundé I, BP 30464, Yaoundé, Cameroon. Email: abossamai@yahoo.fr

Joseph Armathée Amougou

Ministry of Environment, Protection of the Nature and Sustainable Development, Yaoundé, Cameroon. Email: joearmathe@yahoo.fr

Jean Bassahak

Institute of Geological and mining research (IRGM), BP 4110, Yaoundé, Cameroon. Email: jbassahak@yahoo.fr

Stanislas Bessoh Bell

Department of Plant Biology, University of Yaoundé I, BP 812, Yaoundé, Cameroon. Email: bessohbellstanislas@yahoo.fr

Joachim Eisenberg

Institute of Physical Geography, Johann Wolfgang Goethe University, Frankfurt am Main, Altenhöferallee 1, D-60438 Frankfurt, Germany. Email: j.eisenberg@em.uni-frankfurt.de

Joseph Victor Hell

Institute of Geological and mining research (IRGM), BP 4110, Yaoundé, Cameroon. Email: jvhell@yahoo.com

Boniface Kankeu

Institute of Geological and mining research (IRGM), BP 4110, Yaoundé, Cameroon. Email: bonifacekankeu@yahoo.fr

Jürgen Runge

Institute of Physical Geography, Johann Wolfgang Goethe University, Frankfurt am Main, Altenhöferallee 1, D-60438 Frankfurt, Germany. Email: j.runge@em.uni-frankfurt.de

Mark Sangen

Institute of Physical Geography, Johann Wolfgang Goethe University, Frankfurt am Main, Altenhöferallee 1, D-60438 Frankfurt, Germany. Email: m.sangen@em.uni-frankfurt.de

Mesmin Tchindjang

Department of Geography, University of Yaoundé I, BP 30464, Yaoundé, Cameroon. Email: mtchind@yahoo.fr

Scientific background and Capacity building in the framework of a German-Cameroonian research project on long- and short-term landscape dynamics

Jürgen Runge

Department of Physical Geography, Johann Wolfgang Goethe University, Frankfurt am Main, Germany

ABSTRACT: The major objective of the 2003 established and up to 2009 running interdisciplinary Research Unit 510 on 'Ecological and Cultural Change in West and Central Africa' of the German Research Foundation (DFG, Deutsche Forschungsgemeinschaft), was to elucidate correlations between Late Holocene climatic changes, development of land-use strategies and cultural innovations from different proxy data sources, if possible in a high temporal resolution. In the course of the study also older alluvia (>40 kyrs BP) were discovered underlining the further potential of the field sites as proxy data archives. In close cooperation with Cameroonian partners from the University of Yaoundé I, data on climate and landscape history were studied, also adding reflections of recent and future trends in tropical environments (e.g., 'Global Change'). These different aspects were combined for selected areas across the host country, with a strong regional focus on the southern part of Cameroon. The major working areas were situated within river basins in ecologically sensitive transitional zones along the rain forest–savanna boundary. Apart from scientific results it was and it is crucial for Germany's research politics to support intercultural cooperation and academic Capacity building (or 'Capacity Development') of local partners. Therefore, the contributions in this book also act as a showcase for successful international North-South cooperation and Capacity building for empowering African universities. It is problem oriented and applied, and illustrates the way how scientific and interdisciplinary cooperation can work.

1.1 INTRODUCTION

Numerous rain forest and savanna terrains in Western Central Africa are located on flat planation surfaces formed by long lasting surface denudation over Gondwana type basement rocks, often topped by weathering resistant and extended lateritic crusts. There is widespread evidence and recognition that fluctuations in climate during the Quaternary in Central Africa have strongly affected regional environmental conditions such as extension, structure and composition of the rain forest-savanna border by modified rainfall and related hydrological characteristics. Aside from the Precambrian to Palaeozoic perspective on long-term landscape evolution (see contribution by B. Kankeu in this volume), it is also interesting to consider Late Pleistocene human adaptations and Holocene cultural innovations in Sub-Saharan Africa as they still seem to be the least well known of any major area of the Old World

(Brooks and Robertshaw, 1990). There is ongoing limited evidence that may allow a preliminary examination of such ecosystems and human/cultural adaptations closely connected to the Late Pleistocene and Holocene landscape modifications as well as human and cultural history (Lanfranchi and Schwartz, 1990; Maloney, 1998). As it was evidenced by the 'Rain Forest Savanna Contact' (in German 'ReSaKo') subproject for several Cameroonian river basins (e.g., the Ntem River basin with its 'interior delta') extended floodplains and alluvial deposits occur that are characterized by complex sedimentary structures, often interrupted by formerly cut- and fill-features (palaeochannels), which were interpreted in context with former climate induced environmental changes. Because of the linkages between catchment and floodplain, it is obvious that alluvial sediments reflect overall drainage area conditions and therefore they present important sources of information for the palaeo-conditions. A surprisingly old un-calibrated radiocarbon age of 48.232 ± 6.411 yrs BP (fossil wood) in a depth of 340–360 cm at Abong on the Ntem River gave evidence that the sedimentary sequences in Southern Cameroon are containing also the pre-LGM (Last Glacial Maximum) time period what makes them also important for the long-term landscape perspective and climate reconstructions (see contribution by M. Sangen in this volume).

Archaeological findings within Southern Cameroon are still very rare and less is known on how prehistoric people lived as hunter-gatherers in the transitional regions between dense rain forests and open savanna woodland ecosystems. According to Lanfranchi and Schwartz (1990) in Congo–Brazzaville the Maluekien (70–40 kyrs BP) was a relatively dry and cold episode and Middle Stone Age artefacts were frequently discovered in layers of coarse material (stone-lines) within multilayered soil-sediments. The Maluekien was followed by the Nijilien (40–30 kyrs BP) which was again humid. This may be triggered the return of closed forests into the region. Up to now there is no evidence for palaeolithic industries for this time. On the Batéké plateau north of Kinshasa and Brazzaville tropical podzols developed in the Kalahari Sands. These soil features prove striking modifications of the environment (Schwartz, 1988). Shortly before, during and after the maximum of the earth's last glaciation between 30–12 kyrs BP (Léopoldvillien) the climate reverted again to a very dry (less than 50% precipitation) and cold one (reduction of annual mean temperatures in the range of 4° C compared with today's conditions, cf. Runge, 2001) showing an open landscape, highly sensitive to erosion processes. Fluvial systems, streams and also smaller rivers showed high riverbed mobility and because of an assumed strong seasonality of climate, slope sediments (hill wash) and alluvia were deposited to a huge extend. However, these mobile and dynamic sediments have often been re-deposited inside the river channels. Within this open environment isolated, island-like forest 'relicts' around and along rivers ('riparian') might have been a common feature (Runge, 2001).

Late Stone Age industries of the Lupembien (up to 12 kyrs BP) have been discovered in Central Africa (cf. Clist, 1987). From 12 kyrs BP onwards the climate reverted again to humid and warm conditions (Kibangien A) which caused a rapid recolonisation of the low latitudes with dense forest. Stone cultures with microlithes are known for the Tshitolien (12–2.35 kyrs BP) within forested or at least woody biomass dominated landscapes (cf. Lanfranchi and Schwartz (1990). The so-called 'First Millenium Crisis' around 3 kyrs BP (Kibangien B) with strong aridification and savannization triggered the shrinkage of forests while savannas started again to expand (Maley, 2002). It is interesting to recognize that this environmental modification obviously marks the beginning of the Neolithic, which seems—as a not proven hypothesis—to coincide with the establishment of agro-pastoral societies and the 'migration' of pottery making Bantu into Central and Southern Africa (Clist, 1987).

Often it has been argued that the above sketched environmental and cultural models are too deterministic in general, and perhaps also 'eurocentristic' in conception.

When returning to field sites and proxy sources in Southern Cameroon, it has to be stated that not all the streams in the study area contain (big) alluvia—often rivers discharge on Precambrian basement. Therefore, it was of great importance within the project to glean a better understanding of transport and accumulation processes of fluvial sediments inside different catchments. Subsequently, sites had to be identified along river courses where 'suitable' alluvia for palaeoenvironmental interpretation can be expected and where not. The detailed examination of the local neotectonic situation and the over all river network evolution in Southern Cameroon (Lucazeau et al., 2003; Ngako et al., 2003) was fundamental for the ReSaKo project. The long-term landscape perspective contributed directly to the overall palaeoenvironmental objectives of the research (see contribution by B. Kankeu in this volume). The integration of 'neotectonics' helped identifying key-sites for field work where promising proxy data archives for the reconstruction of the Quaternary landscape conditions have been expected (see contribution of J. Eisenberg in this volume).

1.2 'CAPACITY DEVELOPMENT' AND APPLICATION OF RESULTS

The term 'Capacity Development' is a relatively new expression, mainly used in development cooperation, and up recently, it has not been applied to describe relationships between 'academic people'. Capacity Development is a process leading to conditions and capabilities whereby individuals, groups, organizations, and societies contribute to identifying and meeting development challenges in a sustainable manner (Lavergne and Saxby, 2001). Scientific cooperation and collaboration with Third World countries addresses the capacity of individuals (e.g., the Cameroonian partners contributing to this book) as well as organizations and the broader institutional and political context in which they operate (e.g., the Cameroonian universities and their academic system). Capacity Development is also about empowering institutions and building their identity (Ernstorfer and Stockmayer, 2009).

In the case of scientific research the dissemination of results can no longer be restricted to intellectual circles, as it is still often the case in many academic institutions. Apart from basic scientific level of knowledge, it is quite important for a country like Cameroon to carefully examine how research results can serve society and produce sustainable benefits for the livelihoods of people (e.g., poverty reduction).

The contribution of M. Tchindjang in this volume combines the findings from the study of past environments to draw visions and explains facts on 'environmental risks' over a short time scale in different regions and thematic fields within Cameroon. The empirical data base is limited, however, he can track back fluvial discharge of larger streams (e.g., Sanaga River) over several decades, looking on severe droughts in the 1970s and 1980s and their direct and indirect consequences for the savanna and rain forest environment. Human induced deforestation, linked to agriculture, mining or commercial logging also has brought an increase in erosion and degradation of natural conditions. Forest and land management and management of natural resources show explicitly the cross-sector character of conflicting stakeholder groups. The case study shown on deforestation and the efforts for a reforestation in the environs of the capital Yaoundé gives an insight into the complexity of ecological factors on the one hand, on the other hand it highlights societal and socioeconomic conflicts. Illegal logging, uncontrolled mining and corruption are further key topics that hinder the development of Cameroon (cf. Runge & Shikwati, 2011).

1.3 OVERARCHING CONCEPT

In this book quite different features and aspects of landscape evolution are studied on different scales in time and space. At first glance, one could argue that the period from the Palaeozoic to the present is somewhat long to be sufficiently dealt with. Of course, one could reach the conclusion that the dimension of earth history and recent ecological problems don't suite well. On the other hand, when regarding the complexity of earth's ecosystems it seems to be indispensable to integrate as many as possible factors to learn and to understand its characteristics over time, as well as its evolution and dynamics over longer or shorter periods, either in geological or in historical periods.

Chapter (2) by Boniface Kankeu and colleagues outline the geological and tectonic history of Western Central Africa since the Palaeozoic. This contribution gives the necessary introduction to feature the major traces of long-term landscape evolution in Cameroon. It is a prerequisite to follow chapter (3) by Joachim Eisenberg on the geomorphic evolution of two Cameroonian river basins draining into the Atlantic by considering Tertiary and 'younger' neo-tectonic processes. After having explained the features and the making of the Nyong and Ntem catchments, the next chapter (4) by Mark Sangen looks in great detail on the alluvia within these basins and their proxy data potential to document and explain Quaternary features and modifications in southern Cameroonian landscapes, now including conclusions on former climate and vegetation cover. And finally, by arriving at the last chapter (5) by Mesmin Tchindjang and colleagues, current, partly human induced ecosystem modifications and dynamics conclude on 'the big picture' this book intends to draw on Cameroon.

ACKNOWLEDGEMENTS

Between 2003–2009 the work of the Research Unit 510 has been generously supported by grants from the German Research Foundation (DFG) for which I am most grateful.

REFERENCES

Brooks, A.S. and Robertshaw, P. 1990, The Glacial Maximum in Tropical Africa: 22.000–12.000 BP. In *The World at 18.000 BP*, edited by Gamble, C. and Soffer, O., (Low latitudes, UnwinHyman), **2**, pp. 121–169.

Clist, B. 1987, Early bantu settlements in west central Africa: a review of recent research. *Current Anthropology*, **28**, pp. 380–382.

Ernstorfer, A. and Stockmayer, A. 2009, *Capacity Development for Good Governance.* (Baden-Baden: Nomos,), pp. 1–219.

Lanfranchi, R. and Schwartz, D. 1990, *Paysages quaternaires de l'Afrique Centrale Atlantique* (Paris: ORSTOM éditions, Collections didactiques), pp. 1–535.

Lavergne, R. and Saxby, J. 2001, Capacity Development: vision and implications. *CIDA Policy Branch. Occasional Series*, **3**, pp. 1–12.

Lucazeau, F., Brigaud, F. and Leturmy, P. 2003, Dynamic interactions between the Gulf of Guinea passive margin and the Congo River drainage basin: 2. Isostasy and uplift. *Journal of Geophysical Research*, **108** (B8), 2384, pp. 1–19.

Maley, J. 2002, A catastrophic destruction of African forests about 2.500 years ago. Still exerts a major influence on present vegetation formations. *IDS Bulletin*, **33**(I), pp. 13–30.

Maloney, B.K. 1998, *Human activities and the tropical rain forest. Past, present and possible future* (Dordrecht: Kluwer Acad. Publ.), pp. 1–206.

Ngako, V., Affaton, P., Nnange, J.M. and Njanko, Th. 2003, Pan-African tectonic evolutionin central and southernCameroon: transpression and transtension during sinistral shear movements. *Journal of African Earth Sciences*, **36**, pp. 207–214.

Runge, J. 2001, Landschaftsgenese und Paläoklima in Zentralafrika. Physiogeographische Untersuchungen zur Landschaftsentwicklung und klimagesteuerten quartären Vegetations- und Geomorphodynamik in Kongo/Zaire (Kivu, Kasai, Oberkongo) und der Zentralafrikanischen Republik (Mbomou). *Relief, Boden, Paläoklima*, (Berlin, Stuttgart: Gebrüder Bornträger), **17**, pp. 1–294.

Runge, J. and Shikwati, J. 2011, *Geological Resources and Good Governance in Sub-Saharan Africa.* Holistic approaches to transparency and sustainable development in the extractive sector. (Leiden: CRC Press, Taylor & Francis Publ.), pp. 1–265.

Schwartz, D. 1988, Histoire d'un paysage: Le Lousséké. Paléoenvironnements Quaternaires et podzolisation sur sables Batéké. *Editions de l'ORSTOM, Etudes et thèses*, pp. 1–285.

CHAPTER 2

Geological and tectonic history of Western Central Africa since the Palaeozoic

Boniface Kankeu, Jean Bassahak & Joseph Victor Hell
Institute of Geological and Mining Research (IRGM), Yaoundé, Cameroon

Jürgen Runge
Department of Physical Geography, Johann Wolfgang Goethe University, Frankfurt am Main, Germany

ABSTRACT: The complex geological and tectonic history of Western Central Africa since the Palaeozoic provides information which can be used for the better understanding of the interactions between deep lithospheric and surface processes. North of the Archean Congo Craton, the Pan-African rocks and structures are related to the initial formation of Gondwana between 720 and 550 Ma. Post Pan-African formations comprise Cretaceous sedimentary rocks filling pull-apart basins along the Central African Rift System (CARS) and Cenozoic-Quaternary magmatic rocks. Quaternary alluvial sediments occupy topographic depressions. Continuous tectonic activity in Central Africa during the Mesozoic polyphase break-up of the Gondwana supercontinent, in addition to climate and sea level changes concomitant with the northwards drift of the supercontinent are documented. The geotectonic processes observed include normal and strike-slip faulting, extensional stages and the development of associated sedimentary basins, multistage compressional events and uplift/erosion. Cenozoic-Quaternary volcanic and sub-volcanic complexes in Cameroon define a SW-NE straight line (Cameroon Volcanic Line, CVL) and display a "swell and basin" geometry interpreted as the result of rejuvenation of lithospheric Precambrian elements such as fractures, faults and hot spots, and mantle plumes in response to shallow mantle convection. Epicenters of earthquakes reported in Cameroon clearly link neotectonic activity with Pan-African structures in the area. These features are closely related and interpreted as reactivation products of a pre-existing weakness due to a steep basement shear zone. This implies interactions between active tectonics, deep structure, sedimentary basins and polyphase orogeny. We suggest that since the Palaeozoic in Western Central Africa lithospheric structure and evolution have controlled tectonic topography development.

2.1 INTRODUCTION

Numerous rain forest and savanna terrains in Western Central Africa are located on flat planation surfaces formed by long lasting surface denudation over Gondwana type basement rocks, often topped by weathering resistant and extended lateritic crusts. There is widespread evidence and recognition that fluctuations in climate during the Quaternary in Central Africa have strongly affected regional environmental conditions such as extension, structure and composition of the rain forest-savanna border by modified rainfall and related hydrological characteristics. Equatorial ecosystems in Africa can therefore be regarded as highly sensitive to former and recent environmental modifications (Gasse, 2000; Runge, 2001, 2002; Thomas, 2004). For palaeo-environmentally oriented studies the occurrence of thick alluvia trapped inside river basins can be successfully used as a proxy for former climate and vegetation reconstructions in lower latitudes (Thomas, 2000; Runge, 2002; Sangen, 2007).

In the case Gondwanan Africa, the continent's surface is characterized by long lasting denudation with mainly chemical weathering that finally formed extended erosion surfaces (peneplains). It was commonly assumed for a long time that almost no alluvial sedimentation occurring in lower latitudes could be successfully used for palaeoenvironmental reconstruction. However, the opposite is true: the widespread basin and swell topography of Africa (Summerfield, 1985) with locally strong Tertiary and Pliocene 'neo' tectonic movements and heterogeneous bedrock within river catchments, have given rise to a stepped topography with strong downcutting and upwarping. This geomorphological frame bears considerable alluvia inside river basins, which have not yet been well explored (Thomas, 2000; Runge, 2002). Unfortunately, knowledge on geology and tectonic evolution and its influence on geomorphologic features within the topographic history of erosion surface evolution is still poorly understood (Eisenberg, 2009). Runge *et al.* (2006) suggest that neotectonic movements in Western Central Africa since the Pliocene caused the reactivation of Precambrian fault structures which, assisted by intense chemical weathering and deformation, subsequently led to the geomorphic evolution of a macro-relief having smaller basins that could act as "sediment traps" within the rain forest–savanna belt of Western Central Africa. However, as persuasive as such models are, they need to be underpinned with field documentation at a variety of scales, in particular, with respect to the temporal and spatial connection between the Phanerozoic geological evolution and pre-existing basement structures, and its importance for the making of the "younger" Quaternary surface structures and landscapes.

To contribute to a better knowledge of overall landscape evolution, we provide a brief overview of the geological and tectonic evolution of Western Central Africa in terms of the earth's history, focusing on Gondwana and ongoing transformation since the Palaeozoic. The following is a summary of our current understanding of these long term geologic processes drawn largely from the works of Deruelle *et al.* (1991), Maurin *et al.* (1991), Bumby and Guiraud (2005), Guiraud *et al.* (2005) and co-workers (see references).

2.2 PRECAMBRIAN BASEMENT OF WESTERN CENTRAL AFRICA

2.2.1 Origin and significance of the term Gondwana

The term Gondwana, first used by Eduard Suess around 1880, from an Indian word for the name of native people, the Gonds, who lived in central India between the 12th and 17th century. Gondwana meant 'land of the Gonds'.

Gondwana was a supercontinent (Figure 1) located in the southern hemisphere between about 600 Ma, the approximate age of its amalgamation, and Middle Jurassic or 170 Ma, the age of the initiation of its fragmentation. It comprised five of the present-day continents: Africa, South America, India, Australia and Antarctica, as well as the Arabian Peninsula and the large islands of Sri Lanka (formerly Ceylon), Tasmania, New Zealand and Madagascar.

Gondwana was formed during the Pan-African orogeny from 720 to 580 Ma (Unrug, 1996; Caby, 2003) and consisted of (1) several cratonic cores surrounded by

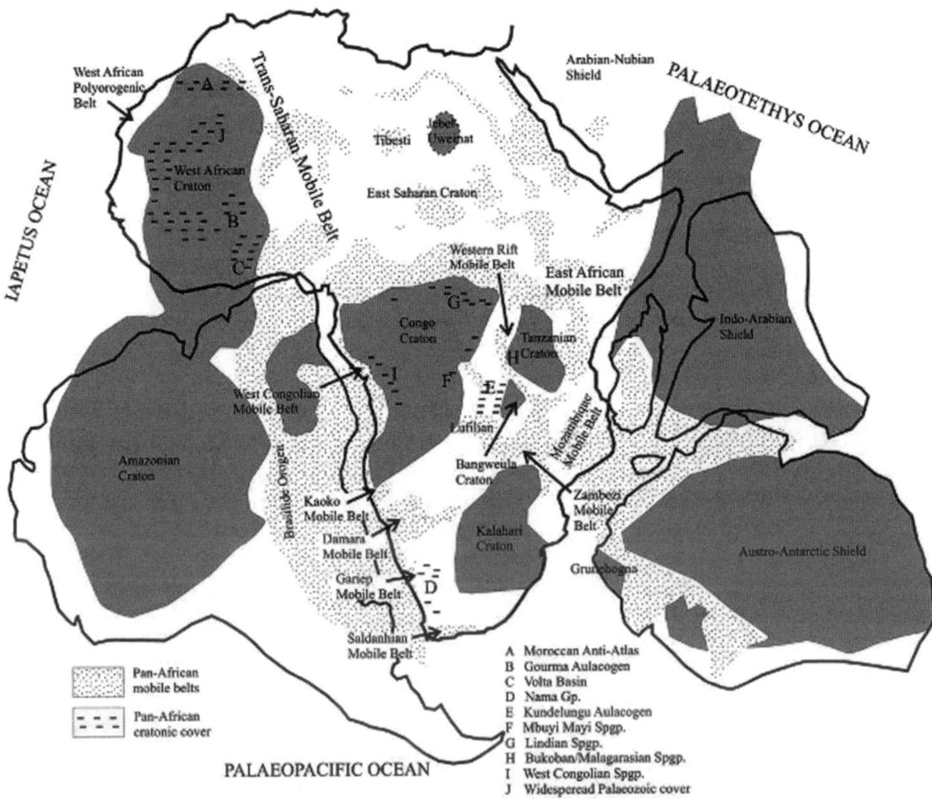

Figure 1. Map showing pre-Pan-African Cratons, Pan-African mobile belts and Pan-African cover rocks within Gondwana (after Petters, 1991; Unrug, 1997; De Wit *et al.*, 1999, Bumby and Guiraud, 2005).

(2) Pan-African fold belts. Deformational and metamorphic events, dated at around 600 Ma, are slightly diachronous. The Pan-African thermo-tectonic event, first recognized by Kennedy (1964), is now regarded as a period of major crustal accretion, where continental, island-arc, and oceanic terranes were brought together to form the crystalline basement of the African continent as part of the late Neoproterozoic supercontinent Gondwana (Toteu *et al.*, 1991; Castaing *et al.*, 1994; Unrug, 1997). Whereas some parts of the Pan-African orogen such as the Arabian-Nubian Shield are typical for accretionary orogens (Windley, 1992), others are characterized by continental collisional tectonics, as the Mozambique Belt (Burke and Sengör, 1986; Stern, 1994).

2.2.2 The Pan-African-Brasiliano fold belts

Comparisons among the Neoproterozoic belts in West Africa, northern Brazil and Central Africa show very strong similarities with respect to lithology, kinematics and metamorphism (Castaing *et al.*, 1993).

The general organization of the Pan-African/Brasiliano chain (Figure 2) is the result of the collision between three major continental domains: the West African and Sao Francisco/Congo Cratons, and a Late Proterozoic mobile domain composed of

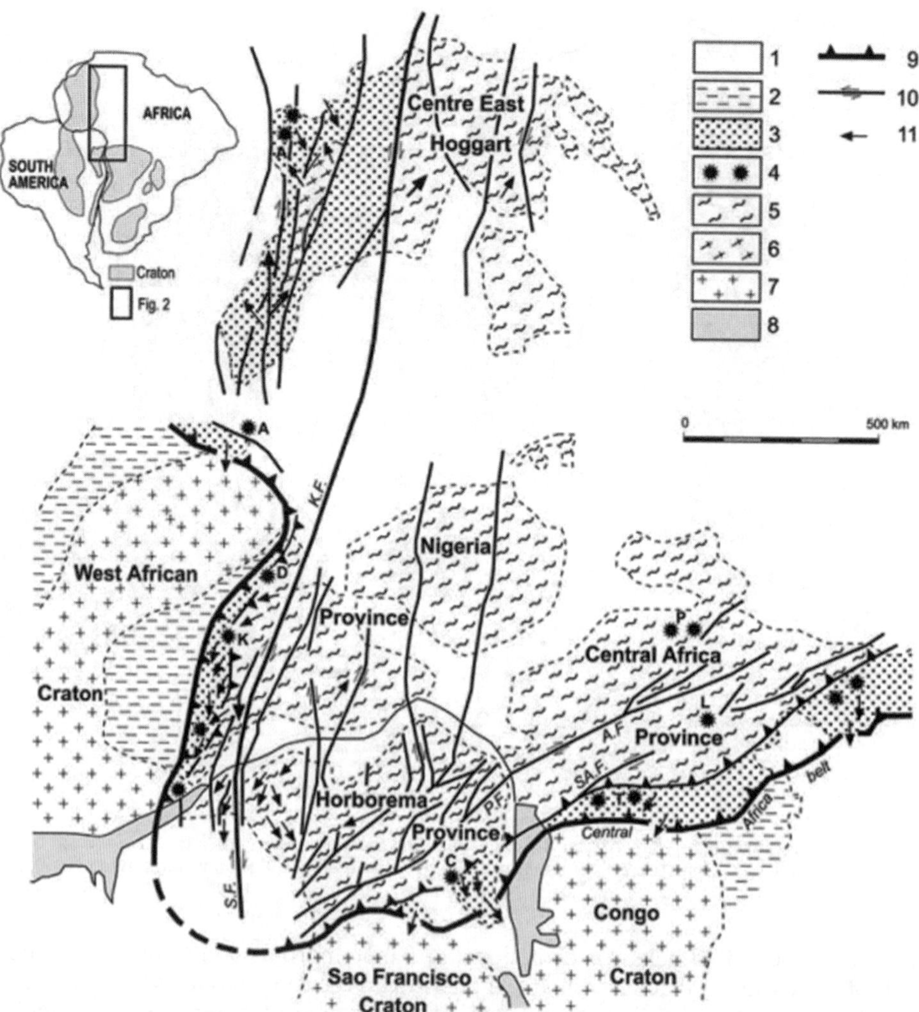

Figure 2. Insert: Main cratonic areas and fold belts (white space) in Africa and South America and localization of Figure 2. General organization of the Late Proterozoic Pan-African/Brasiliano belt as the result of the collision between three major continental domains during setting up of Gondwana (modified from Castaing *et al.,* 1993). 1 = Phanerozoic cover, 2 = Proterozoic cover, 3 = Pan-African/Brasiliano metasedimentary fold belts including basement granulite (6) in the western Hoggar, 4 = main basic and ultrabasic massifs underlining the suture zone, 5 = undifferentiated mono-and polycyclic gneiss and metasediment, migmatites, granite and charnockite, 6 = Ouzzal and Iforas Eburnian, 7 = 2 Ga Cratons, 8 = sea, 9 = major shear zone, 10 = shear zones, 11 = tectonic nappe-transport direction. A.F. = Anaga-Adamaoua fault, P.F = Pernambuco fault, S.F = Sobral fault, SA.F = Sanaga fault, K.F. = Kandi fault.

a polycyclic basement province (Hoggar, Nigeria, Borborema and Central Africa) in its inner part, surrounded by a fringe of metasedimentary (Pharusian, Atacora, sergipano and Central Africa) thrust belts.

Three domains can be recognized in the chain (Black *et al.*, 1979; Cahen *et al.*, 1984; Castaing *et al.*, 1994): (1) the external "nappes" or frontal part of the belt thrusted over the Craton, derived from passive-margin sedimentary deposits having medium to high pressure metamorphism, (2) the intermediate domain having basic and ultrabasic massifs including suture-zone rocks with eclogitic assemblages (protoliths with an age of ca. 800 Ma) and mainly monocyclic metasedimentary rocks and orthogneiss, (3) the internal domain composed of high-grade, commonly anatectic gneiss (anatexis at ca. 600 Ma), including orthogneiss (Eburnian plutons), elongated granulite belts, reworked Archaean basement, and outcropping intrusive granite (ca. 600 Ma).

The tectonic evolution can be divided into three main periods: the first characterized by oblique collision inducing thrusting towards the cratonic forelands; the second with anatexis and doming; and a third period having wrench faulting with associated post-foliation folding under late shortening. In the course of orogenic activity, numerous Pan-African granitoids and major deformational zones originated all across the belt.

2.2.3 The Pan-African central Africa fold belt

Cameroon and the Central African Republic overlay the margin of the Pan-African belt towards the Congo Craton (Figure 2), which is represented by the Central African Fold Belt (CAFB). This belt consists of marginal thrusts, which transported Pan-African orogenic material southwards onto the orogenic foreland of the Congo Craton (Castaing *et al.*, 1994). In contrast, central and northern Cameroon, the northern Central African Republic and southern Chad, which represent the more internal domains of the orogen, are dominated by strike slip tectonics.

The formation of the Pan-African–CAFB north of the Congo Craton can be interpreted as convergence and collision (Toteu *et al.*, 2001; Ngako *et al.*, 2003) between the stable Archaean Craton to the south and one or two Palaeoproterozoic plates, i.e., the so-called "central Sahara plate" including Hoggar.

The most notable structural features of the CAFB are represented by:

- Narrow external and frontal thrust zones along the Congo Craton characterized by horizontal tectonics, which favored the emplacement of high-grade metamorphic terranes. These thrust belts (Figure 3) are generally interpreted as Pan-African collisional zones north of the Congo Craton and the Oubanguides belt (Vellutini *et al.*, 1983; Poidevin, 1985; Boudzoumou and Trompette, 1988; Maurin *et al.*, 1991);
- Large internal domains mostly composed of polycyclic gneiss that derive from Archean or Palaeo-Proterozoic (Eburnean-Trans Amazonian, ca. 2 Ga) plutonic rocks, deformed and recrystallised during the Pan-African-Brasiliano thermo-tectonic events. The entire terranes are crosscut by syn-kinematic (ca. 650–600 Ma) and post-kinematic (ca. 580–530 Ma) granitoids (see reviews by Caby, 1989; Boullier, 1991);
- Large mylonitic shear zones, which extend from the Borborema province to Central Africa. In the Borborema province, the shear zones correspond to a network of anastomosing transpressive steep mylonitic bands of several kilometers in width. In Central Africa, these lineaments are now regarded as Pan-African strike-slip shear zones having a subordinate component of compression, separating external Pan-African domains of compression along the northern margin of the Congo Craton from internal domains dominated by high-angle strike-slip and transpressional deformation (Kankeu and Greiling, 2006; Kankeu *et al.*, 2009). Apart from their importance to Pan-African evolution, these structures are inferred to have influenced subsequent tectonic evolution.

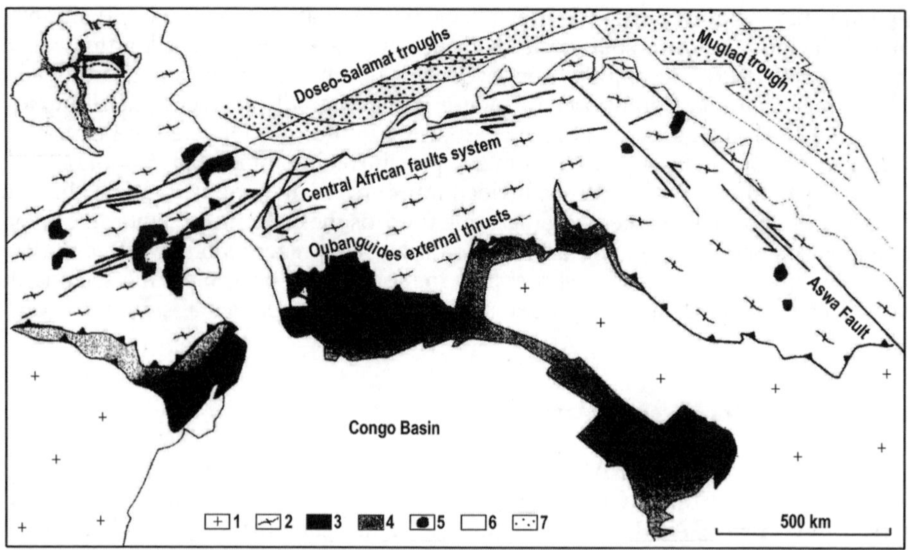

Figure 3. Pan-African Oubanguides belt (redrawn from Poidevin, 1985) showing the location of the rift system in southern Chad and in Sudan (after Genik, 1992). 1 = Archean-Lower Proterozoic basement; 2 = undifferentiated reworked gneisses; 3 = Lower Proterozoic units: 4 = Upper Proterozoic units; 5 = late Pan-African granites; 6 = Phanerozoic cover: 7 = Early Cretaceous depocentres.

2.3 PALAEOZOIC EVOLUTION

Only gentle deformation affected Central Africa during the Palaeozoic. Eroded material identified as Pan-African molasses are found in within-belt Palaeozoic grabens. The Mangbei-type semi-grabens in northern Cameroon (Balche, Nigba) are located at the intersections between NS secondary and EW shear zones. Main lithologic units in these grabens include: sandstones, conglomerates and continental tholeites represented by acidic, intermediate and basic lavas (Béa *et al.*, 1990). The principal palaeogeographic and geomorphologic characteristics of Western Central Africa were the permanency of large scale terrestrial denudation and formation of planation surfaces.

2.4 MESOZOIC

The Mesozoic is marked by the break-up of Gondwana and the extensive development of rifts. These processes were dominated from the Late Carboniferous onwards. As a result of this extension, basins developed and filled with thick sequences of lacustrine to marine deposits, both at the margins and within the continental African plate. Rifting can be considered to have occurred in three distinct pulses:

- From the Late Carboniferous to Middle Jurassic when East and West Gondwana separated, and the Central Atlantic Ocean began to open,
- From the Late Jurassic to Early Cretaceous the South Atlantic opened progressively northwards, and intraplate extension led to the development of the West, Central and East African rift basins,

- The final episode of rifting took place between the Late Eocene to Early Miocene (Guiraud and Bellion, 1995; Wilson and Guiraud, 1998) and opened the Dead Sea–Red Sea–Gulf of Aden basins, with rifting in the East African Rift continuing to the present day.

2.4.1 Late Carboniferous to Middle Jurassic

The first stage of rifting (Late Carboniferous–Middle Jurassic) affected the northern and eastern margins of Africa, Arabia, and the Central Atlantic province (Figure 4). This period of rifting can be considered as the initial stage of the break-up of Gondwana at about 180 Ma, separating West Gondwana (Arabia–Africa and South America) from East Gondwana (Antarctica, Australia, India, Madagascar and New Zealand). Initiation of this rifting in the south was most likely controlled by the presence of a mantle plume beneath southeastern Africa and Antarctica, which had the local effect of initiating rifting within the Kalahari-Grunehogna Craton. Contemporaneous with rifting was the deposition of widespread sandstones, known as "Continental intercalaire" (Kilian, 1931; Lefranc and Guiraud, 1990), between the Late Carboniferous and early Late Cretaceous.

2.4.2 Late Jurassic to Early Cretaceous

The second stage of Gondwana rifting occurred from the Late Jurassic (180 Ma) to Early Cretaceous and had a broad impact. The equatorial regions of the Atlantic Ocean continued to rift, accomplished under dextral transtensile conditions (Guiraud

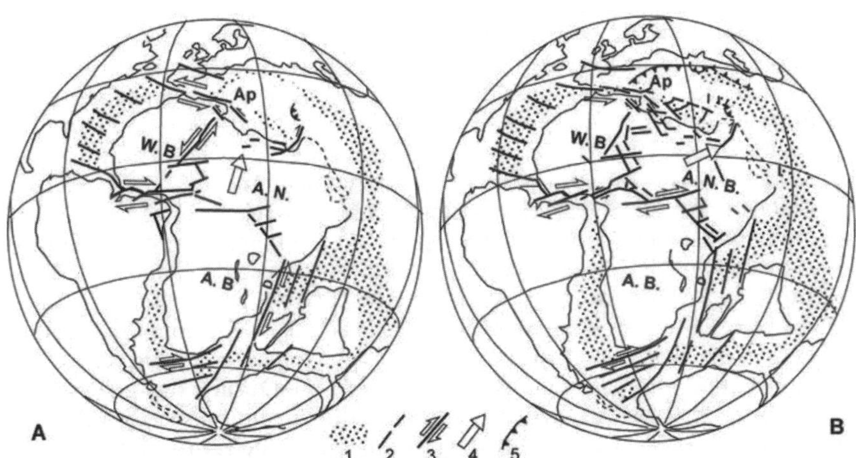

Figure 4. Subdivision of Africa into major blocks during the Early Cretaceous. (A) Late Barremian (≈ 120 Ma). (B) Early Albian (≈ 100 Ma). Reconstructions according to "Terra Mobilis" model (Scotese and Denham, 1988). A.B. = Austral block; A.N.B. = Arabian-Nubian block: W.B. = Western block: Ap = Apulian plate: D = Davie Ridge. 1 = Oceanic crust; 2 = Fault or strike-slip fault, thick lines indicate block boundaries; 3 = Major strike-slip fault, or transfer fault, with relative motion; 4 = Relative motion of the Arabian-Nubian block during the Early Cretaceous; 5 = Thrust.

and Maurin, 1992). In addition to rifting around the edge of the present plate, the Central African Rift, East African Rift and West African Rift systems also began to open (Guiraud and Maurin, 1992; Guiraud *et al.*, 2005), which reflects an aborted break-up within the African plate itself.

2.4.3 Cretaceous

The Cretaceous corresponds to a crucial period for Central Africa. Active rifting episodes over large domains led to the break-up of Western Gondwana and the opening of the South and Equatorial Atlantic oceans. From Aptian times, peaking in the Late Cretaceous, strong warming in the global climate resulted in wide transgressions over continental subsiding basins.

2.4.3.1 Early Cretaceous

Diachronous opening of the South and Equatorial Atlantic oceans and polyphased development of the Central African Rift System

A pivotal stage in the break-up of western Gondwana occurred during the Early Cretaceous illustrated by the crustal separation of the African and South American plates (latest Albian), opening of the South and Equatorial Atlantic oceans, and the polyphase development of the Western and Central African Rift System (WCARS). The system comprises a complex array of linked rift basins that straddle the continent from Nigeria to Kenya. Large parts of the rift systems have no present-day surface expression and were discovered largely by regional geophysical studies (Browne and Fairhead, 1983). The basic geometry of the rift system is that of a series of major fault zones diverging from the Gulf of Guinea into Africa. The rift systems comprise two parts: (1) The Central African fault zone extending from the Gulf of Guinea in an ENE direction crossing Southern Chad and the Central African Republic into western Sudan. In Sudan the fault is linked to an orthogonal system of extensional basins striking NW-SE through the Sudan and Northern Kenya, (2) the Benue Trough and Gongola Rifts form a series of basins that pass to the NE into the Lake Chad region and link with a series of NNE-trending basins in Chad, Eastern Niger and southern-most Algeria.

Rifting is polyphase (Guiraud and Maurin, 1991, 1992). Although rifting became evident by the Late Jurassic in the very oriental branch of the CARS (Reeves *et al.*, 1986/87; Bosworth, 1992), essentially the whole of rift development took place during the Early Cretaceous. This development was contemporaneous with the diachronous opening of the South and Equatorial Atlantic Oceans (Popoff, 1988; Binks and Fairhead, 1992). Two stages of rift development and fracturing (Figure 5) have been identified. Rifting was initiated during the earliest Neocomian in response to a submeridian extensional regime. A second rifting phase commenced during the middle-late Aptian generating pull-apart basins: (1) During the Neocomian to the early Aptian (Cretaceous Syn-Rift I Stage), the opening of the South Atlantic (Dingle *et al.*, 1983) was accompanied by rifting along the future continental margins from Angola to Cameroon. Contemporaneous subsidence of E-W-trending half grabens in the Upper Benue, Northern Cameroon, Southern Chad, and in Sudan was governed by a regime of approximately N-S-trending crustal extensions (Figure 5A) (Maurin and Guiraud, 1990). (2) During the middle-late Aptian and Albian (Cretaceous Syn-Rift II Stage) the progressive opening of the Equatorial Atlantic caused dextral transform movement and subsidence of pull-apart basins

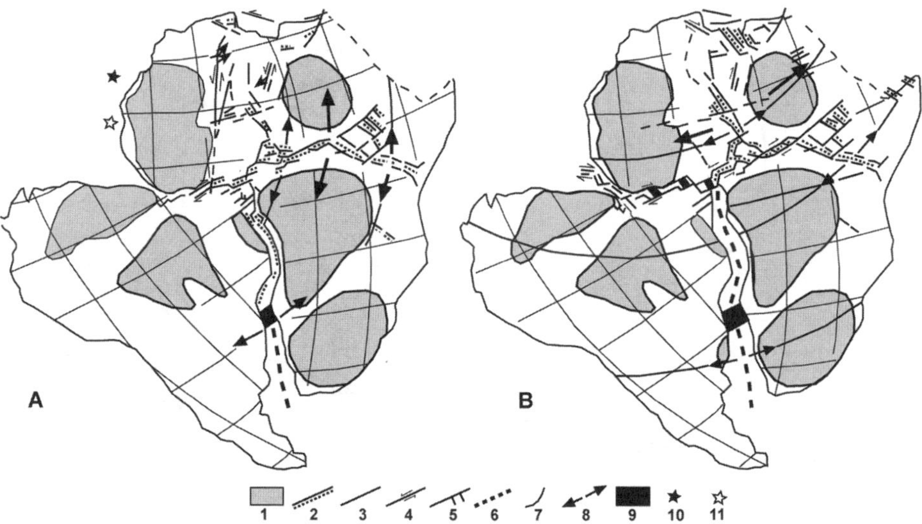

Figure 5. Two stage development of African rifts during Early Cretaceous. (A) Barremian.
(B) Albian. (After Fairhead, 1988; Unternehr *et al.*, 1988; Guiraud and Maurin, 1991). 1 = Craton;
2 = Rift; 3 = Active fault; 4 = Major strike-slip fault; 5 = Compressional zone at the end of wrench fault;
6 = Oceanic crust; 7 = Flow line; 8 = Extension direction; 9 = Walvis-Rio Grande Ridge; 10 = Pole of
rotation, after Fairhead (1988) for Fig. 3A, Klitgord and Schouten (1986) for Fig. 5B; 11 = Pole of
rotation proposed for the Fabian-Nubian block.

along the equatorial fracture zones (Mascle *et al.*, 1988). At the same time, the large
NW-SE-trending troughs of eastern Niger, Kenya and Sudan subsided in a tensional
(strike-slip and extension) regime (Fairhead, 1988). Also transtensional subsidence
of half grabens and pull-apart basins in the area of the Benue Trough (Benkhelil,
1988) and southern Chad (Genik, 1992) testified to wrench movements along the
Central African fault system, governed by a regime of an approximately N50°E-
trending crustal extension.

In terms of plate tectonics, the WCARS is considered by Guiraud and Maurin
(1991, 1992) as a more or less continuous system of weak zones dividing the continent
into three internal, relatively rigid, blocks (Figure 4); the Western block, the Arabian
Nubian block and the Austral block. Decoupling of these blocks from each other
along the WCARS is described by Guiraud and Maurin (1992) as follows: (1) During
the Neocomian to the early Aptian (Figure 4A), decoupling is evident by a northward
displacement of the Arabian-Nubian block involving crustal extension (first rifting
phase) along its common boundary with the Austral block and sinistral strike-slip
movements along the Trans-Saharian fault system. (2) During the middle-late Aptian
and Albian (Figure 4B), the Arabian-Nubian block tended to migrate towards the
northeast; this involved crustal extension (second rifting phase) along its common
boundary with the Western block and dextral strike-slip along the Central African
fault system separating it from the Austral block.

Latest Albian to Middle Santonian (Late Rift to Sag Basin Stage) is characterized by a decrease in tectonic activity and a marine transgression over Western
Central Africa. The subsidence is considered to be due to NE-SW directed extension (Bosworth, 1992; McHargue *et al.*, 1992; Guiraud, 1993; Janssen, 1996), sometimes associated with thermal relaxation. Mesozoic and Cenozoic fault-bounded

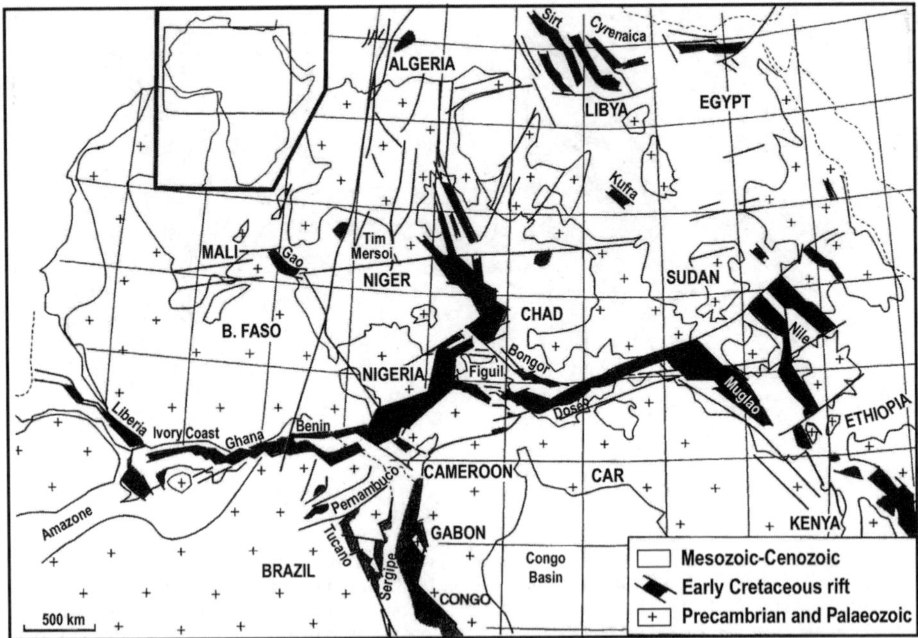

Figure 6. Structural map showing the geometry of the Western and Central African Rift System (WCARS).

sedimentary basins (Benue, Doba, Mbere and Djérem and Bake, Figure 6), more or less rich in oil and gas resources, are related to this evolution.

Pan-African basement structures and rift relationships

The WCARS is developed along an ENE-trending fault zone generated during a period of dextral strike-slip displacement in the Late Cretaceous. The fault zone is a major crustal structure, which is clearly visible on regional gravity and magnetic maps (Cornachia and Dars, 1983). The major Cretaceous fault system that traverses Cameroon is superimposed on the Adamawa Pan-African shear zone system. This is clear in Cameroon (Dumont, 1987; Ngangom, 1984; Kankeu *et al.*, 2009), but difficult to prove for Sudan due to the lack of outcrop of the fault zone for several hundred kilometers. The evolution of WCARS is related to the tectonic evolution of the Atlantic Ocean (Figure 7), and in particular the equatorial fracture zone. Between 74 Ma and 65 Ma, the differential opening of the equatorial Atlantic implies a dextral shear along the fracture zones (Fairhead, 1988). This period of dextral shear coincides with a contractional deformation of the Benue Trough, previously opened by sinistral strike-slip motion during Atlantic rifting. It also coincides with renewed dextral displacement along the WCARS and the related extension within the basins of Sudan. Thus, we have an example of a plate tectonically derived stress causing failures in the African continental crust. This occurred by the reactivation of a pre-existing weakness provided by a steep basement shear zone. Contrasting types of fault reactivation patterns are proposed from pure extension to transtension depending on the orientation of the pre-existing faults with respect to the palaeostress field.

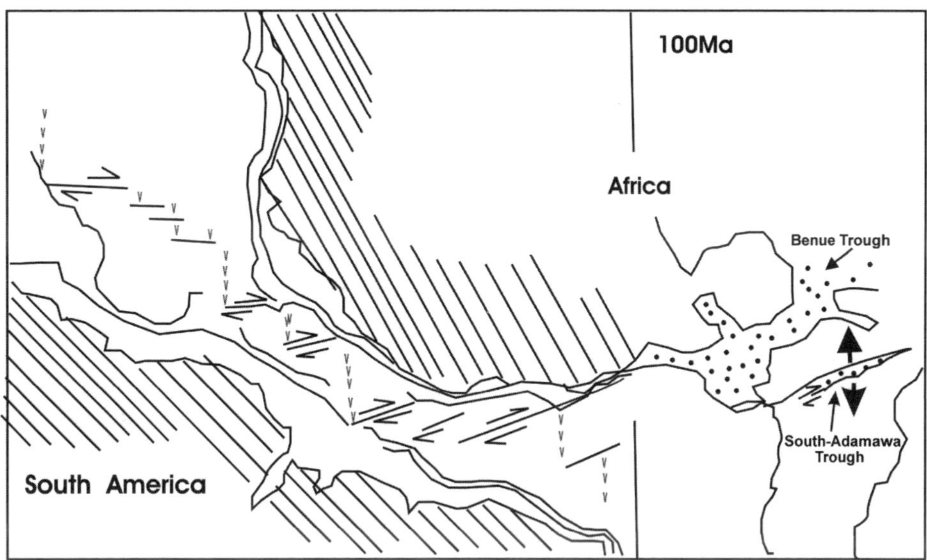

Figure 7. Map of the Atlantic opening, on the left, state of the equatorial Atlantic, extended to the East as far as Adamawa, showing possible correlation between the opening of the South Atlantic and stretching (extension) defined in the Meiganga region (modified from Dumont, 1986; Mascle *et al.*, 1986).

2.4.3.2 Upper Cretaceous

During the upper Cretaceous, there is a change in the regional stress field reflected by the onset of basin inversion and/or folding during Late Santonian times, and which affected both the E-W striking northern African–Arabian plate margin and Central African rifts (Guiraud and Bosworth, 1997). Evidence for inversion tectonics in the basins is provided by large-scale fold and flower structures, mainly concentrated along the Central African fault system. These deformations are related to a compressive event, which corresponds to the first stage of the collision of the African–Arabian and Eurasian plates in response to changes in the opening directions and rates of the South, Central, and North Atlantic Oceans (Binks and Fairhead, 1992; Fairhead *et al.*, 2003; Guiraud *et al.*, 2005). More generally, the Santonian event corresponds to a global event. The African-Arabian plate then strongly rotated in an anticlockwise direction and this resulted in the beginning of the collision with the Eurasian plate (Olivet *et al.*, 1984). The shortening is followed by a period of surface denudation (etching) and planation or leveling by climate controlled strong chemical weathering that formed the post-Cretaceous erosion surface (Le Marechal and Vincent, 1971).

Also noticeable was the development of lateritic crusts and enrichment of Bauxite over Western Central Africa. Major transgressions took place during Late Cenomanian to Lutetian times. The high sea levels, associated with warm climates, favored the development of carbonate platforms along Central Africa (Dercourt *et al.*, 1993; 2000).

Following this event, a new stress field appeared in Africa–Arabia. Tensional activity favored breccia and slump sedimentation, and magmatism. Around the end-Cretaceous, a brief compressional pulse occurred during which most of the Santonian structures were amplified.

2.5 CENOZOIC EVOLUTION

2.5.1 Palaeocene to Middle Eocene

During these times, magmatic activity was reduced, but persisted along the Cameroon Line (Wilson and Guiraud, 1998).

2.5.2 Early Late Eocene (37 Ma)

By the end Bartonian–Early Priabonian a major plates scale compressional event occurred. Transpression occurred along the major intraplate fault zones. Stress field analysis shows a general NNW-SSE to N-S shortening, resulting from a new stage in the collision of the African–Arabian and Eurasian plates (Guiraud and Bellion, 1995). Like the Late Santonian event, it resulted from changes in the rates and directions of the opening of the Atlantic oceans. The E-W trending Central African Fold Belt also began to slowly develop.

2.5.3 Late Eocene to recent

During this stage, the geological evolution of West Central Africa was dominated by tectonic and magmatic activities, and rapid and intense changes in the global climate. These phenomena resulted in frequent changes in palaeoenvironments.

At the plate scale, tectonic events led to the development of Oligocene rifts such as the Red Sea–Gulf of Aden–East Africa Rift System that resulted in (1) the separation of the Arabian plate and (2) the development of the Alpine Belt along the northern African–Arabian plate margin, which was polyphased and generated deformation in the intra-plate Central Africa domains. From the earliest Miocene, the collision between the African-Arabian and Eurasia plates episodically intensified. Another change in plate dynamics occurred around 22 Ma and generated minor or more localized compressional deformation during the Burdigalian. During the Tortonian (8.5 Ma), the African plate motion changed again. Some major intraplate fracture zones were rejuvenated again. Finally, an early Pleistocene event occurred in Africa-Arabia. Major intra-plate fault zones were rejuvenated and regional uplifts intensified.

Large intra-plate magmatic provinces initiated or developed in the Jos Plateau (Nigeria) and along the Cameroon Line (Wilson and Guiraud, 1992; 1998).

Cameroon Volcanic Line—Nature and origin

One special macro-morphological and geological feature in Western Central Africa is the presence of a unique within-plate volcanic province that straddles a continental margin (Fitton and Dunlop, 1985), generally referred to as the Cameroon Volcanic Line (CVL) (Figure 8). Stretching for over 1500 km, the CVL is marked by a striking NE-SW alignment of recent volcanic and sub-volcanic complexes, appearing in both Central Africa and the adjacent South Atlantic Ocean (Figure 9). The CVL splits into two branches in its north-eastern part, one crossing the Upper Benue valley in Nigeria, the other running eastward to the Adamawa or Ngaoundéré Plateau. The line is made up of an oceanic (Annobon, Sao Tomé, Principé and Bioko) and a continental (Mounts Etindé, Cameroon, Manengouba, Bambouto, Oku, Ngaoundéré, Bui and Mandara Mountains) sector.

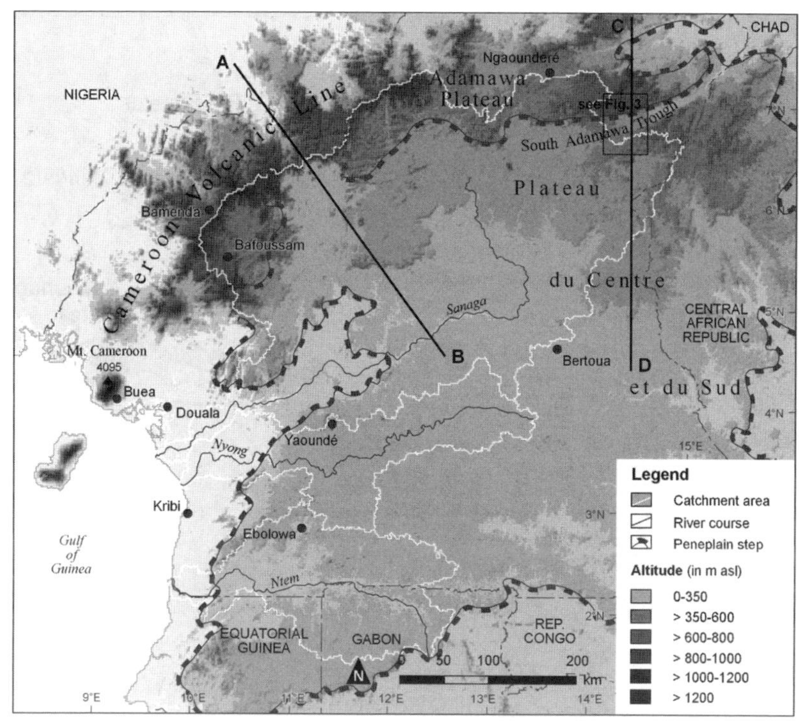

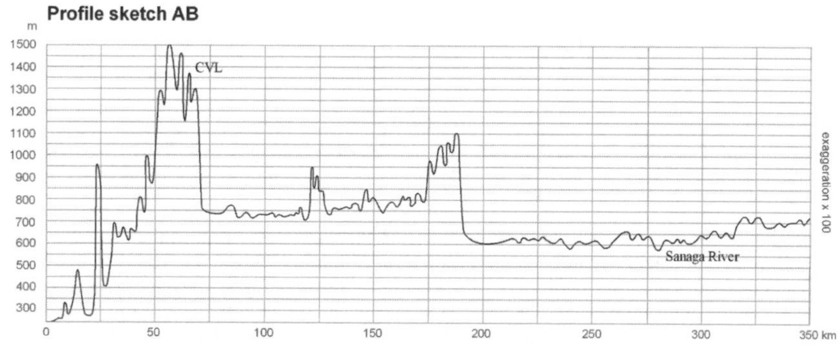

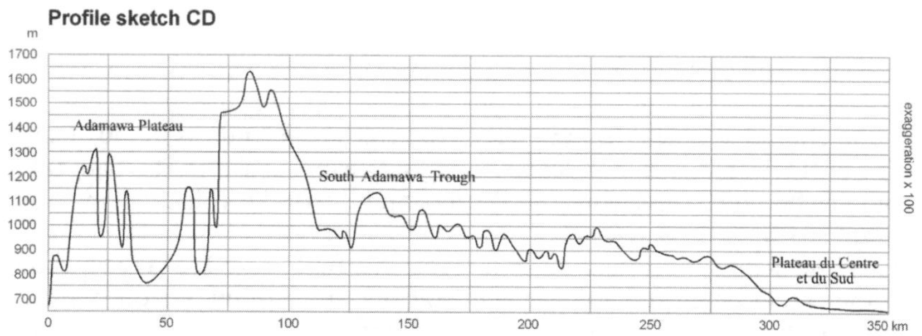

Figure 8. Map of central and southern Cameroon (SRTM data) highlighting the macromorphological setting.

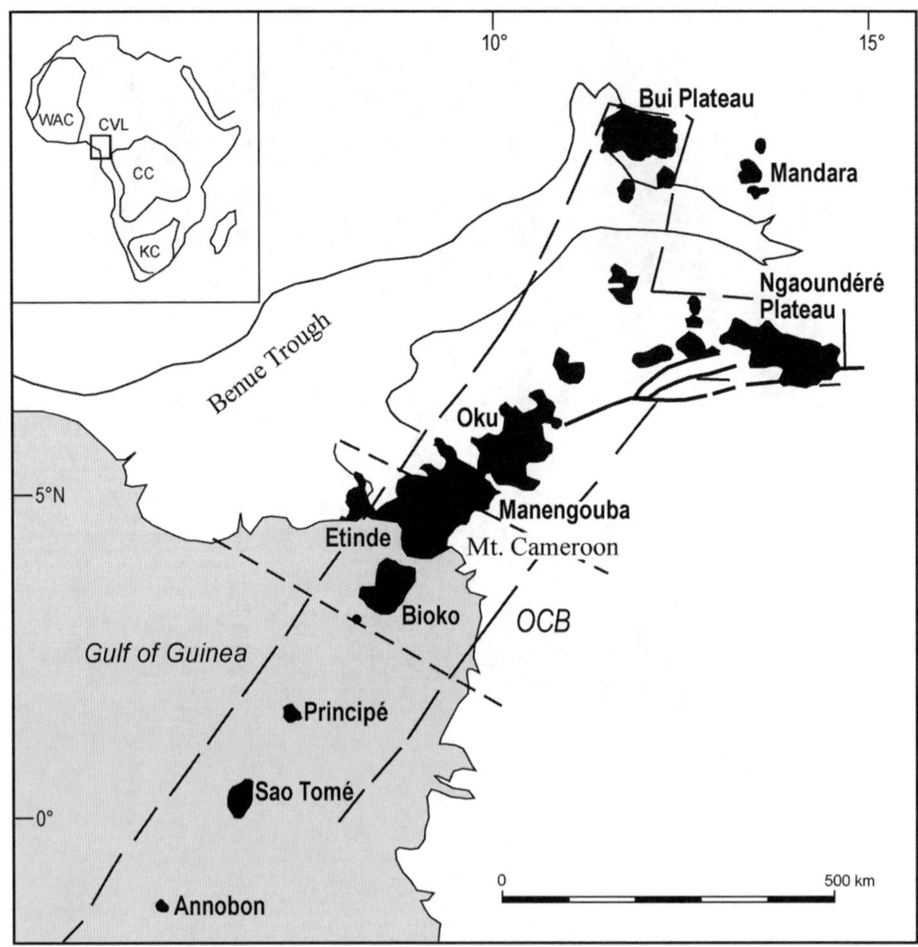

Figure 9. Map of the Cameroon Volcanic Line (CVL) in the Gulf of Guinea. Main Cenozoic volcanic centers are indicated in black. Oceanic sector volcanoes are Annobon, Sao Tomé and Principé. Volcanoes of the ocean/continent boundary are Bioko, Etinde and Mount Cameroon. Volcanoes of the continental sector are Manengouba, Bambouto, Oku, Ngaoundéré, Mandara Mountains and Biu Plateau. Inset shows the CVL within the main tectonic units of Africa (WAC = West African Craton; CC = Congo Craton; KC = Kalahari Craton; OCB = Ocean Continent Boundary).

The basement consists mainly of formations of the Pan-African Mobile Belt of Central Africa overlain in some places by Palaeozoic volcano sedimentary formations. In the Gulf of Guinea the bedrock of the volcanic islands, is found at shallower depths, from the Pagalu strato-volcano lying on oceanic crust, to the Sao Tomé volcanic complex built on Cretaceous sandstones, and the Bioko volcano lying on a deep sedimentary basin.

Magmatic activity started 80 Ma ago; the first 40–50 Ma were characterized by the emplacement of about 60 subvolcanic ring complexes on both sides of the Cameroon Line in its continental part (Moreau *et al.*, 1994) but have not been recorded in the Gulf of Guinea islands. The plutonic complexes are small in size (mostly 5–10 km in diameter) and except for four complexes in North and South-west

Cameroon, they are mainly constituted of granites or syenites, to which subordinate intermediate and basic rocks are sometimes associated (Kambou *et al.*, 1989; Njonfang *et al.*, 1992; Njonfang and Moreau, 2000). From 30 Ma to present, intense volcanic activity affected the southern part of the CVL and the Ngaoundéré area, spreading southwestward into the Gulf of Guinea. Volcanic rocks are usually aligned and in places associated with the plutonic complexes. Tertiary-Quaternary tectonic movements are associated with the emplacement of subvolcanic and volcanic complexes. In the southern part of the Cameroon Line, a series of grabens and horsts are bounded by N20°–30° trending normal faults. A general basement uplift having a N20° axis is related to the magmatic activity; this uplift, at least, begins with the emplacement of an orogenic complex (Benkhelil, 1988).

The origin of the CVL is still controversial, and there are several possible scenarios:

- Vertical movements, which created either a series of horsts and grabens bounded by SW-NE normal faults (Deruelle *et al.*, 1991) or (failed) continental rifting (Guiraud and Maurin, 1992; Wilson and Guiraud, 1992), probably directly produced by a thermal anomaly in the asthenosphere.
- The movement of the African plate over a hotspot trail (Morgan, 1983; Van Houten, 1983). The similarities in size and shape between the CL and the bordering Benue Trough to the NW led Fitton (1980) to postulate that it resulted from the anticlockwise rotation of the African lithosphere over an asthenospheric hotspot at about 80–65 Ma. Recently, Meyers *et al.* (1998) have proposed that due to convection in the lower mantle, the CL as well as other linear volcanic chains of West Africa, may have formed as mantle upwellings or "hotlines" due to heat transfer and/or shear along discontinuity.
- Reactivation of ancient basement fractures (Moreau *et al.*, 1987; Fairhead, 1988) by Cenozoic large-scale left-lateral transcurrent shearing/movement. The proposed Riedel model (Figure 10) implies a rejuvenation of the N70° Adamawa fracture zone, the CVL appearing as a tension gash, the network of the N-S Pan-African and E-W trends superimposed to R' and P directions, respectively.

The Cenozoic-Quaternary CVL occurs exclusively in the Pan-African domains where it is associated with domal uplift and reactivation of pre-existing Pan-African shear zones and transcurrent faults. Climates registered several periods of strong cooling during this time, resulting in marine regressions and major changes in the continental environments (e.g., terrestrial denudation and alluviation of rivers).

2.6 RECENT TECTONIC ACTIVITY

One crucial element of recent evolution in West Central Africa is the strong seismicity recorded in Cameroon and its possible relationship with lithospheric structures. Over 60 felt earthquakes have been reported in Cameroon (Figure 11), for the period 1615 to 2005 (Nangé, 2007). The epicenters of these events do not occur hazardly but are associated with prominent tectonic features. These include (1) the Mt. Cameroon region, (2) the Central Cameroon or Adamaoua Shear Zone (CCSZ), (3) the Sanaga Shear Zone (SSZ) and the northern margin of the Congo Craton.

The earliest documented earthquake felt in the Mt. Cameroon region occurred in 1852 and between this time and 1910 at least 9 earthquakes have been felt. In 1986,

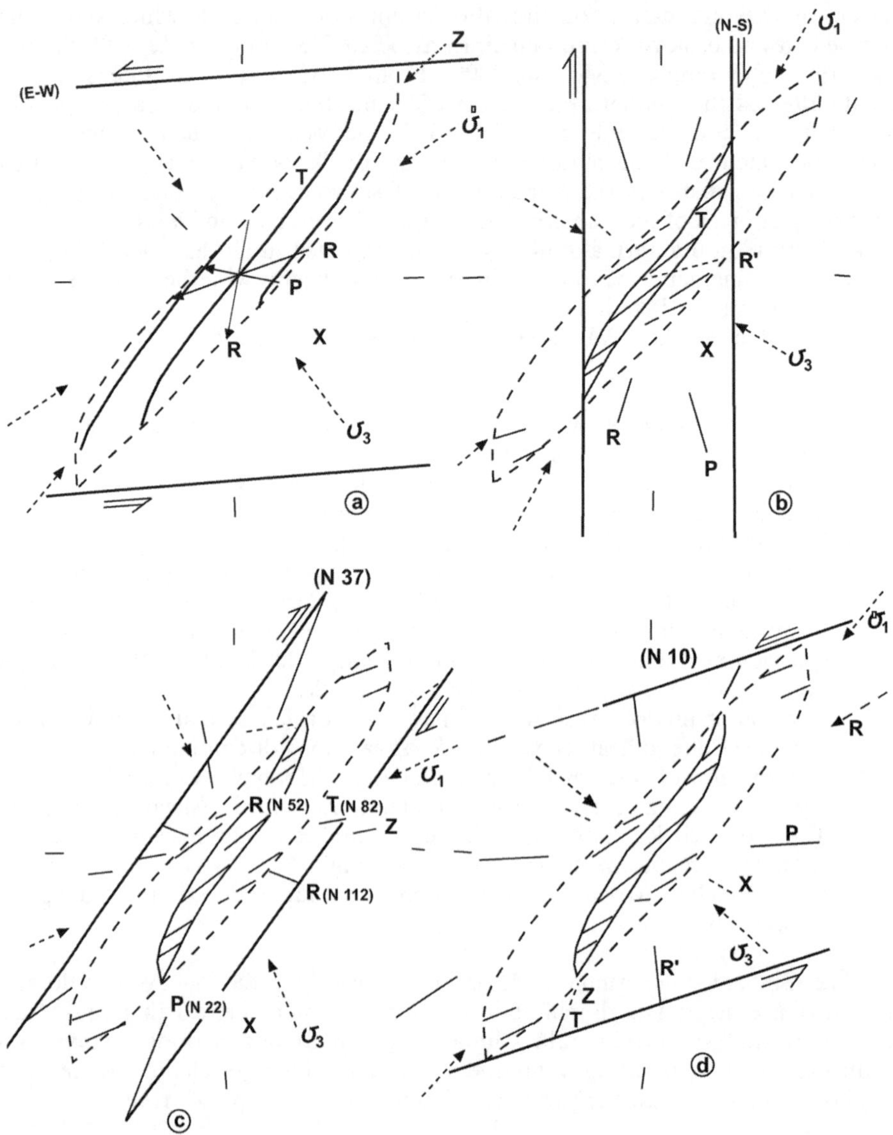

Figure 10. Models of shearing along the main trend found on satellite imagery and the autocorrelation graph. Following the Riedel model on all figures (a, b, c ,d): T = Tension gashes; R and R′ = Conjugate shear faults; P = symmetrical to R, X and Z = Maximum stretching and shortening axes; δ_3 and δ_1 = Tensional and compressive principal stress axes (a) sinistral E-W shear, (b) dextral N-S shear, (c) dextral N37° shear, (d) sinistral N70° shear (Moreau *et al.*, 1987).

a total of 8 earthquakes were registered in the NW of Mt. Cameroon. In 1989 and 1990, 3 earthquakes were felt at Buea, and the 1999/2000 Mt. Cameroon eruption was accompanied by a swarm of earthquakes. Recently felt events reported from the Adamaoua Shear Zone include the 1969 Yoko event, the 1983 earthquake centered around Magba and the Tibati earthquake of 1987.

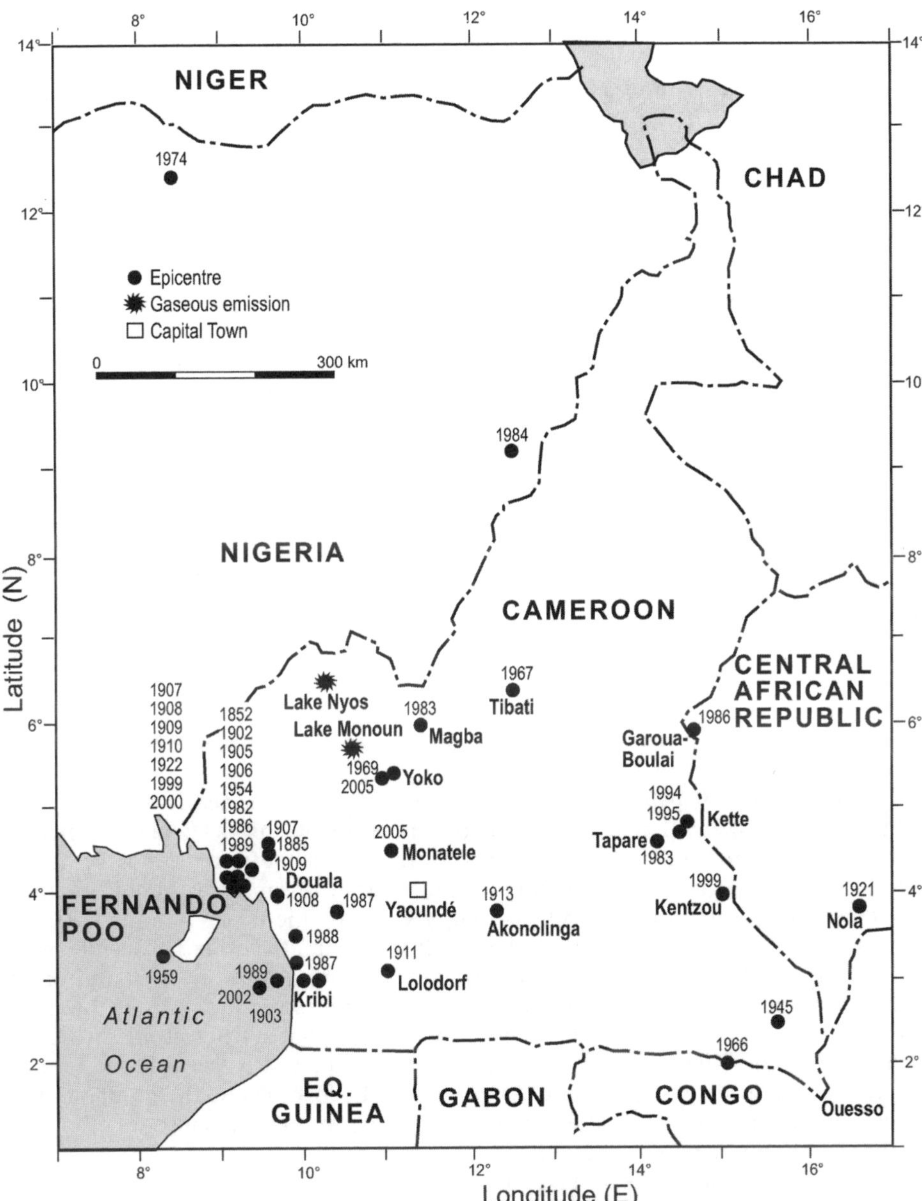

Figure 11. Seismic map of Cameroon (simplified from Nangé, 2007).

Felt seismic events reported from the SSZ include the 1986 Garoua Boulai (east Cameroon) earthquake, the 2005 Monatélé (north of Yaoundé) earthquake and some of the earthquakes reportedly felt at Edéa (coastal region). From 1987 to 1989, the area of Kribi (south of Edéa) was subject to a series of earthquakes. The latest event reported in this area occurred in 2002. Others felt events associated with the geophysical northern margin of the Congo Craton include the 1983 Tapare earthquake, the

1999 earthquake at Kenzou in the Eastern province, and the March 1911 Lolodorf earthquake. In 1913, an earthquake having its epicenter near Akonolinga was widely felt.

The distribution of epicenters of both historic earthquakes and recent seismic events in Cameroon, correlate with the lithospheric Pan-African structures (Figure 11), strongly suggesting that neotectonic activity is responsible for seismicity in the area.

2.7 SUMMARY, DISCUSSION AND CONCLUSIONS

Structurally, the most striking features of the Pan-African/Brasiliano domains as a whole are the large steep shear zones, which extend from the Borborema province to Central Africa and Sudan. Since the Palaeozoic, the Pan-African belt of Central Africa was repeatedly deformed as a consequence of rifting, initiated in the interior of a residual Gondwana continent, and collision events that took place throughout the Mesozoic and Cenozoic. The complex polyphase evolution included: (1) normal faulting and strike-slip structures coeval with the initial opening of the Cretaceous West Central Africa Rift basins; (2) compressional structures coeval with the Late Santonian event registered by the African-Arabian plates, and which occurred in response to changes in the opening direction and rate of the South, Central and North Atlantic Oceans; (3) N–S extension related to the Cenozoic-Quaternary alkaline volcanism of the Cameroon Volcanic Line and significant northward motion of the African-Arabian plate; (4) historic earthquakes and recent seismic events in the region (Cameroon), which show correlations with the Pan-African structures, providing direct evidence for neotectonic activity that may be responsible for seismicity in the area. This evolution implies interactions between active tectonics, deep structures, sedimentary basins and polyphase orogeny.

Topographically, general structure in Western Central Africa is represented by either domes or swells (Adamawa, Cameroon Line) or by basins (Benue, Mbéré and Djérem, Chad and Congo). In Western Central Africa the four islands of Bioko, Principe, Sao Tomé and Annobon, and the four central volcanoes, Mount Cameroon, Manengoumba, Bambouto and Oku, occupy swells of the CVL. Also included are two large seamounts, one between Bioko and Principé and the other between Principé and Sao Tomé. They define a SW–NE straight line and display a "swell and basin" geometry reminiscent of the overall horst and graben structure. All ten began to rise ca. 30 Ma ago. It has been suggested (Morin, 1989) that active tectonics (since the beginning of the Oligocene, ca. 34 Ma) could have played a key role in elevating the volcano-capped swells on the Cameroon line. Unfortunately, the chronology of block faulting in relation to the successive volcanic eruptions and episodes of erosion surface-formation remains uncertain.

Burke and Gunnel (2008) have shown that tectonic controls on denudation and geomorphic history of Africa are of considerable importance at the plate scale. They describe the topographic evolution of Africa tied to the history of the African Surface and use this to illustrate how the continent's distinctive tectonics is reflected in a unique geomorphic history over the past 180 million years. The African Surface is defined as the end product of composite cycles of erosion that started in the Middle Jurassic (ca. 180 Ma) or Early Cretaceous (ca. 125 Ma) with continental breakup events when the Central Atlantic and Indian Oceans began to open. During Late Jurassic and Early Cretaceous times (150–100 Ma), the shoulders of the newly formed, rifted, continental margin around Africa-Arabia were eroded to low elevations. By ca. 100 Ma, in late Albian times, the whole continent had become a low lying surface with little relief, probably not exceeding mean elevations of 0.5 km asl (above sea level) according to Segalen (1967) and Guiraud and Bosworth (1977). This low lying surface persisted with various local and regional perturbations until ca. 30 Ma, when upward flexure of Africa's active swells began. The African Surface is both the continent's most extensive and oldest major erosion surface.

The data presented in this paper show that there was almost continuous tectonic activity in Central Africa. Even so, there was a relatively quiet tectonic setting, which provided, in most of the area of Afro-Arabia situated away from intra continental rifts and rifted margins, an environment for the development, over tens of millions of years (between 100 Ma and 30 Ma), of extensive, deeply weathered erosion of the Africa surface.

The Congo Craton has remained nearly rigid since Pan-African times. The northern border of the Congo Craton and the narrow Pan-African zone, have largely accommodated these dynamic processes. It is suggested that since the Palaeozoic, lithospheric structure and evolution have controlled the development of tectonic topography: vertical movement along major Pan-African shear zones controlled the location, type and evolution of Mesozoic sedimentary basins along the WCARS. Likewise, within plate Cenozoic magmatic activity along the CVL is largely determined by reactivation of Pan-African transcurrent faults. Fault reactivation may be caused by pure extension or transtension (simple shear with a component of extension across the zone). Episodes of structural inversion, lead to differential vertical movements at both regional and local scales. This is associated with significant changes in tectonic topography, landscape evolution and rapid geomorphic evolution of the main fluvial system. Moreover, subsequent thick alluvia trapped inside river basins can be successfully used as a proxy for reconstructions of former climate and vegetation in lower latitudes of Central Africa (see contribution by M. Sangen in this volume).

REFERENCES

Béa, A., Cochemé, J.J., Trompette, R., Affaton, P., Soba, D. and Sougy, J. 1990, Grabens d'âge Paléozoïque inférieur et volcanisme tholéiitique associé dans la région de Garoua, Nord Cameroun. *Journal of African Earth Sciences*, **10**, pp. 657–667.

Benkhelil, J. 1988, Structure et évolution géodynamique du bassin intracontinental de la Bénoué (Nigéria). *Bulletin des Centres de Recherches Exploration—Production Elf-Aquitaine*, **12**(1), pp. 29–128.

Binks, R.M. and Fairhead, J.D. 1992, A plate tectonic setting for the Mesozoic rifts of Western and Central Africa. In Geodynamics of Rifting, Volume II. Case History Studies on Rifts: North and South America and Africa-Arabia, edited by Ziegler, P.A., *Tectonophysics*, **213**, pp. 141–151.

Black, R., Caby, R., Moussine-Pouchkine, A., Bayer, R., Bertrand, J.M., Boullier, A.M. and Fabre, J. 1979, Evidence for late Precambrian plate tectonics in West Africa. *Nature*, **278**, pp. 223–227.

Bosworth, W. 1992, Mesozoic and early Tertiary rift tectonics in East Africa. *Tectonophysics*, **209** pp. 115–137.

Boudzoumou, F. and Trompette, R. 1988, La chaine Pan-Africaine ouest Congolienne au Congo (Afrique tquatoriale): un socle polycyclique char & sur un domaine sub-autochtone formé par l'aulacogéne du Mayombe et le bassin de l'ouest-Congo. *Bulletin de la Société Géologique de France*, **8**, pp. 889–896.

Boulier, A.M. 1991, The Pan-African Trans-Saharan Belt in the Hoggar Shield (Algeria, Mali, Niger): a review. In *The West African Orogen and Circum-Atlantic*

Correlatives, edited by Dallmeyer, R.D. and Lecorche, J.P. (New York: Springer), pp. 85–105.

Browne, S.E. and Fairhead, J.D. 1983, Gravity study of the Central African Rift System: a model of continental disruption. 1. The Ngaoundéré and Abu Gabra Rifts. *Tectonophysics,* **94**, pp. 187–203.

Bumby, A.J. and Guiraud, R. 2005. The geodynamic setting of the Phanerozoic basins of Africa. Journal of African Earth Sciences, **43**, pp. 1–12.

Burke, K. and Gumel, Y. 2008, The African Erosion Surface: A Continental-Scale Synthesis of Geomorphology, Tectonics, and Environment Change over the Past 180 million Years. *The Geological Society of America,* **201**, p. 66.

Burke, K. and Sengör, C.,1986, Tectonic escape in the evolution of the continental crust. *AGU Geodynic,* **14**, pp. 41–53.

Caby, R. 1989, Precambrian terranes of Benin-Nigeria and northeast Brazil and the Late Proterozoic South Atlantic fit. *Geological Society of America Bulletin, Special Publication,* **230**, pp. 145–158.

Caby, R. 2003, Terrane assembly and geodynamic evolution of central—western Hoggar: a synthesis. *Journal of African Earth Sciences,* **37**, pp. 133–159.

Cahen, L., Snelling, N.J., Delhal, J. and Vail, J.R. 1984, *The geochronology and evolution of Africa,* (Oxford: Clarendon Press), p. 512.

Castaing, C., Feybesse, J.-L., Thièblemont, D., Triboulet, C. and Chèvremont, P. 1994, Palaeogeographical reconstructions of the Pan-African/Brasiliano orogen: closure of an oceanic domain or intracontinental convergence between major blocks. *Precambrian Research,* **69**, pp. 327–344.

Castaing, C., Triboulet, C., Feybesse, J.-L. and Chèvremont, P. 1993, Tectonometamorphic evolution of Ghana, Togo and Benin in the light of the Pan-African/ Brasiliano orogeny. *Tectonophysics,* **218**, pp. 323–342.

Cornachia, M. and Dars, R. 1983, Un trait structural majeur du continent africain. Les linéaments centrafricains du Cameroun au golfe d'Aden. *Bulletin de la Société Géologique de France,* **1**, pp. 101–109.

De Wit, M.J., Thiart, C., Doucour, M. and Wilsher, W. 1999, Scent of a Supercontinent: Gondwana ores as chemical tracers—tin, tungsten and the Neoproterozoic Laurentia–Gondwana connection. *Journal of African Earth Sciences,* **28**, pp. 35–51.

Dercourt, J., Ricou, L.E. and Vrielynck, B. 1993, *Atlas Tethys, Palaeoenvironmental Maps,* (Paris: Gauthier-Villars).

Dercourt, J., Gaetani, M., Vrielynck, B., Barrier, E., Biju–Duval, B., Brunet, M.F., Cadet, J.P., Crasquin, S. and Sandulescu, M. 2000, Atlas Peri-Tethys, Palaeogeographical maps. (Paris: CCGM/CCMW), *24 maps and explanatory notes,* **I–XX**, p. 269.

Deruelle, B., Moreau, C., Nkombou, C., Kambou, R., Lissom, J., Njofang, E., Ghogomu, T.R. and Nono, A. 1991, The Cameroon Line: a review. In *Magmatism in Extensional Structural Settings, the Phanerozoic African Plate,* edited by Kampunzu, A.B. and Lubala, R.T. (Berlin: Springer-Verlag), pp. 274–327.

Dingle, R.V., Siesser, W.G. and Newton, A.R. 1983, *Mesozoic and Tertiary Geology of Southern Africa,* (Rotterdam: Balkema), p. 375.

Dumont, J.D. 1987, Etude structurale des bordures Nord et Sud du plateau de l'Adamaoua: influence du contexte Atlantique. *Geodynamique,* **2**, pp. 55–58.

Eisenberg, J. 2009, Morphogenese der Flusseinzugsgebiete von Nyong und Ntem in Süd-Kamerun unter Berücksichtigung neotektonischer Vorgänge. PhD. Thesis (Frankfurt a. Main: Johann Wolfgang Goethe Universität), pp. 1–219.

England, P. and Houseman, G. 1984, On the geodynamic setting of kimberlite genesis. *Earth and Planetary Sciences Letters,* **167**, pp. 89–104.

Fairhead, J.D. 1988, Mesozoic plate tectonic reconstructions of the central South Atlantic Ocean: The role of the West and Central African rift system. *Tectonophysics*, **155**, pp. 181–191.

Fairhead, J.D., Guiraud, R. and Stone, V. 2003, Mesozoic Plate Tectonic controls on Rift Basin Development on North Central Africa: a major Cretaceous Basin Systems. In *AAPG Annual Meeting 2003: Energy-Our Monumental Task*, (Salt Lake City 12–14 April, Abstract), p. 4.

Fitton, J.G. 1980, The Benue Trough and the Cameroon Line—a migrating rift system in West Africa. *Earth and Planetary Science Letters*, **51**, pp. 132–138.

Fitton, J.G. and Dunlop, H. 1985, The Cameroon line, West Africa, and its bearing on the origin of oceanic and continental alkali basalt. *Earth Planetary Sciences Letters*, **72**, pp. 23–38.

Gasse, F. 2000, Hydrological changes in the African tropics since the Last Glacial Maximum. *Quaternary Science Reviews*, **19**, pp.189–211.

Genik, G.J. 1992, Regional framework and structural aspects of rift basins in Niger, Chad and the Central African Republic. In Geodynamics of Rifting, Volume II. Case History studies on Rifts: North and South America, Africa-Arabia, edited by Ziegler P.A., *Tectonophysics,* **213**, pp. 169–185.

Guiraud, M. 1993, Late Jurassic rifting–Early Cretaceous rifting and Late Cretaceous transpressional inversion in the upper Benue Basin (NE Nigeria). *Bulletin du Centre Recherches Exploration—Production Elf-Aquitaine,* **17**, pp. 371–383.

Guiraud, R. and Bellion, Y. 1995, Late Carboniferous to Recent geodynamic evolution of the West Gondwanian cratonic Tethyan margins. In *The Ocean Basins and Margins, the Tethys Ocean,* edited by Nairn, A.E.M., Ricou, L.-E., Vrielynck, B. and Dercourt, J. (New York: Plenum Press), pp. 101–124.

Guiraud, R. and Boswort,W. 1997, Senonian basin inversion and rejuvenation of rifting in Africa and Arabia: synthesis and implications to plate-scale tectonics. *Tectonophysics,* **282**, pp. 39–82.

Guiraud, R., Boswort, W., Thiery, J. and Delplaque, A. 2005, Phanerozoic geological evolution of Northern and Central Africa: an overview. *Journal African Earth Sciences*, **43**, pp. 83–143.

Guiraud, R. and Maurin, J.C. 1991, Le rifting en Afrique au Crétacé inférieur: synthèse structurale, mise en évidence de deux phases dans la genèse des bassins, relations avec les ouvertures océaniques péri-africaines. *Bulletin de la Société Géologique de France,* **162**(5), pp. 811–823.

Guiraud, R. and Maurin, J.C. 1992, Early Cretaceous Rifts of Western and Central Africa: an overview. In *Geodynamics of Rifting,Volume II. Case History studies on Rifts: North and South America, Africa-Arabia,* edited by Ziegler, P.A., *Tectonophysics,* **213**, pp. 153–368.

Janssen, M.E. 1996, Intraplate deformation in Africa as a consequence of plate boundary changes. Inferences from subsidence analysis and tectonic modelling of the Early and Middle Cretaceous period, PhD. Thesis, (Holland: Free University Amsterdam), p. 161.

Kambou, R., Nzenti, J.P. and Soba, D. 1989, Apport à la connaissance des complexes anorogéniques d'âge tertaire de la Ligne du Cameroun: le complexe pluto-volcanique de Tchégui (Nord-Cameroun). *Comptes Rendus de l'Académie des Sciences Paris,* **308**, pp. 1257–1260.

Kankeu, B. and Greiling, R.O. 2006, Magnetic fabrics (AMS) and transpression in the Neoproterozoic basement of Eastern Cameroon (Garga Sarali area). *Neues Jahrbuch für Geologie Paläontologie—Abhandlungen*, **239**, pp. 263–287.

Kankeu, B., Greiling, R.O. and Nzenti, J.P. 2009, Pan-African strike-slip tectonics in eastern Cameroon—Magnetic fabrics (AMS) and structures in the Lom basin and its gneissic basement. *Precambrian Research,* **174,** pp. 258–272.

Kennedy, W.Q. 1964, The structural differentiation of Africa in the Pan-African (= 500 million year) tectonic episode. *8th Annual Report of Research Institute of Geology* (Leeds University), pp. 48–49.

Kilian, C.A. 1931, Des principaux complexes continentaux du Sahara. *Comptes Rendus Sommaire de la Societé Géologique de la France,* **9,** pp. 109–111.

Klitgord, K.D. and Schouten, H. 1986, Plate kinematics of the Central Atlantic. In *The Western North Atlantic Region. The Geology of North America,* edited by Tucholke M.B.E. and Vogt, P.P., *Geological Society of America Bulletin,* pp. 351–378.

Le Franc, J.P. and Guiraud, R. 1990, The Continental intercalaire of Northwestern Sahara and its equivalents in the neighbouring regions. In *African continental Phanerozoic sediments,* edited by Kogbe, C.A. and Lang, J., *Journal of African Earth Sciences,* **10,** pp. 27–77.

Le Marechal, A. and Vincent, P. 1971, Le fossé Crétacée du Sud Adamaoua (Cameroun). *Cahier ORSTOM. Series sciences Géologiques,* **3,** pp. 67–83.

Mascle, J., Blarez, E. and Marinho, M. 1988, The shallow structures of the Guinea and Ivory Coast-Ghana transform margins: their bearing on the Equatorial Atlantic Mesozoic evolution. *Tectonophysics,* **188,** pp. 193–209.

Mascle, J., Marinho, M., and Wanneson, J. 1986, The structure of the Guinean continental margin: implications for the connection between the Central and the South Atlantic Oceans. *Geologische Rundschau,* **75/1,** pp. 587–599.

Maurin, J.C., Boudzoumou, F., Djama, L.M., Gion, P., Michard, A., Mpenda, B.J., Peucat, J.J., Pin, C. and Vicat, J.P. 1991, La chaine protérozoïque ouest-congolienne et son avant-pays au Congo: nouvelles données géochronologiques et structurales, implications en Afrique Centrale. *Comptes Rendues de l'Académie des Sciences* (Paris), Série II, **312,** pp. 1327–1334.

Maurin, J.C. and Guiraud, R. 1990, Relationships between tectonics and sedimentation in the Barremo-Aptian intracontinental basins of Northern Cameroon. In *African Continental Phanerozoic Sediments,* edited by Kogbe C.A. and J. Lang. *Journal of African Earth Sciences,* **10**(1/2), pp. 331–340.

McHargue, T.R., Heidrick, T.L. and Livingstone, J. 1992, Tectonostratigraphic development of the Interior Sudan Rifts, Central Africa. In *Geodynamics of Rifting, Volume II. Case History studies on Rifts: North and South America, Africa-Arabia,* edited by Ziegler, P.A., *Tectonophysics,* **213,** pp. 187–202.

Meyers, J.B., Rosendahl, B.R., Harrison, C.G.A. and Ding, Z.-D. 1998, Deep-imaging seismic and gravity results from offshore Cameroon Volcanic Line and speculation of African hot-lines. *Tectonophysics,* **284,** pp. 31–63.

Moreau, C., Demaiffe, D., Bellion, Y. and Boullier, A.-M. 1994, A tectonic model for the location of Palaeozoic ring complexes in Aïr (Niger, West Africa). *Tectonophysics,* **234,** pp. 129–146.

Moreau, C., Regnoult, J.M., Deruelle, B. and Robineau, B. 1987, A new tectonic model for the Cameroon Line, Central Africa. *Tectonophysics,* **139,** pp. 317–334.

Morgan, W.J. 1983, Hotspot tracks and the early rifting of the Atlantic. *Tectonophysics,* **94,** pp. 123–139.

Morin, S. 1989, *Hautes terres et bassins de l'ouest du Cameroun: Etude géo-morphologique.* Ph.D. Thesis (Talence: University of Bordeaux 3), p. 1190.

Nangé, M.J. 2007, The seismicity of Cameroon. *Sciences et Développement, MINRESI, Cameroun.* **5,** pp. 24–25.

Ngako,V., Affaton, P., Nnange, J.M. and Njanko, J.T. 2003, Pan-African tectonic evolution in central and southern Cameroon: transpression and transtension during sinistral shear movements. *Journal of African Earth Sciences,* **36**, pp. 207–214.

Ngangom, E. 1984, Etude tectonique du fossé de la Mbére et du Djérem, Sud Adamaoua. Cameroon. Bulletin des Centres de Recherches Exploration—Production Elf-Aquitaine, **7**, pp. 339–347.

Njonfang, E. and Moreau, C. 2000, The mafic mineralogy of the Pandé massif, Tikar plain, Cameroon: Implications for a peralkaline affinity and emplacement from highly evolved alkaline magma. *Mineralogical Magazine,* **64**, pp. 525–537.

Njonfang, E., Kamgang, P., Ghogomou, T.R. and Tchoua, F.M. 1992, The geochemical characteristics of some plutonic-volcanic complexes along the southern part of the Cameroon Line. *Journal of African Earth Sciences,* **14**, pp. 255–266.

Olivet, J.L., Bonnin, J., Beuzart, P. and Auzende, J.M. 1984, Cinématique de l'Atlantique nord et central. *Rapports scientifiques et techniques du Centre national pour l'exploitation des océans,* **54**, p. 108.

Petters, S.W. 1991, Regional geology of Africa. *Lecture Notes in Earth Sciences* (Springer-Verlag), **40**, pp. 722.

Poidevin, J.L. 1985, Le Protérozoïque supérieur de la République Centrafricaine. Musée Royal d'Afrique Centrale, Tervuren. *Annales des Sciences Géologiques,* **91**, p. 75.

Popoff, M. 1988, Du Gondwana à l'Atlantique sud: les connexions du fossé de la Bénoué avec les bassins du Nord-Est Brésilien jusqu'a l'ouverture du golfe de Guinée au Crétacé inferieur. In *The West African Connection, Journal of African Earth Sciences,* 7(2), pp. 409–431.

Reeves, C.V., Karanja, F.M. and MacLeod, I.N. 1986/87, Geophysical evidence for a failed Jurassic rift and triple junction in Kenya. *Earth and Planetary Science Letters,* **81**, pp. 299–311.

Runge, J. 2001, Landschaftsgenese und Paläoklima in Zentralafrika. Physio-geographische Untersuchungen zur Landschaftsentwicklung und klimagesteuerten quartären Vegetations- und Geomorphodynamik in Kongo/Zaire (Kivu, Kasai, Oberkongo) und der Zentralafrikanischen Republik (Mbomou). *Relief, Boden, Paläoklima,* (Berlin, Stuttgart: Gebrüder Bornträger), **17**, pp. 1–294.

Runge, J. 2002, Holocene landscape history and palaeohydrology evidenced by stable carbon isotope (δ^{13}C) analysis of alluvial sediments in the Mbari valley (5°N/23°E), Central African Republic. *Catena,* **48**, pp. 67–87.

Runge, J., Eisenberg, J. and Sangen, M. 2006, Geomorphic evolution of the Ntem alluvial basin and physiogeographic evidence for Holocene environmental changes in the rain forest of SW Cameroon (Central Africa)-preliminary results. *Zeitschrift für Geomorphologie. N.F., Supplement.-Band,* **145**, pp. 63–79.

Sangen, M. 2007, Physiogeographische Untersuchungen zur holozänen Umweltgeschichte an Alluvionen des Ntem-Binnendeltas im tropischen Regenwald SW-Kameruns. *Zentralblatt für Geologie und Paläontologie, Teil1,* **1/4**, pp. 113–128.

Scotese, C.R. and Denham, C.R. 1988, User's Manual for Terra Mobilis: Plate Tectonics for the Macintosh. In *Earth in Motion Technologies,* (Houston, Texas) (unpubl.).

Segalen, P. 1967, Les sols et la géomorphologie du Cameroun. Cahiers ORSTOM, Série Pédologie, **5**, pp. 137–187.

Stern, R.J. 1994, Arc assembly and continental collision in the Neoproterozoic East African orogen: Implications for the consolidation of Gondwanaland. *Annual Revue Earth and Planetary Sciences,* **22**, pp. 319–351.

Summerfield, M.A. 1985, Tectonic background to long term landform development in tropical Africa. In *Environmental Change and Tropical Geomorphology,* edited by Douglas, I. and Spencer, T., (London: Allan and Unwin), pp. 281–294.

Thomas, M.F. 2000, Late Quaternary environmental changes and the alluvial record in humid tropical environments. *Quaternary International,* **72**, pp. 23–36.

Thomas, M.F. 2004, Landscape sensitivity to rapid environmental change—a Quaternary perspective with examples from tropical areas. *Catena,* **55**, pp. 107–124.

Toteu, S.F., Bertrand, J.M., Penaye, J., Macaudière, J., Angoua, S. and Barbey, P. 1991, Cameroon: a tectonic keystone in the Pan-African network. In *The Early Proterozoic Trans-Hudson Orogen of North America,* edited by Lewry, J.F. and Stauffer, M.R., *Geological Society of Canada, Special Paper,* **37**, pp. 483–496.

Toteu, S.F., Van Schmus, W.R., Penaye, J. and Michard, A. 2001, New U-Pb and Sm-Nd data from north-central Cameroon and its bearing on the pre-Pan African history of central Africa. *Precambrian Research,* **108**, pp. 45–73.

Unrug, R. 1996, Geodynamic map of Gondwana supercontinent assembly. *Publication Council of Géoscience, Pretoria, South Africa and Bureau de recherches géologiques et minières,* (Orléans, France).

Unrug, R. 1997, Rodinia to Gondwana: the geodynamic map of Gondwana supercontinent assembly. *GSA Today,* **7**, pp. 1–6.

Unternehr, P., Curie. D., Olivet, J.L., Goslin, J. and Beuzart, P. 1988, South Atlantic fits and intraplate boundaries in Africa and South America. *Tectonophysics,* **155**. pp. 169–179.

Van Houten, F.B. 1983, Sirte Basin, North-Central Libya: Cretaceous rifting above a fixed mantle hot spot? *Geology,* **11**, pp. 115–118.

Vellutini, P.J., Rocci, G., Vicat, J.P. and Gion, P. 1983, Mise en évidence de complexes ophiolitiques dans la chaine du Mayombe (Gabon-Angola) et nouvelle interprétation géotectonique. *Precambrian Research,* **22**, pp. 1–21.

Wilson, M. and Guiraud, R. 1992, Magmatism and rifting in West and Central Africa, from Late Jurassic to Recent times. *Tectonophysics,* **213**, pp. 203–225.

Wilson, M., and Guiraud, R. 1998, Late Permian to Recent magmatic activity on the African–Arabian margin of Tethys. In *Petroleum Geology of North Africa,* edited by Mac Gregor, D.S., Moody, R.T.J. and Clark-Lowes, D.D., *Geological Society London, Special Publication,* **132**, pp. 231–263.

Windley, B. 1992, Proterozoic collisional and accretionary orogens. In *Proterozoic crustal evolution,* edited by Condie, K.C. (Amsterdam: Elsevier), pp. 419–446.

CHAPTER 3

Geomorphic evolution of the Nyong and Ntem River basins in Southern Cameroon considering neo-tectonic influences

Joachim Eisenberg

Institute of Physical Geography, Johann Wolfgang Goethe University, Frankfurt am Main, Germany

ABSTRACT: Since the opening of the South Atlantic, the geomorphologic evolution of the catchment areas of the Nyong and the Ntem has been influenced by tectonic processes. Suitable sediment traps, for use as proxy data archives, were identified and used for palaeo environment reconstruction. Geomorphologic forms were investigated by remote sensing and field work with regard to the evolution of the catchment areas. Four geomorphologic formations were recognized: (1) Peneplains and inselbergs, (2) 'demi-oranges' and 'bas-fonds', (3) anastomosing river courses, and (4) a peneplain step having inselberg ranges and V-shaped valleys. In relation to these geomorphologic formations, the resultant data was interpreted from a morphodynamic-genetical viewpoint. Additionally, a lineament analysis was conducted along the peneplain step. Four cluster regions were identified. They are defined by different lineaments reflecting structural directions primarily related to the Pan-African orogeny. Indications relating lineaments to Archaic structures, such as the Ntem faults, could not be completely confirmed. In the synthesis, an attempt to extrapolate results obtained from individual sites of the study region to the region as a whole is made.

The results give an overview of tropical geomorphology and its different forms as they relate to the evolution of the Ntem and Nyong catchments. The combination of remote sensing and field work allows the elucidation of the directions of geological structures, and it was found that remobilization has taken place mostly along the N-S, E-W, and NE-SW directions. In addition, in the Nyong's floodplain, a NNE-SSW direction is documented, which originates in the context of neo-tectonic offsets. The collision of Africa and Eurasia has led to increasing uplift since the Miocene and is the cause of geomorphological modifications, including incision, denudation, and extensional fractures resulting from the rise of the Congo Craton. It is likely that the vicinity of the active Cameroon Volcanic Line and the compression of the Congo Craton by the East African, South Atlantic, and Southwest Indic Rift, triggered the tectonic events. The repeated renewal of the central African hinterland by uplift led to a re-arrangement of the drainage network. The probable sub-recent river capture of the Nyong catchment by the Sanaga River underlines the fact that evolution is continuous and on-going within the Cameroonian hinterland.

3.1 INTRODUCTION

3.1.1 Objectives of the study

Triggered by the opening of the South Atlantic, the drainage network of Southern Cameroon was fundamentally modified. Rifting accompanied by uplift and graben formation connected with faulting at the edge of the graben, were the base for formation of the recent landscape morphology.

Much geomorphological research has been carried out dealing with former palaeoclimates triggering different processes, among them Segalen (1967) and Martin (1967, 1970, 1973), who discussed the evolution of Cameroonian soils and peneplains, Kadomura (1986) and his team, who carried out research on the Last Glacial Maximum (LGM) in Cameroon and its forms, and Kuete (1990), who was part of Kadomura's research unit. His thesis gives deep insights into Quaternary formation processes on the South Cameroonian plateau.

Discussions concerning (neo-)tectonics are mainly associated with recent volcanic activity along the Cameroon Volcanic Line (Ateba and Ntepe, 1997; Suh *et al.*, 2001) and there is only rudimentary knowledge about earthquake activity along the hypothetical border of the Congo Craton (Ateba *et al.*, 1992; Nfomou *et al.*, 2004) and the Sanaga Shear Zone north of the studied basins (c.f. Kouankap, 2006). The study areas are situated on the northwestern swell of the Congo Craton. Certain authors postulate the absence of tectonic activity, due to the thick and rigid cratons (c.f. Bremer, 1989; Wirthmann, 1994; MacKenzie, 2001), however, these have to be scrutinized with regard to several forms suggesting a (neo-)tectonic overprint onto Precambrian structures. Summerfield (1996:14) refers to the fact that little research has been done so far on correlating tectonics and the evolution of the African drainage network.

The main goal of this paper is to discuss morphological forms at several places in the Nyong and Ntem river basins and the processes that triggered their evolution, focusing especially on tectonic events. Local geomorphic evolution will be extrapolated to the respective landscape in order to draw a regional sketch of the whole study area.

In the framework of the ReSaKo project, it is searched for information about geomorphic evolution; typical landscapes and forms are selected for, among other things, further research involving identification of sediment archives for palaeo environmental studies (see contribution of M. Sangen, this volume).

3.1.2 Methods—GIS, field work, and laboratory

Based on multispectral (Landsat 7) and radar data (JERS-1-SAR, SRTM) as well as vector data from the Digital Chart of the World (DCW) and field data, several processing steps were conducted using a Geographical Information System (GIS).

Using SRTM data, the boundaries of the Nyong and Ntem basins were edited manually at a scale of 1:50,000. The resultant outlines were used for Digital Landscape Models (DLM) of the basins by clipping the SRTM data. The DLM provided the basic viewing map and was used to obtain hypsometric curves for the different catchment areas. Hypsometric curves reflect the landscape (Strahler, 1952; Guarnieri and Pirrotta, 2008:267) as seen in Figure 1. The LU1 (Landscape Unit 1) represents a landscape with deep incised valleys and steep slopes. Contrary to LU1, LU2 represents parallel isohypses belonging to a homogeneous landscape having plateau-like (peneplain) or steep relief. LU2.1 is a special form characterized by a peneplain having inselbergs surmounting the plane.

The term lineament was used following the definition of O'Leary *et al.* (1976) as a sub-linear to linear physio-geographical formation of some hundred meters up to several kilometers in length including river courses, scarps, ridges, and also vegetation changes (density or height) as well as combinations of these features. An overview of the lineaments was given in three different scales (SRTM—1:200,000; Landsat 7—1:60,000; JERS-1—1:30,000) in a region coincident with the smallest dataset

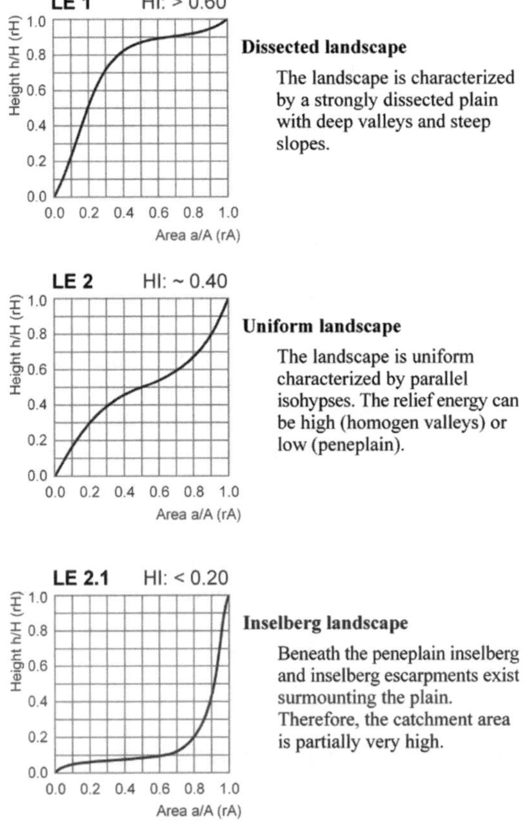

Dissected landscape

The landscape is characterized by a strongly dissected plain with deep valleys and steep slopes.

Uniform landscape

The landscape is uniform characterized by parallel isohypses. The relief energy can be high (homogen valleys) or low (peneplain).

Inselberg landscape

Beneath the peneplain inselberg and inselberg escarpments exist surmounting the plain. Therefore, the catchment area is partially very high.

Figure 1. Examples of different landscape units drawn as hypsometric curves.

of JERS–1. The lineaments were divided into 18 categories due to their azimuth of 10° each.

Additionally river courses were acquired in detail using both Landsat data and topographic maps. According to Bellisario *et al.* (1999) river courses highlight geological structures. These authors have used linear river courses to identify below ground strike-slip faults. It becomes apparent that rivers of a higher order (Strahler, 1957) are aligned to old geological structures and, of a lower order, to sub-recent tectonic activity. However, Ciccacci *et al.* (1987) presume that rivers of a higher order can also reflect the remobilization of older structures. For the main rivers, the whole courses were digitized using a GIS. Afterwards they were subdivided in each linear unit of the GIS' line feature. Two graphs were drawn into one diagram showing both the number of linear features of one azimuthal direction and the sum of all features' length reflecting the same direction. Thereby the structural influence of a river can be highlighted: if the sum of lineaments is higher than the whole length structural influence is low, hence, if the sum of lineaments is lower than the whole length, a structural influence is inferred. Additionally, the drainage density (river course kilometers/square kilometers of the respective unit) was calculated for areas of 25 km² each.

The main courses of the Nyong and Ntem Rivers and also a line connecting source and mouth of the rivers were clipped with the DLM to obtain sketch profiles.

In the floodplain, about 160 cores having a depth of up to 420 cm were taken using an Edelmann-corer by Eijkelkamp. Soil profiles were dug and newly constructed road embankments were documented. Geomorphological parameters e.g., residues of lateritic crusts, outcrops and their weathering status were documented. Inclination and topo-sequences were measured using a Breithaupt inclinometer and a Thommen classic altimeter.

Along wide Precambrian outcrops, found mostly at cataracts, linear structures occurring as small steps, fissures, and partially eroded quartz veins were categorized. Their orientation was recorded using a Breithaupt geologist's compass. Clasts were collected for further laboratory work.

In conjunction with sedimentological studies (see contribution of M. Sangen, this volume) certain mineralogical studies were performed. Rock fragments were cut at *Institut des Recherches Géologiques et Minières* and at Frankfurt University in the mineralogical unit. Thin sections and an element distribution map were made from the most interesting parts of rocks showing fissures and iron coated fragments. Several rocks were characterized by a dark-colored patina whose composition was determined by X-ray diffraction.

Due to missing material, direct datings of the coatings were not possible. However, processes suggestive of wetter or drier climates were related to what was known from long-term palaeo-climatological research in the study area. Additionally, the size of morphological forms was correlated with the duration of evolution. An evident bias towards either climate-based or tectonic-based interpretations makes a detailed temporal placement difficult, and for this reason a balanced approach considering both was used (cf. Watchman and Twidale, 2002).

3.1.3 Tectonics and geomorphology in Southern Cameroon

Tectonics

A look at small-scale topographic maps highlights the river network in the study area, which is mainly linked to Precambrian structures. They were often reshaped, covered, and leveled (Segalen, 1967; Toteu *et al.*, 2004). Krenkel (1939:217) spoke of a "unentwirrbar erscheinende[n] Chaos [von ...] Gebirgsstöcken und Bergkuppen" ["an inextricable chaos of massifs and summits"] in Cameroon. Also nowadays the geological-geomorphological history of Cameroon, especially of the mobile belt north of the Congo Craton, is not completely understood (Toteu *et al.*, 2004:74).

The collision of the West African craton with the East Saharan Palaeoproterozoic block and the central African palaeo-continent during the pan-African orogeny led to the formation of the central African mobile belt ranging from Cameroon to Congo as part of a network connecting all cratons of Africa (Bumby and Guiraud, 2005; Ngako *et al.*, 2003; Toteu *et al.*, 2004). In Southern Cameroon, the Yaoundé series is a meta-sedimentary thrust fault having an E-W trending thrust front thrusted over the Congo Craton. It is the western extension of the Central African Oubanguide series (Feybesse *et al.*, 1998; Maurizot, 2000; Milesi *et al.*, 2006; Owona *et al.*, 2011b; Toteu *et al.*, 1990).

The Nyong and Lokundje series in the South Cameroonian hinterland were thrust-faulted south-eastwards over the Congo Craton during the Eburnean orogeny in the Palaeoproterozoic with a NE-SW stretching thrust front (Feybesse *et al.*, 1998; Owona *et al.*, 2011a; Toteu *et al.*, 2004).

Due to the thrust faults, the exact border of the Archean Congo Craton can be determined only hypothetically by means of the thrust cleavage. Also, during the Pan-African orogeny, large shear zones developed like the Central Cameroon Shear Zone (CCSZ), the Sanaga Shear Zone (SSZ) both trending N70°E, and the South western Cameroon Shear Zone (SCSZ) having a NNE-SSW orientation (Ngako *et al.*, 2003; Toteu *et al.*, 2004; Figure 2).

After the Variscan orogeny around 200 Ma, a long planation period took place leveling the Gondwana continent to the Gondwana Surface. Residuals of this surface are the Mandara escarpment and parts of the Adamawa Plateau in the North of Central

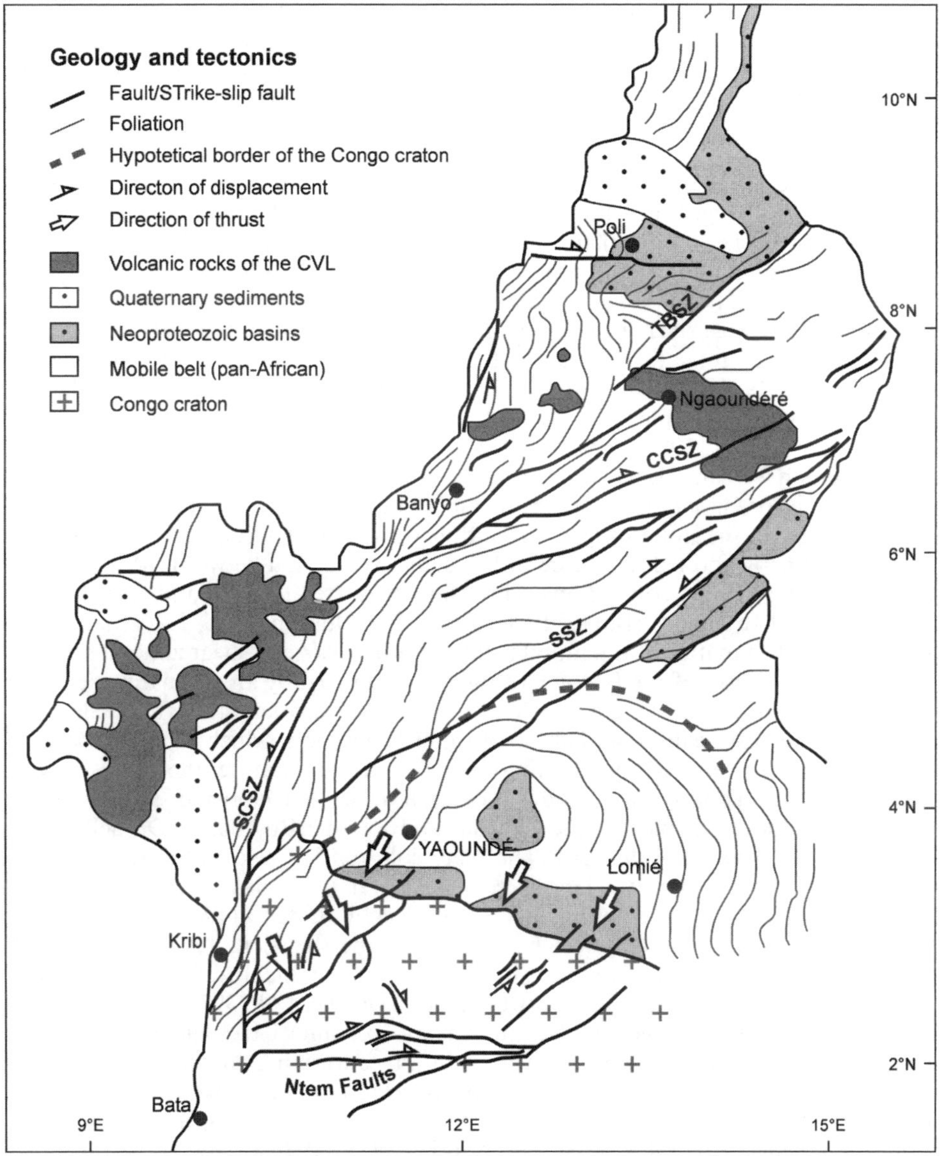

Figure 2. Geological overview showing principal fault structures of the central African hinterland. TBSZ: Tcholliré-Banyo Shear zone; CCSZ: Central Cameroon Shear zone; SSZ: Sanaga Shear zone; SCSZ: South Cameroon Shear zone (modified from Feybesse *et al.*, 1998 and Toteu *et al.*, 2004).

Cameroon (Segalen, 1967). Since Jurassic times the break-up of Gondwana took place along spreading ridges initiated by the formation of triple junctions triggered by both remobilization of Precambrian structures and domal uplifts (Burke and Whiteman, 1973; Summerfield, 1991). In the Lower Cretaceous, the South Atlantic spreading ridge was formed (Robert, 1987:101; Summerfield, 1985), and in the Upper Cretaceous, at around 95 Ma, South America and Africa were entirely separated (Reyment and Tait, 1972). A triple junction located below the Niger delta is responsible for the evolution of the recent West African coast. The Benoue Trough represents one ridge; an aulacogen having an orientation N45°E, which stopped opening due to the faster drift of North-west Africa in relation to South Africa. It was filled with Cretaceous sediments, which became metamorphosed 80 Ma ago (Burke and Whiteman, 1973:747; Keller *et al.*, 1995; Nwajide, 1987; Robert 1987:104). The southward rift led to the opening of the South Atlantic. The westward rift is characterized by a transform fault and the respective Chain-, Romanche- and St. Paul's faults along the West African coast. Next to Cameroon the Kribi fault zone, a transform fault, meets the continental crust, and recently this has triggered earthquakes at the coast and nearby hinterland (Rosendahl and Groschel-Becker, 1999, 2000; Ateba *et al.*, 1992; Nfomou *et al.*, 2004).

The uplift led to denudation processes leveling the Post-Gondwana Surface represented by parts of the Adamawa plateau in Cameroon. It also set up the drainage direction of the large rivers running off the uplifted areas mainly towards the centers of the African cratons, in contrast to the rivers draining directly into the Atlantic (Summerfield, 1991:424). Renewed uplift took place during the Eocene and Early Oligocene triggering the planation to the African Surface I (Burke and Whiteman, 1973; Burke, 1996; Robert, 1987; Segalen, 1967), and also the first intrusive activity along the Cameroon Volcanic Line (CVL) (Suh *et al.*, 2001). Lee *et al.* (1994:121) counted around 17 plutons along the CVL having an age between 65 to 30 Ma (Figure 3).

Triggered by the alpine orogeny around 30 Ma ago, rift velocity of the African Plate reduced distinctly and led to a more or less static connection between the lithosphere and underlying magma convections. The Afar rift started to open and also the CVL began its eruptive history (Burke, 1996; Fitton and Dunlop, 1985; Lee *et al.*, 1994). It was probably a matter of a remobilization of a pan-African shear zone. It seems to be an eastward offset of the Benoue Trough characterized by an alternating sequence of horsts and grabens having a N70°E and N130-140°E orientation (Déruelle *et al.*, 1987). The highest peak is the Fako of Mount Cameroon, which erupted about six times in the 20th century, the last time in 2001. In the Atlantic, the basin between Mount Cameroon and Bioko Island was covered by around 7000 m of sediment since Cretaceous times. It is part of the Douala- and Rio del Rey basin (Déruelle *et al.*, 1987:198; c.f. Murawski, 1980:30). The plate collision led also to the formation of the widespread basin and swell landscape in Africa. In Central Africa, the continental margin underlay an uplift resulting in a further planation period (African Surface II), forming the coastal surface since the Early Miocene (Segalen, 1967; Summerfield, 1985). At the same time, the formation of the Antarctic ice shield additionally increased terrestrial erosion as a result of the lower sea level (Burke, 1996:397).

The term neo-tectonics implies an attempt to connect tectonic activity with a temporal focus. However, the reactivation of Precambrian structures enhances the time span enormously. When talking about neo-tectonics these mixed forms also have to be taken into consideration. Embleton refers to Obruchev (1948; cited in Embleton 1987:2; c.f. Vita-Finzi, 1986:14; Stewart and Hancock, 1994:371) who used the term for tectonic processes occurring since Tertiary times until the Quaternary (Neogen). Panizza and Castaldini (1987:174 et seq.) connect morpho-neo-tectonics with the forms shaped by sub-recent tectonic activity. Regarding the principle of uniformitarianism

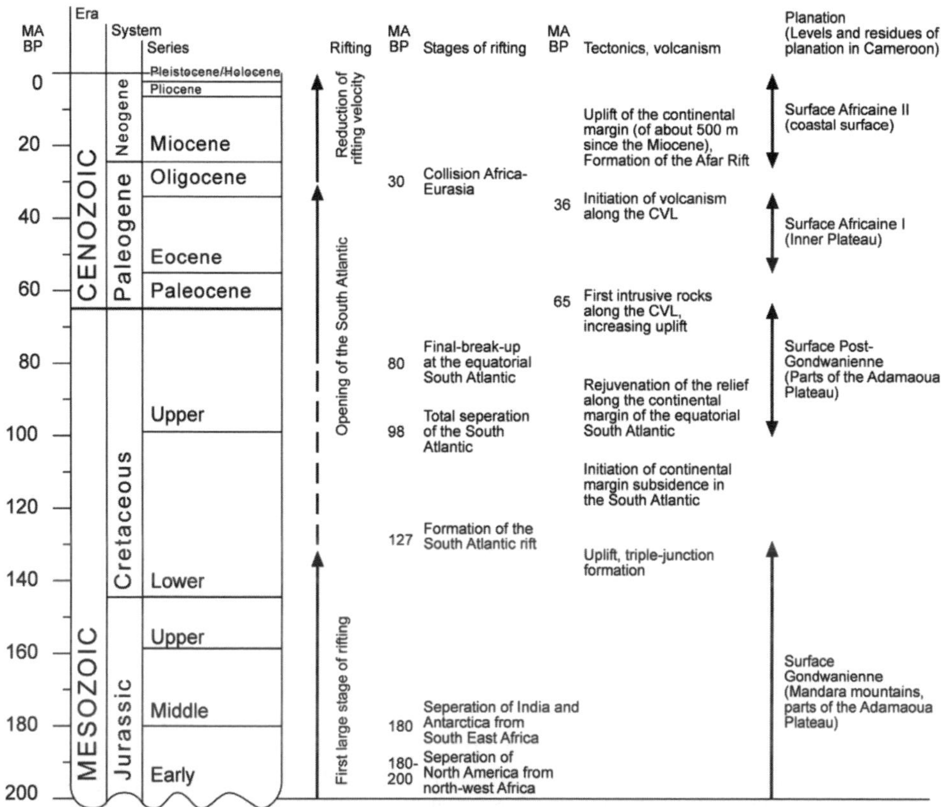

Figure 3. Correlation of rifting, tectonics, and planation periods of Central Africa (after Burke, 1996; Burke and Whiteman, 1973; Reyment and Tait, 1972; Robert, 1987; Segalen, 1967; Summerfield, 1985, 1991).

in geology, such forms in a region without tectonic activity can be used as indicators for activity in the past and probably in the future, too. In geomorphology different methods for the verification of neo-tectonics will be used, depending on the scale of observation: at the regional scale focus is on morphometric measurements and analysis of the river network, at the local scale, focus is on identification of forms shaped by tectonic activity. An important tool is the geomorphologic-neo-tectonical mapping of the study area (Stewart and Hancock, 1994; Panizza and Castaldini, 1987).

In Africa, neo-tectonic impetus is mainly connected with the young East African Great Rift Valley and its south-westward prolongation at the Okavango delta (Fairhead and Girdler, 1969; McCarthy, 1993; McCarthy *et al.*, 1993; Summerfield, 1991, 1996:3). Neo-tectonics at the Central African continental margin will be mentioned with focus on the CVL (Déruelle *et al.*, 1987; Fairhead, 1985; Fitton and Dunlop, 1985; Suh *et al.*, 2001; Ubangoh *et al.*, 1997). Fairhead (1985) refers to a seismological map of Africa edited by Krenkel (1921), which is still current, though, imprecise for regional works due to the global scale. More than 3000 micro-earthquakes were linked to volcanic activity of CVL by Ubangoh *et al.* (1997). Ateba and Ntepe (1997), too, measured earthquake swarms having epicenters 60 km below Mount Cameroon and interpreted them as recent activity in and below the crust. The Sanaga Shear Zone is located closer to the study area. Kouankap (2006) reported an earthquake in 2005 of magnitude

4.6 along the shear zone resulting in cracks in walls and pavements in Yaoundé, which was 65 km from the epicenter. Nfomou *et al.* (2004) used an earthquake in the Atlantic Ocean close to Kribi to draw a risk map of the coast and its hinterland. They assume the Sanaga Shear Zone and the Southwestern Cameroon Shear Zone to be geologic-tectonical units, which were repeatedly remobilized (c.f. Burke, 1996; Ngako *et al.*, 2003; Skobelev *et al.*, 2004). Toteu *et al.* (2004) delineate a reactivation for the whole northern border of the Congo Craton since pan-African times.

The activity trigger is probably the isostatic equilibrium of the northwestern swell of the Congo Craton due to thick sediment layers of the Congo drainage on the craton. Lucazeau *et al.* (2003) assume an uplift of the swell of around 500 m since the Miocene (c.f. Karner *et al.*, 1997; Lavier *et al.*, 2001). In addition, long lasting denudation of the landscape and accumulation on the continental margin led to uplift until recent times (Summerfield, 1996).

Geomorphology

Basic work on geomorphology in Cameroon was done by Segalen (1967). He distinguishes four different peneplain levels according to the pediplanation model of King (1949, 1953). He attributes their evolution to uplift and long lasting planation sequences. Depending on the uplift, a new erosion level was reached from which back wearing took place. The theory of an initial uplift is reminiscent of Davis' geographical cycle (1899).

Burke (1996) considers the complicated Cameroonian topography to be the result of intrusive and effusive activity and, herein lays the difficulty of a classification of different surfaces. However, Segalen (1967) allocates parts of the Adamawa Plateau and the Mandara Mountains between 1300 to 1400 m a.s.l. to the Gondwana Surface and parts of the Adamawa Plateau between 1000 to 1200 m a.s.l. to the Post-Gondwana Surface. The Inner Plateau having heights between 600 to 800 m a.s.l., is part of the African Surface I and the coastal surface is part of the African Surface II. Martin (1967, 1970, 1973) focuses in detail on the Sanaga region, distinguishing two different levels of the Inner Plateau. The lower surface cut into the upper one until the Tertiary. For recent times, he assumes morphological stability. Another location in the hilly Fébé region between the Inner Plateau and the coastal surface, was examined by Embrechts and de Dapper (1987). They identified several pedimentation cycles due to arid periods in the Quaternary. A comparison of the work of Embrechts and de Dapper (1987) and Martin (1973) highlights the differences due to activity and stability on the same surface in the younger geological past (c.f. Burke and Gunnell, 2008).

Geomorphological forms in Cameroon were described as part of the Tropical African Geomorphology Research Project initiated by Kadomura (1977b, 1986, 1989) during the 1970 s and 80 s dealing with palaeoclimatic change and its modifications in the landscape during the LGM and recent times. Kadomura (1977a) describes the Inner Plateau as undulating relief having several hills between 40 to100 m high and poorly drained floodplains in between. He distinguishes between inselbergs characterized by large outcrops and *demi oranges*, a special form having convex slopes and a continuous weathering mantle covering the hill. Kuete was part of Kadomuras working group. In his thesis, Kuete (1990) synthesized research on several localities into a complex view of the Inner Plateau. Another interdisciplinary project was the Tropenbos Cameroon programme (TCP) of the Cameroonian Ministry for Ecology and Forest (MINEF) and the Tropenbos Foundation (1992), which dealt with integrated management of Cameroonian rain forest (Gemerden and Hazeu, 1999). Part of the project involved a geomorphological survey considering slope length and steepness, relief intensity, number of interfluves, and altitude range. However, this project was primarily a descriptive one to identify locations for ecological survey.

Olivry (1986) described a large part of the Cameroonian hydrological network, and only briefly dealt with geomorphic evolution.

3.1.4 Characteristics of the Nyong and Ntem River catchments

In Southern Cameroon two catchment areas of about 25,000 km² exist draining the Inner Plateau and the coastal surface and flowing into the Gulf of Guinea. The basins extend between 9°30' to 13°30'E and 1°30' to 4°30'N. Beside draining the Cameroonian area, they also drain parts of Gabon and Equatorial Guinea. Explaining their geomorphic evolution leads to geomorphic understanding of the area as a whole. As we shall see further on, the coincidence of both catchment areas makes them suitable for comparable research due to their forms and evolution, with but small differences in climate and geology (Figure 4).

Nyong

The second largest river basin of Cameroon after the Sanaga, is the Nyong draining an area of about 27,800 km² and having a length of 690 km. The river has its source at a height of 690 m a.s.l. at the eastern end of the basin on the Inner Plateau and has a mean incline of 1‰ (Olivry, 1986). The incline varies along the longitudinal profile with high values at the peneplain step and also along the coastal surface between Eséka and the mouth, and low values crossing the Inner Plateau. Its largest tributary is the Soo, which has a catchment area of 3120 km² and a length of 233 km; it flows into the Nyong close to Mbalmayo (3°31'N, 11°30'E) coming from the South.

In the upper and middle catchment, the river crosses mica schists, quarzites and gneisses of the Yaoundé series. Fifteen kilometers downstream of Mbalmayo, the

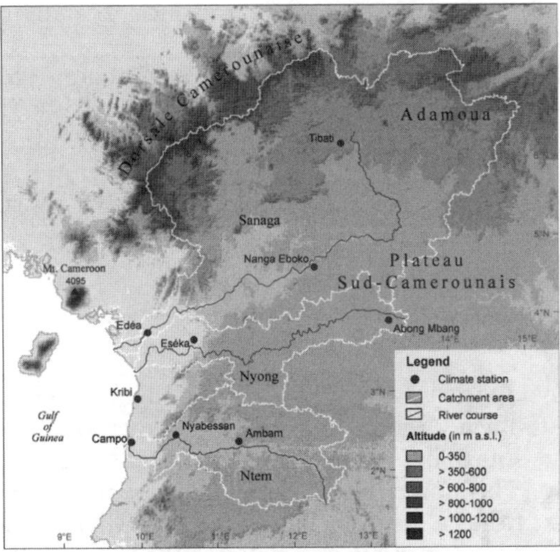

Figure 4. Overview of the study area showing limits and topography of the Sanaga, Nyong and Ntem Rivers (topography based on SRTM data).

Nyong flows about 50 km along the thrust front of the Yaoundé series forming the border between the neoproterozoic series and the Congo Craton (c.f. Feybesse *et al.*, 1998; Maurizot, 2000; Toteu *et al.*, 2004). Close to the peneplain step, the craton's northwestern swell is characterized by neoarchaic amphibolites and quarzites. The coastal surface is formed by rocks of the thrust fault and the Congo Craton. The orientation of the river's course is characterized by linear sections and abrupt knick points suggestive of a structural origin. It is likely that the knicks relate to the crossing of the NNE-SSW oriented Southwestern Cameroon Shear Zone (SCSZ). However, the exact trend of this zone is not mapped in small-scale geological maps of the region. At Dehané (3°29'N, 10°4'E) the river enters a region of Quaternary sediments accompanied by a distinct reduction of the structural trend. The mean yearly discharge of the Nyong River at Dehané amounts to 425 m³/sec (Olivry, 1986).

Olivry (1986:105f.) describes three locations where there are possible river captures to the detriment of the Nyong River: a tributary of the Sanaga Catchment, the Téré River, and two tributaries of the Congo Catchment, the Doumé and Lobo Rivers.

The TREES project (1997) characterized the upper catchment of the Nyong River as swamp grassland. In more detail, Villiers and Santoir (1995) refer to the region as "dépressions marécageuses périodiquement inondées". Actually, the floodplain at and below its source is flooded the whole year (pers. obs. 2007). The vegetation is composed of aquatic grasses (*Echinochloa stagnina*, Santoir, 1995), *Raphia* sp., and swamp forest of around 25–30 m in height (*Sterculia subviolacea* sp., *Macaranga* sp., Villiers and Santoir, 1995; c.f. Kuete, 1990:208). Along the Nyong River, *Sterculia subviolacea* is only known from this place, therefore it is called as endemic basin by Villiers and Santoir (1995). Below Makok (3°58'N, 12°42'E), the vegetation changes progressively to a grass covered floodplain of *Echinochloa pyramidalis,* instead of swamp forest (Kuete, 1990:208; Villiers and Santoir, 1995). Segalen (1967) describes the floodplain as alluvial accumulation having Gleysols distinct from the surrounding Tertiary peneplain. Downstream from Akonolinga (3°46'N, 12°14'E), the floodplain becomes narrower within the evergreen and semi-deciduous forests (Villiers and Santoir, 1995). Sometimes it is missing completely. The coastal plain is covered by evergreen lowland rainforest which undergoes a transition to mangrove vegetation just before the mouth at the Atlantic (Letouzey, 1985).

The tropical humid climate (A2h; Lauer and Frankenberg, 1987) is characterized by changes in precipitation and rainfall due to different altitudes a.s.l. and the distance to the coast. In the whole catchment, there are two distinct rainy and dry seasons. A long rainy season runs from August to November, a shorter one from April to June. The long dry season lasts from November to April, and the shorter one from June to August. Rainfall increases from 1674 mm at Abong Mbang close to the source, to 2836 mm at Kribi located about 35 km south of the mouth.

Ntem

The Ntem basin drains approximately 31,000 km² (Olivry, 1986:201) covering three countries: Equatorial Guinea, Gabon, but mostly, Cameroon. The river's length is about 460 km having its source in Gabon at a height of 700 m a.s.l. and its mouth into the Atlantic at Campo. The Lower Ntem represents the border between Equatorial Guinea and Cameroon. It has a mean incline of 1.52‰ (Olivry, 1986), however, there is a distinct knick point at the peneplain step between the Inner Plateau and the coastal surface marking a divergence in the river's ordinary longitudinal profile. The largest tributary is the Kom followed by the Mvila, Biwomé and Ndjo'o Rivers (Olivry, 1986).

The source and upper catchment are located on the Inner Plateau. As a result of planation, the Archaic basement was capped, and gneisses, granites, charnockites, and quartzite chains crop out, which have to be crossed by the river. After receiving its primary tributaries, the Ntem crosses the peneplain step and flows onto the coastal plain with its neo-Archaic and Palaeoproterozoic metamorphic rocks, before reaching the Atlantic Ocean. The mainly linear river course highlights the structural influence of the hydrological network (Segalen, 1967; c.f. Maurizot, 2000). On the coastal plain, the Ntem divides into two branches forming a large island called Dipikar about 340 km^2 in size. Below the confluence, the Ntem enters Cretaceous and Quaternary deposits. The mean yearly discharge of Ntem River at Nyabessan (still above the peneplain step) amounts to 440 m^3/sec (Olivry, 1986). Further data from gauging stations at the mouth are not available.

The vegetation of the periodically inundated floodplain in the upper catchment is composed of swamp forest (Segalen, 1967; compare satellite imagery with those of Nyong upper catchment). The swamp forest occurs also in the middle catchment along with evergreen and semi-deciduous forests (Villiers and Santoir, 1995). The TREES project (1997) focused on secondary forest species due to resulting from recent anthropogenic influences. They do not mention the swamp forest, probably because of missing data and their small-scale resolution (1:500,000).

The tropical humid climate (A2h; Lauer and Frankenberg, 1987) is characterized by two rainy and dry seasons. A very humid season having high precipitation lasts from September to November, a less humid rainy season lasts from March to July. The long lasting dry season runs from November to March, and the shorter one from June to September (Olivry, 1986; c.f. Wiese, 1997). Rainfall increases from 1670 mm at Ambam, where the Nyong crosses the border between Gabon and Cameroon, to 2696 mm at Campo at the mouth. As a result of missing data, there is no information available to make precise climatological statements respecting the headwaters in Gabon.

3.2 RESULTS AND INTERPRETATION

3.2.1 Regional view on South Cameroonian geomorphological features

Drainage density

Drainage density (Dd) was calculated for three different landscape units in each basin as a function of their representativeness of the respective landscape as source region, peneplain step, and coastal hinterland. The drainage density was calculated for a quadratic surface of 25 km^2. Each landscape unit consisted of 400 km^2.

The lowest mean density was found in the headwaters of the Nyong and Ntem River basins (1.08 Dd, resp. 1.14 Dd) increasing towards the peneplain step (1.59 Dd, resp. 1.61 Dd) and the mouth (1.77 Dd, resp. 1.82 Dd), respectively.

At first glance, it would appear that Dd increases due to higher rainfall at the coast. However, the drainage network is different in the three units according to the structural settings. The headwaters are characterized by the planation surface with *demi-orange* relief intersected by large hydromorphic floodplains. Hence, huge amounts of water are stored in the floodplain without a direct impact on hydrological structure. The river network changes at the peneplain step where several linear units hint to the structural framework of the basement. The highest Dd exists in the coastal hinterland characterized by a flat landscape, which is strongly defined by the geological structures. The sampled unit within the Nyong catchment lies at the transition from basement to

Quaternary deposits. This transition becomes visible also within the unit as evinced by pixels having higher Dd along the basement and lower Dd along the Quaternary deposits (Figure 5). In the Ntem Catchment, the transition is located closer to the coast; hence, it will not become obvious in the discussed landscape unit. When we compare both basins it is clear that rainfall affects Dd. The closer the basin is situated to the equator, the higher the Dd in the respective basin.

Longitudinal river profiles and hypsometric curves

The Nyong River has a mean incline of 1‰ composed of several features differing from an ordinary longitudinal profile (Figure 6). Starting from the source until the peneplain step, the river overcomes 70 m on a section of around 480 km. The incline of 0.07‰ represents the planation surface characterized by a flat and undulating relief. The river crosses the peneplain step across four knick points, which are the *Chutes de Mpoume, Chutes de Makai, Chutes de Njok*, and *Chutes de Mbombo Ngouima*, with a mean incline of 6.25‰. Also, the Nyong's tributaries have several knick points at the height of the main step, reflecting resistance of the basement rocks. A last knick point is represented by the *Chutes de Dehané*, below which the Nyong is navigable for 40 km until its mouth.

The hypsometric integral (HI) of 0.47 reflects the nearly horizontal trend of the hypsometric curve of the Nyong in its upper catchment. A distinctive knick point of the curve represents the peneplain step. It is very difficult to allocate the curve to a landscape unit (LU). It is appropriate to consider the curve as a compilation of two units divided at the afore-mentioned knick point. The upper one corresponds to

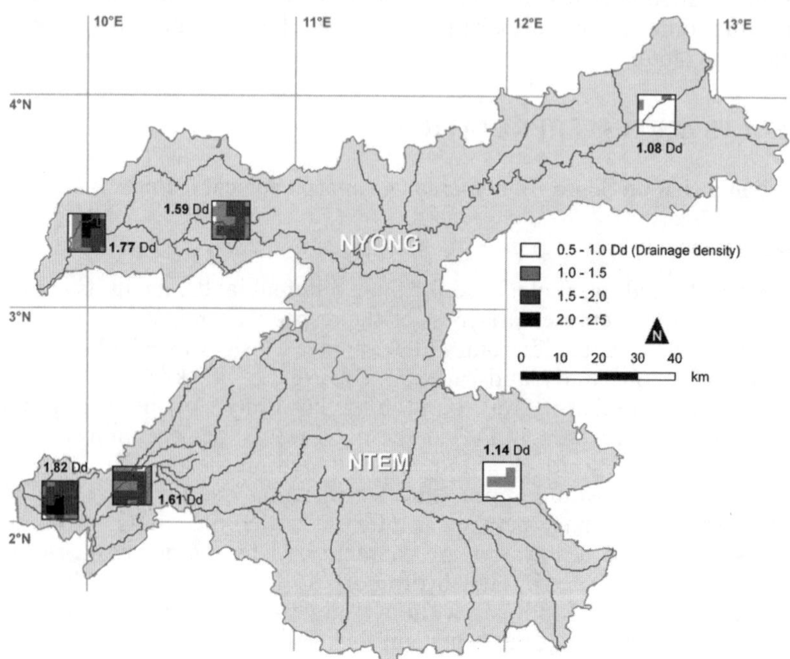

Figure 5. Drainage density (Dd) of the upper, middle, and lower catchments of the Nyong and Ntem, in comparable units (based on respective regional topographical maps).

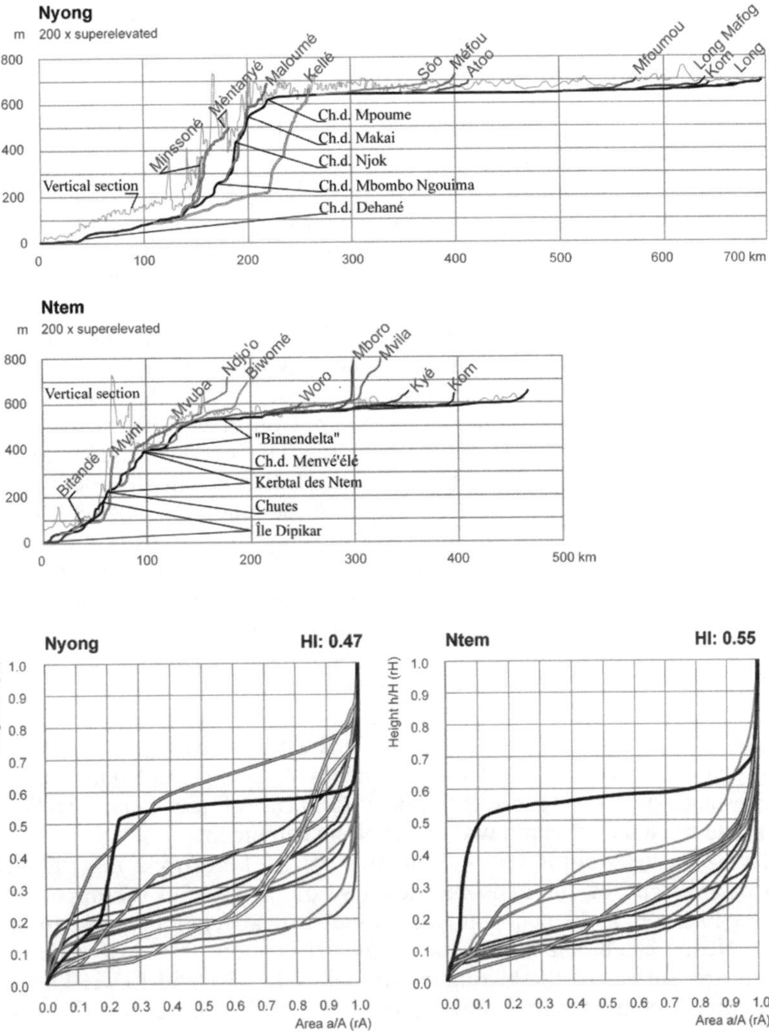

0.26 Kellé	0.24 Atoo	0.23 Bitandé	0.16 Woro
0.26 Minssoné	0.14 Mfoumou	0.23 Mvini	0.17 Mboro
0.35 Mèntanyé	0.31 Long Mafog	0.32 Mvuba	0.19 Kyé
0.57 Maloumé	0.24 Kom	0.36 Biwomé	0.15 Kom
0.26 Soo	0.38 Long	0.29 Ndjo'o	0.22 Quellfluss Ntem
0.13 Méfou	0.29 Quellfluss Nyong	0.18 Mvila	

Figure 6. Longitudinal river profiles and respective hypsometric curves of the Nyong, Ntem and selected tributaries, respectively, as well as longitudinal landscape profiles connecting the rivers' source and mouth (based on SRTM data).

LU 2.1, reflecting the planation surface and its surmounting inselbergs; the lower one corresponds to LU 1, hinting to the accentuated relief of the coastal surface before crossing the Tertiary and Quaternary deposits. The tributaries support this view and reflect the landscape more in detail. In the upper Nyong catchment, they correspond to LU 2 progressing to LU 2.1 the nearer their confluence is located to the peneplain step, with values around 0.38 (Long) to 0.13 (Méfou). Below the step, the tributaries correspond to LU 1. The closer the confluence is to the Nyong's mouth, the more they correspond to LU 2 (Figure 1).

The mean incline of the Ntem River amounts to 1.52‰. In the upper catchment, the profile is almost similar to the Nyong profile reflecting the planation surface of the Inner Plateau, although already 160 km below the source a first knick point appears at the confluence with the main tributary, the Kom. It is probable that it is induced by the increasing amount of water, and therefore, increasing erosional activity. However, the step occurs just before the confluence, like the mouth of a hanging valley of the receiving stream, suggesting a tectonic origin. The same form occurs at the confluences with Mboro and Woro Rivers below.

The tributaries draining the catchment south of the main valley are marked by smooth and regular longitudinal profiles. They have their source 50 to 100 m higher than the receiving stream, reaching their middle catchment after a short section and then flow after a long flat section into the Ntem. Conversely, the northern tributaries are characterized by a more accentuated profile. Having their source 200 to 250 m higher than the receiving stream, they cross this difference in altitude by one or two distinct steps in their profile before flowing into the Ntem. Just before the peneplain step, the Ntem fans out to an interior delta showing small steps in the watercourses. The delta terminates in the *Chutes de Menvé'élé*. They are the onset of a deep V-shaped valley along which the main river crosses the peneplain step about 450 m height having a mean inclination of 4.5‰. A few knick points become obvious below the *Chutes de Menvé'élé*, and referred to simply as *Chutes* in the topographic map of the region. Moreover, Olivry (1986) did not refer to names within the lower course.

On the coastal surface, the two branches of the Ntem show two distinct knick points at the same height of 80 and 30 m a.s.l. The northern branch crosses the second step just before the confluence of both branches. At 5 km, the Ntem crosses Quaternary deposits before flowing into the Gulf of Guinea. A flat undulating relief is nearly absent in the coastal hinterland within the Ntem catchment.

The HI of the Ntem River amounts to 0.55 (c.f. Leturmy *et al.*, 2003) reflecting the horizontal trend of the hypsometric curve of the river in the upper catchment. Due to the location of the peneplain step, which is relatively more close to the mouth, as in the Nyong catchment, the value of the HI is about 0.07 higher. However, the curve is nearly similar to the Nyong composed of LU 2.1 and LU 1. The tributaries of the upper catchment have a low HI of around 0.15 (Kom) to 0.22 (source river of the Ntem) reflecting the LU 2.1. In contrast, the tributaries of the lower catchment around the peneplain step and below are accentuated with reference to LU 2. The nearer the confluence of the tributary and main river is to the Ntem's mouth, the lower the HI.

Discussion

Longitudinal profiles and hypsometric curves point out the landscape's character. The high HI of both river courses is a result of the step between the Inner Plateau and coastal surface. If the hypsometric curve is split into two curves, the relief features of the landscape become visible. A detailed view on the hypsometric curves of the tributaries highlights the character of different locations along the receiving streams.

Both basins are characterized by uniform and flat undulating planation surfaces in the headwaters, with low relief and aquatic to sub-aquatic floodplains. The network of the valleys is structured by elongated *demi-oranges* with a convex declivity into the floodplain. In the lower catchment, the landscape alters from the peneplain step to the coast. Directly below the step, the landscape is characterized by an upland relief consisting of rough hills and strongly incised river courses in the basement. The terrain becomes flatter towards the coast with a distinct shift between basement and sedimentary rocks in the hinterland.

The drainage density accentuates the morphology. Ruhe (1952) describes increasing drainage density with increasing age in a periglacial landscape in the retreat area of glaciers. Distinct from his system, the low drainage density of the Inner Plateau highlights the rather old age of the planated terrain. It seems that the influence of the parent rock on the drainage network decreases with increasing age of a landscape unit; this is seen at the peneplain step and coastal surface directly below the step, with rectilinear river courses tracing the structures of the geology. However, also in the headwaters, linear river courses occur beneath meanders as prominent hydro-geomorphological features. The dependence of the river network on the geological structures is more or less omnipresent.

3.2.2 Zoom in—local view of Southern Cameroon

Nyong River on the Inner Plateau

The term Inner Plateau was defined primarily by Segalen (1967:137) referring to heights between 600 and 800 m a.s.l. However, in its upper catchment, the Ntem River flows for the most part below the 600 m a.s.l. mark, unlike the Nyong. Nevertheless, Segalen assigns the Ntem to the Inner Plateau. Whereas both rivers cross the peneplain step, this distinct knick point in the rivers' gradient curve should be taken as edge of the Inner Plateau in the West.

A further distinct step occurs between the Nyong and Ntem catchments. The upper catchment of the Nyong River is bordered by the Sanaga catchment in the North and the Boumba and Dja catchments in the East, in this way, the upper catchment acquires its own plateau-like character in the North of the Inner Plateau. The river orients along a smooth syncline (Kuete, 1990), which is also obvious in the longitudinal profiles of the Nyong's tributaries.

The Nyong catchment and the Inner Plateau are not totally congruent (Figure 7). The dendritical system orients towards the syncline; however, some tributaries do not reach the Nyong. The DLM (having an altitude resolution of a few meters) identifies a barrier hindering the Téré River draining into the Ntem; rather, it flows into the Sanaga, crossing a deeply incised valley, which contrasts with the flat planation surface of the Inner Plateau. The transition between wide floodplain and small valley is abrupt and coincides with a distinct bulge into the Inner Plateau (4°25'N, 12°35'E). The upper catchment of the Téré has a size of about 950 km². Below the barrier, the floodplain is as large as above without corresponding to a source region as can be seen at other tributaries of the Nyong River in the region. Similar phenomena become obvious not only at the Téré catchment but also at further locations at the northern, eastern, and southern rim of the Nyong's catchment, but they do not reach the same size as that of the Téré's. If the sizes are added up it results in 1213 km² for three regions draining all into the Sanaga, 156 km² for two regions draining into the Congo, and 97 km² for six regions draining into the Nyong. Overall, an area of 1446 km² exists having a dendritical network oriented towards the Nyong River without draining into its catchment.

The source region of the Nyong is characterized by marshy depressions (Villiers and Santoir, 1995) having stagnant waters and swamp forests situated between "demi-oranges" and a few bulging inselbergs. The vegetation is mainly composed of *Raphia* sp. and *Sterculia subviolacea*—trees about 25–30 m tall, having a bright grey

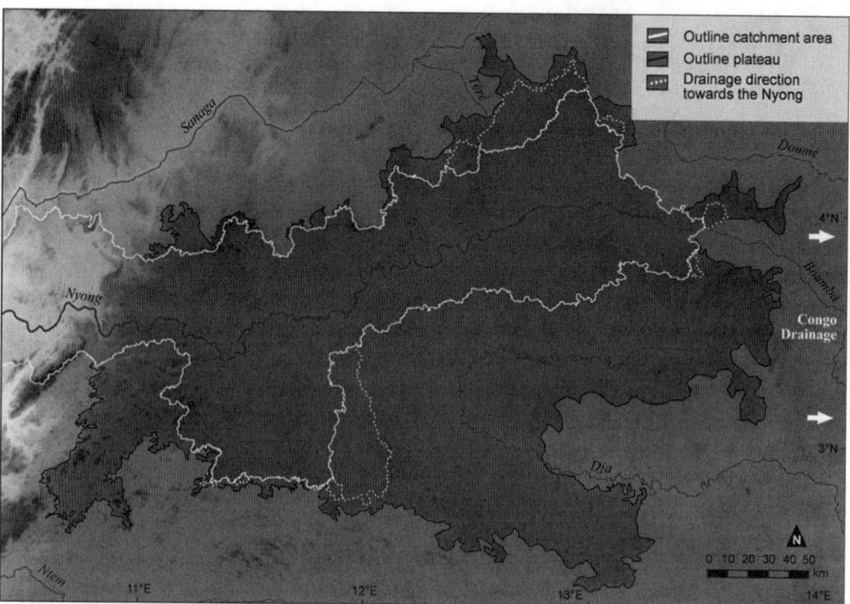

Figure 7. Overview of the upper Nyong catchment (white line), entire plateau (black line), and areas having a dendritic network oriented towards the Nyong but draining into another receiving stream (based on SRTM data).

trunk; they appear in the Landsat imagery as a dark blue to green cloudy texture (dark grey cloudy texture of Figure 8). The main channel of the Nyong River appears as dark to black-colored meanders in the center of the floodplain, which has some light red colors (light grey; Figure 8) reflecting *Echinochloa stagnina* (Santoir, 1995), an aquatic grass; this grass is a pioneer species of more or less unperturbed water bodies with or without low current. At Abong Mbang, the width of the floodplain varies between 800 and 1500 m. The southern river bank has been altered by fishermen and laundresses, and shows less vegetation and pisolites (iron concretions) sedimented in the bank. A coring at the river bank shows a pisolitic horizon at 60 cm depth marking the transition between the underlying saprolite and the covering hill wash sediments. An attempt to recover material in the Nyong's floodplain failed due to the high water content. However, the depth of the river bed was determined to be 4 m at about 20 m away from the bank. Only some pisolites remained in the core, pointing to their deposition in the near-shore river bed, too. There is less stagnant water in the tributaries the further they are from the confluence, hence further corings could have been made in the floodplain of two tributaries showing sandy to clayey dark brown sediments over saprolitic material. The thickness changes between 50 to 80 cm, and represents mainly shallow sediments, which form the tributaries' floodplains (Figure 8).

Downstream from Makok (3°58'N, 12°42'E), the vegetation changes to semi-aquatic grasses (primarily *Echinochloa pyramidalis*, Villiers and Santoir, 1995) and there is high moisture penetration only during the rainy season. Additionally, the cyan and green colors of the Landsat image (Path 185, Row 057, 18.05.2000) indicate grasses. The width of the floodplain reaches up to 4500 m. The more uniform the colors, the less anthropogenic influences can be detected. The black-colored curvaceous line represents the main river channel. Dark grey and dark blue areas highlight abandoned channels and inundated areas hinting to the rainy season during which the image was

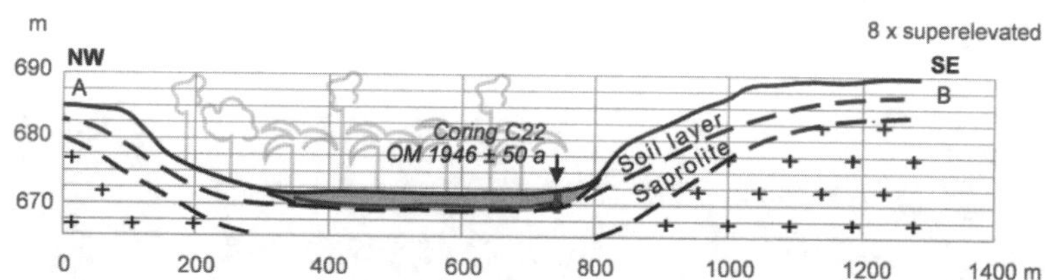

Figure 8. Abong Mbang region, transect AB in the West of the region, and photo of current vegetation at the upper catchment of the Nyong (map based on Landsat 7 imagery).

taken. Beneath meandering river sections, straight courses occur oriented in an ENE and NNW direction. In contrast to the Nyong's floodplain, here tributaries are covered with swamp forest reflecting the same morphology as at the headwaters at Abong Mbang. The greyish green floodplain is set off against the reddish green forest signature of the watershed's semi-deciduous forest, a site of higher humidity. Sometimes the swamp forest invades the main floodplain as a gallery forest. Another Landsat image taken during the dry season (Path 185, Row 057, 16.01.2002) shows symmetrical brown areas reflecting burned grasses in the floodplain. A coring was performed in the alluvial plain at Ayos, it contained sandy to clayey brownish dark sediments having an age of 679 ± 28 cal. yrs BP at a depth of 280 cm, suggestive of sub-recent processes of river migration. Moreover, a cut through a *demi-orange* contiguous to the floodplain was inspected; it showed a regolith layer of about 6 m thickness composed of 4 m saprolite material and a further 2 m reddish soil layer covering the mica-schist basement rocks (c.f. Figure 48). The basement descends just before the floodplain.

A comparable landscape could be observed at Akonolinga consisting of semi-aquatic grasses in the floodplain, characterized by cyan and green colors of the false color Landsat image (light grey, Figure 9). The northern slope up to Akonolinga is steep, similar to the *demi-orange* morphology of Ayos. The southern bank is build by

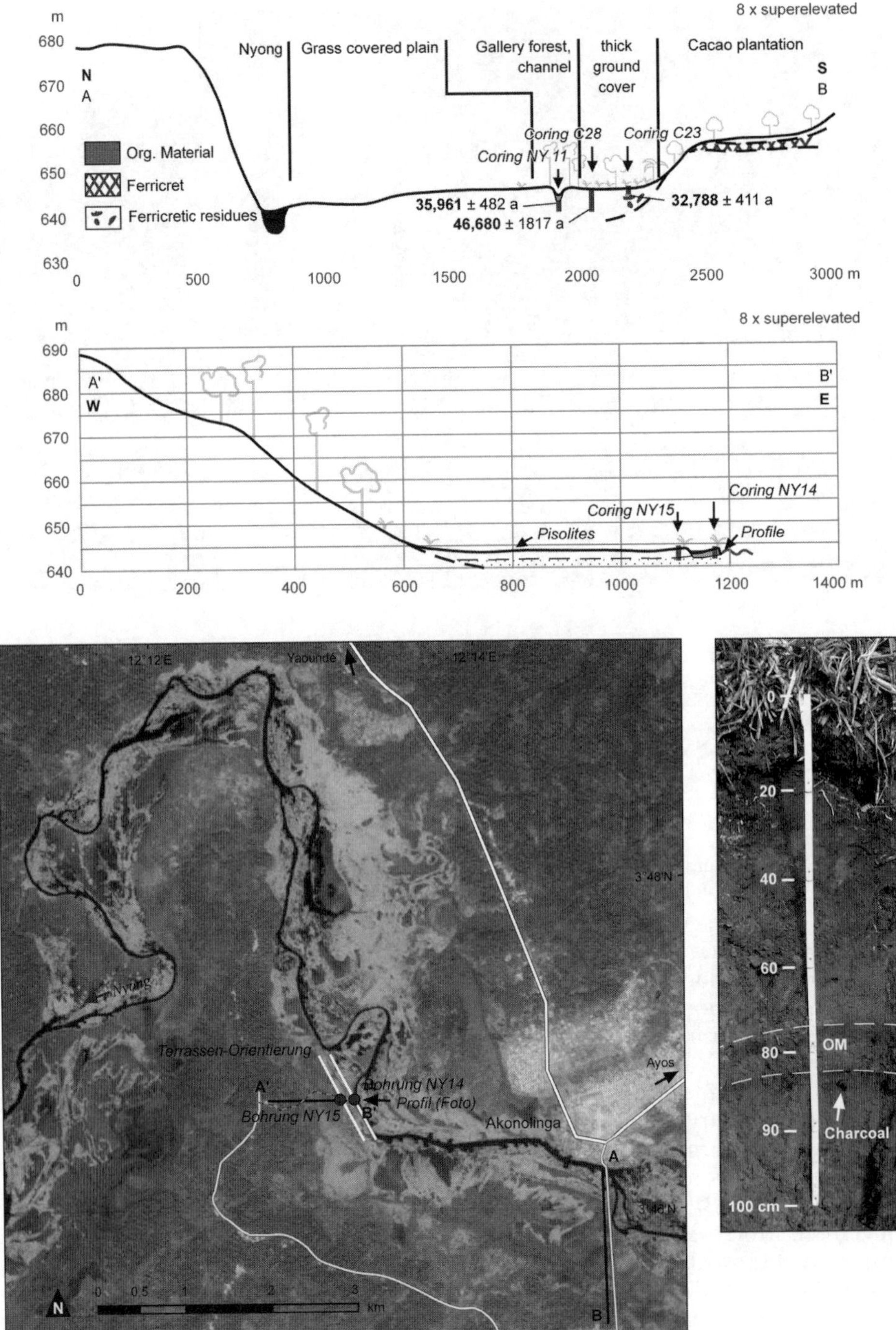

Figure 9. Akonolinga region, transects AB in the East and A'B' in the West, and photo of the river bank at the end of transect A'B' (map based on Landsat 7 imagery, photo by M. Sangen).

a terrace 11 m above the Nyong's floodplain. A coring on the terrace reveals a 10 cm thick soil cover over a lateritic crust, which could not be penetrated by the corer. Kuete (1990:206) describes the crust as an iron crust having particles of the basement's strong weathered mica schist, which contrasts with a deep weathered oxidation soil at the other valley border. The terrace is part of a catena along the floodplain and three additional corings were made. Between the southern bank and the Tetar, a small tributary channel of the Nyong, sediments were found in a depth of 420 cm having a maximum radiocarbon age of 46.7 cal. kyrs BP (C28); this age was not found elsewhere in the Nyong catchment. A further coring was done directly in the dried-out channel. It seems that it was incised into the Nyong's floodplain, which had already been subject to long lasting soil formation characterized by oxidation mottles in a clayey environment. The core reveals a sandy layer at the base becoming finer with oxidation mottles until a depth of 180 cm. The oxidation layer is superimposed by a sandy layer, which probably represents the basis of the incised channel. Once more, the granulation becomes finer up to clay and the color becomes a dark brown at the top. Sediments for dating could only be recovered at a depth of 280 cm, these show an age of 36 cal. kyrs BP (NY11). A last coring was made (C23) close to the southern bank; it was characterized by a clayey texture with oxidation mottles. The core was limited to a depth of 420 cm by a lateritic crust residue. At 150 cm, a layer of organic sediments could be dated to an age of 32.8 cal. kyrs BP. It is assumed that the residue had fallen into the floodplain before sediments covered it. Nearby, a large ferricretic block was observed at the surface, also close to the bank, in a small area of swamp forest inundated by a source. About 3 km west of Akonolinga, two terraces were identified directly at the Nyong. They have a sub-linear to linear appearance in the field and on the satellite image, respectively. The upper one has a height of 250 cm (NY15) and is build by a substrate of sand at the base, turning successively to clay at the top with oxidation mottles, similar to the site south of Akonolinga. The lower terrace having a height of 220 cm, consisted of sand at the base but thereafter mainly of brownish dark, loamy material (NY14). The terraces height increases smoothly to its fronts; therefore, they are covered by taller vegetation because there is less inundation.

The landscape changes once more at 27 km west of Akonolinga. The floodplain becomes narrower, sometimes characterized only by the river course, supported by the satellite data. Cyan and green colors (light grey, Figure 10) representing grassy vegetation decline and red colors of secondary forest occupy the river's surroundings (Path 185, Row 057, 16.01.2002; medium grey color, Figure 10). The change is accompanied by modification of the river's course for about 30 km SSW, after a *de novo* modification to the West. Kuete (1990:467) describes quartzite chains having a meridional orientation, which are crossed by the Nyong at this location. However, the geological map (Maurizot, 2000) does not refer to these chains, probably due to the small scale which was used. In contrast to the aforedescribed landscape, the river forms a wide meander to the West close to Edjom. The southern border is marked by a geological transition of mica schists to quartzitic epi-schists. A resistant net of grasses cover greyish clay soils alternating with water covered surfaces. The remains of charred tree trunks testify to anthropogenic burning activities during the dry season. The satellite image shows a floodplain 3 km in width. It is characterized by several units divided by linear sections passing parallel to the river (Figure 10). Sediment of a coring taken in the floodplain reveals an age of 16.3 cal. kyrs BP (260–280 cm) and 13 cal. kyrs BP (220 cm).

At the level of Nkolmvondo (3°23'N, 11°54'E) the Mvondo and Essaa Rivers flow into the Nyong. About 3 km before its mouth, both rivers form a wide branch, which is incongruous to their small catchment area. In this region, further tributaries of the Nyong have a distinctly smaller stream and floodplain than the Mvondo and Essaa. Together both rivers form an island having a large floodplain before their confluence with the Nyong (Figure 11). The soil of the floodplain is characterized by sand at the bottom and clayey layers, up to 60 cm at the top. Organic sediments for dating were not found. The plain is covered by grasses and woody shrubs. Black charred trunks of dead trees testify to anthropogenic burning activities. The Mvondo River is described as lacustrine having stagnant waters and local people say there is an abundance of fish. Towards the mouth, aquatic grasses cover the water so that all you can see is a small open channel. The Mvondo has a nearly linear border at the West formed by magmatitic gneisses oriented towards the NNW-SSE.

Until its confluence with the Soo River (3°22'N, 11°27'E), the large floodplain along the Nyong is suspended.

Along the sub-aquatic floodplain, the Nyong's channel is easy to identify. The river has formed meanders, however, they occur erratically and are characterized in part by linear to sub-linear sections suggestive of a structural origin. Figure 12 contains all flow directions of the Nyong from its source to its confluence with the Soo

Figure 10. Edjom region and photo of the floodplain taken in the East of the region (map based on Landsat 7 imagery).

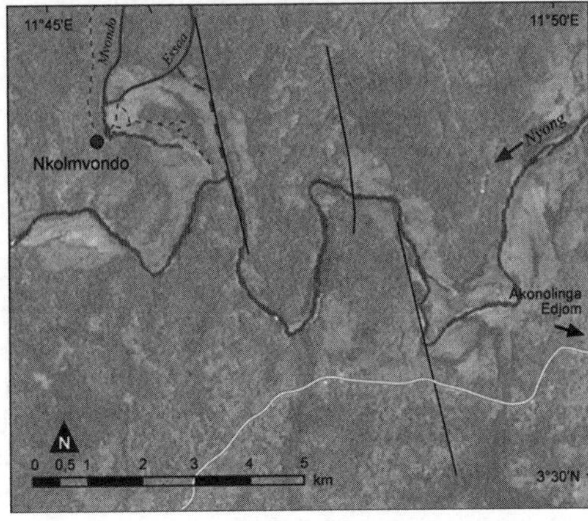

Figure 11. Nkolmvondo region, faults after Maurizot (2000) (map based on Landsat 7 imagery).

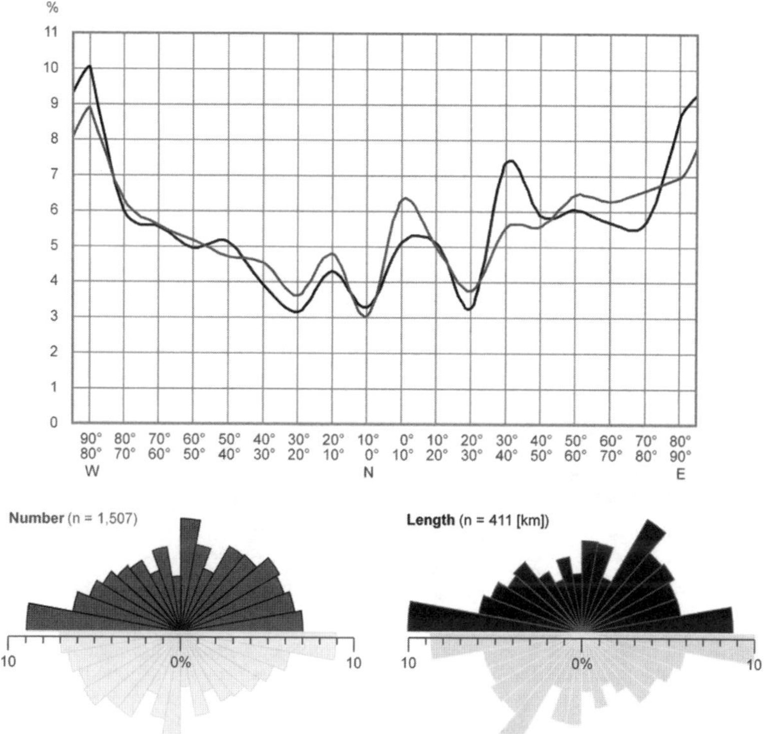

Figure 12. Flow directions of the Nyong in its upper catchment, subdivided into number (grey) and length (black) of the river sections (flow directions were edited using Landsat 7 imagery).

River, highlighting three sub-peaks at N/NE (N10–20°W, N0–10°E, and N30–40°E) and the main peak in a E-W orientation (N80–90°E, N80–90°W) (Figure 12).

Ntem River on the Inner Plateau

The Ntem River is oriented along the Ntem faults (c.f. Feybesse *et al.*, 1998:166), which extend from Southern Cameroon to the Central African Republic. The Kom River, the largest tributary of the Ntem, and the Ayina River, a tributary of the Ogooué, also align along this fault direction (c.f. Santoir, 1995). The Ntem catchment consists of a wide syncline, limited in the North by the plateau drained by the Nyong and Dja Rivers, and in the South by a plateau drained by the Mbini River. The Nyong catchment's border in the West is difficult to identify due to a lack of a distinct step. Landsat imagery shows a floodplain of up to 1650 m width having a cloudy texture and darkish green to red colors. Also on the DLM, a channel occurs only partially, but clearly points to the Ayina/Ogooué catchment in the East. On the image a dark curved feature 5 km below the watershed points to the distinctive water course of the Ayina River. At the level of Ayina's wide floodplain, the dendritic river network orients towards the Kom River against the flow direction of Ayina, and covering an area of 1602 km² in total. The same phenomenon can be observed at the Ntem's source river. Two areas of 31 km² and 10 km², which are drained by the Ogooué, have their dendritic network oriented towards the Ntem (Figure 13).

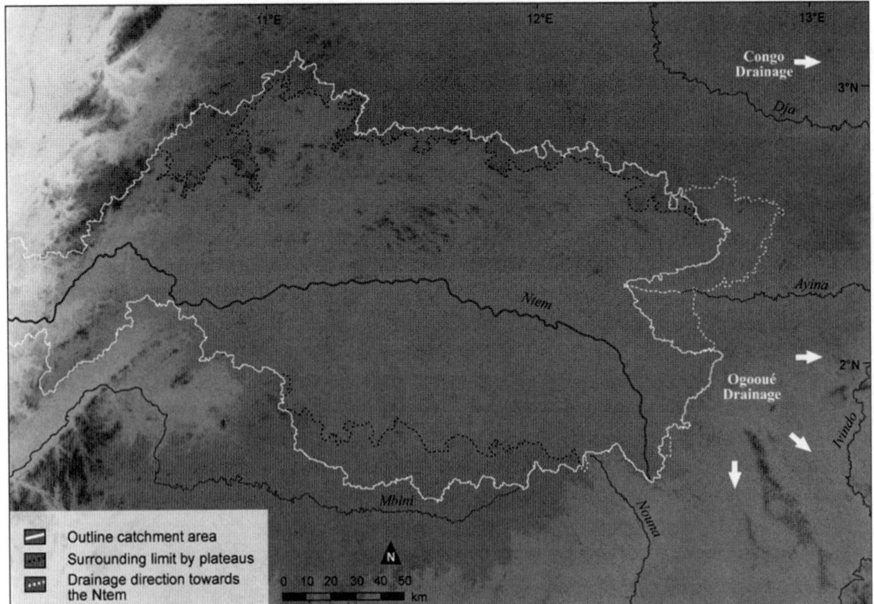

Figure 13. Overview of the upper Ntem catchment (white outline), and areas having a dendritic network oriented towards the Ntem but draining into another receiving stream (white dotted line), (based on SRTM data).

Kuete (1990:240) describes the whole upper catchment area of the Ntem as an unfinished planation surface having irregularly eroded hills. In fact, the main river course subdivides the catchment into two geomorphological units. The northern unit is characterized by inselbergs made of the strongly heterogeneous basement (banded gneiss, gabbro, granite); they are about the height of the North Yaoundé region (around 1050 m a.s.l.), these inselbergs break the monotony on the flat landscape. The step to the northern plateau passes in a WSW-ENE direction parallel to the Ntem's main course and this continues also in the Ebolowa region (see below), which is characterized by several inselbergs of different height. The southern unit is more homogeneous and has only a few inselbergs surmounting the slightly rising planation surface to the South; these reach maximum heights of 790 m a.s.l. The transition to the plateau in the South is not as obvious as in the North. The plateau is mainly drained by the Mbini River and only in parts by the Ntem. The hydrologic network of these parts is dendritic and oriented towards the Mbini, and it shows an abrupt directional change towards their northern receiving waters.

The Ntem has its source above a distinct step between the Inner Plateau in the North and a wide planation surface in the South, drained by the Ogooué. Along 1 km, height varies 200 m. The source is located at 700 m a.s.l. The hydrological network of the Ntem and Ogooué tributaries is different. Thirteen kilometers downstream, the Ntem's floodplain has a width of 600 m, and also the tributaries are characterized by large floodplains. The rivers draining into the Ogooué have finely dissected the landscape (Figure 14).

The cloudy green to red texture of the Landsat imagery (Path 185, Row 58, 02.03.2001) of the Ntem's floodplain hints to a swamp forest as described by Olivry (1986) for the upper catchment. Further 10 km downstream, the floodplain widens up to 1500 m and a main channel manifests until the confluence with the Kom River. From there on, the width of the floodplain reduces slightly until it is abandoned

(2°10'N, 11°59'E). The diagram from which the obviously structural affected river course was drawn, shows the main peak in a E-W orientation (N80–90°W, N80–90°E) (Figure 15).

The Ntem's largest tributary is the Kom River, they have their confluence 140 km downstream of the main river's source. The Kom catchment area is separated from the

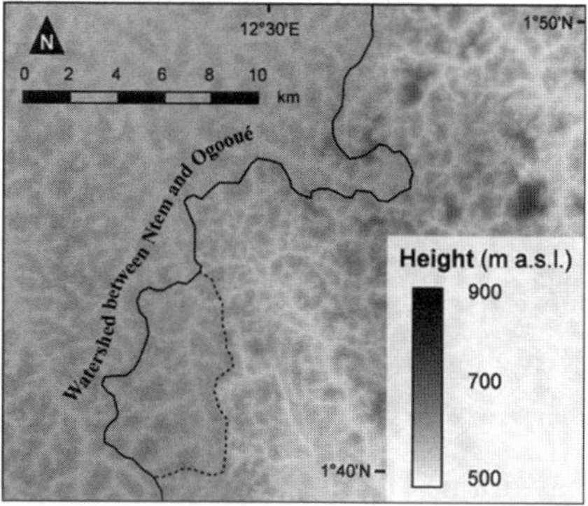

Figure 14. Watershed between Ntem and Ogooué catchments (based on SRTM data).

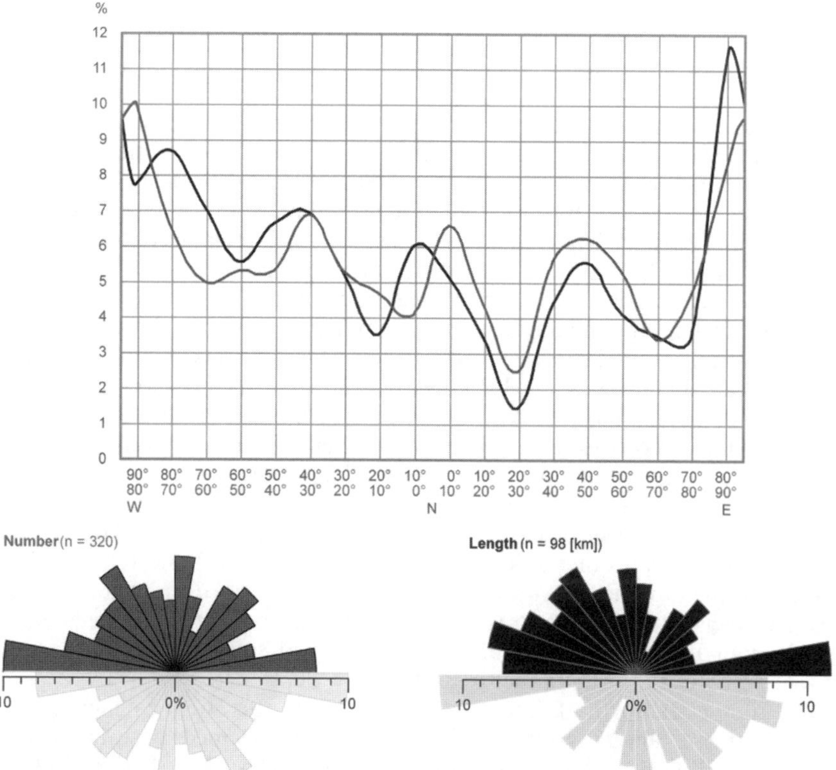

Figure 15. Flow directions of the Ntem in its upper catchment, subdivided into number (grey) and length (black) of the river sections (flow directions were edited using Landsat 7 imagery).

Ntem's by a linear chain of inselbergs of 120 m maximum, surmounting the plateau. Calculating the catchment area at their confluence, the Ntem has a value of 7128 km² and the Kom of 6576 km²—almost equal areas. The longitudinal profile of the Ntem shows a knick point at the confluence with the Kom. If the source region of the Ntem was replaced by the Kom, the whole longitudinal profile is much more congruent with the ordinary profile of a river in the tropics probably in relation to the structural connectivity with the Ntem faults.The Kom area could not be visited during field studies. However, the observation of Landsat and DLM data shows a homogenously planated landscape having inselbergs and *demi-oranges* surmounting the surface up to 400 m. The differences in height are most obvious in the Nlobo catchment, a river flowing into the Kom just before its confluence with the Ntem. The Nlobo is characterized by wide hydromorphic valleys in its upper catchment draining the Mvangan plateau, which is limited by the Nyong plateau in the West, North and East. The hydrological net joins at Mvangan (2°39'N, 11°44'E) following a geological structure to N10°E (c.f. Kuete, 1990:249). At a scale of 1:50,000, the satellite imagery shows a somewhat dark meandering line, the linear character of which becomes obvious at a scale of 1:200,000. At its lower catchment, the Nlobo crosses a gneissic chain causing strong relief.

The Mvangan plateau drained by hydromorphic valleys was formed over Archaic charnockite. The transition to the mostly linear feature is exactly accompanied by a geological transition to gneisses (c.f. Maurizot, 2000). The floodplain in the Kom's source region has a width of 350 m open up to 1500 m, 44 km below its source and close to the watershed with the Ayina (see above). Downstream, width progressively decreases. At the ferry of Akoabas (2°17'N, 12°03'E), the floodplain has a width of 500 m and on satellite imagery it appears greenish red, which is only slightly different to the reddish brown of the surrounding rain forest. The Kom's tributaries also show a width of about 500 m just before their mouth formed between the *demi-oranges* of the region. Beside sporadic meanders, the Kom's river course displays linear units suggestive of a structural origin that are mainly to N80–90°W and N80–90°E (Figure 16).

Some kilometers north of the Ntem a profile of several meters was excavated along the Trans-African Highway. At the base, gneissic rocks were formed by spheroidal weathering with fissures filled by saprolitic material (Figure 17). The saprolite also occurs sometimes as a small fringe covering the rocks. The material is covered by a ca. 100 cm thick layer, probably accumulated in the framework of road construction. Further onwards, the rocks descend below the surface. At this location, the saprolite is covered by a light clayey layer which delimits a horizon having several pisolites (iron concretions) in a much more darkish brown, clayey matrix. Pisolite content increases slightly until the top of the layer. The pisolitic layer is covered by yellow-brown, silty sediments. Only at the edge is the cover cut and the pisolites occur at the top. Another profile shows the same layer sequence. The surface at the top of the profiles represents the undulated planation surface of the region having smooth hills and depressions.

The pisolitic layer sometimes runs parallel to the surface. However, it mainly runs discordant to the top; the transition to the sediments is formed by an accentuated line having deep incisions, which contrasts with the surface. In this region, the Ntem passes slowly through a small floodplain consisting mainly of sandy sediments. Large basement rocks crop out directly in the river, at the low depth here.

Ten kilometers north of Ebolowa, the border of the Ntem catchment is represented by a step of about 150 m in height. The town is framed by several inselbergs. Southeast of Ebolowa, there is an inselberg called Ako'Akas. Kuete (1990:243ff.) describes it as a granitic rock having distinct indentations likely related to different planation phases in the region. Large karren were formed out of the monolith,

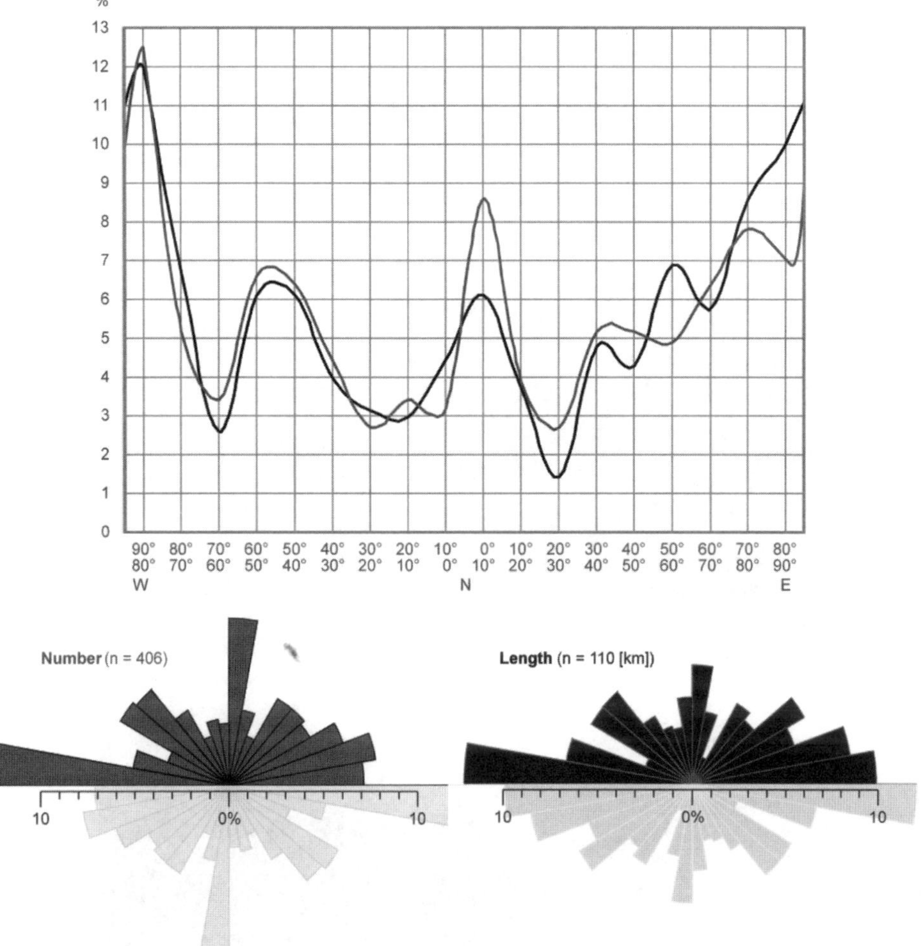

Figure 16. Flow directions of the Kom in its upper catchment, subdivided into number (grey) and length (black) of the river sections (flow directions were edited using Landsat 7 imagery).

the reflection of pseudocarst processes resulting from high rainfall and related weathering (Figure 18).

Peneplain step

The peneplain step between the Inner Plateau and the coastal surface does not consist of a clearly formed step along a few kilometers. A much wider spectrum of forms characterize the step. Between Kribi in the North and the mouth of the Ntem at Campo in the South, it runs nearly parallel to the coast 35 km inland. North of Kribi, the parallelism stops due to the Tertiary and Quaternary sediments of the Sanaga and Nyong Rivers. The rise up to the Inner Plateau of around 500 m occurs along 30–65 km. Sometimes in front of the step, sometimes on the step, NE-SW (N30–40°E) stretching escarpments occur, reaching heights of 1000 m a.s.l. The valleys inbetween are prominent. The lower course of the Mbini in the South runs through a basin of

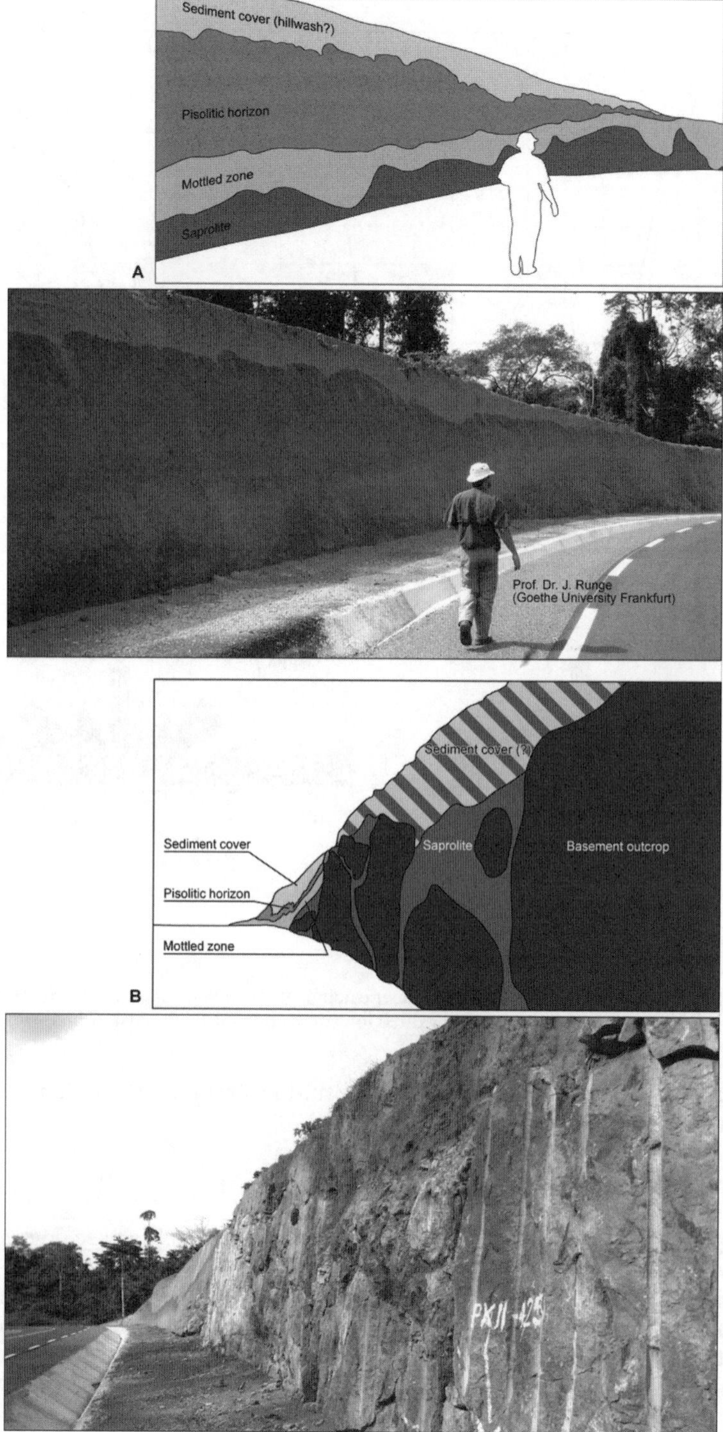

Figure 17. Two exposure sites along the Trans-African Highway roadside, close to Ambam.

Figure 18. Karren at the Ako'Akas inselberg.

15 km in width and 100 km in length (Benito rift) limited by escarpments in the North and South. The escarpment dividing the Mbini and Ntem is strongly dissected and extends towards the Inner Plateau in a wide arch. On the plateau, the Ntem's river course orients along this form. North of the Ntem, the Nkolebengue escapment forms the watershed and is characterized by a lower grade of dissection. Both ridges are built by coarse-grained charnockitic material.

In the South, the Nyong catchment is limited by the Ngovayang escarpment, which is a very prominent ridge built of amphibolits and gneisses of the Neoproterozoic basement extending in the same direction as the Nkolebengue escarpment. Small rivers drain the area between these two escarpments. The Kienke River, particularly with its source at the peneplain step and it flow into the Atlantic at Kribi, has modeled terraces into the basement, a fact that becomes obvious in the DLM. East of the ridges, the step is formed as a distinct edge, hence it is frayed following the basement structures (Figure 19).

Directly above the peneplain step, the planation surface is characterized by several inselbergs surmounting the surface up to 300 m. In the Nyong catchment, the occurrence of inselbergs decreases with distance to the step. In the Ntem catchment, the occurrence of inselbergs was already been described (see above).

Lineament mapping

Mapping faults and main structural lineaments directly in the field is difficult (Kuete, 1990:312). Haman (1976) remarks how lineament mapping helps highlight the primary structural orientations of a region. For this reason, lineament mapping using DLM-, SAR-, and multi-spectral data was conducted to obtain a structural geomorphologic sketch of the study area (Table 1).

Three thousand six hundred and forty-one lineaments were mapped and the following were distinguished: linear depth contours, footslopes, river courses, contour

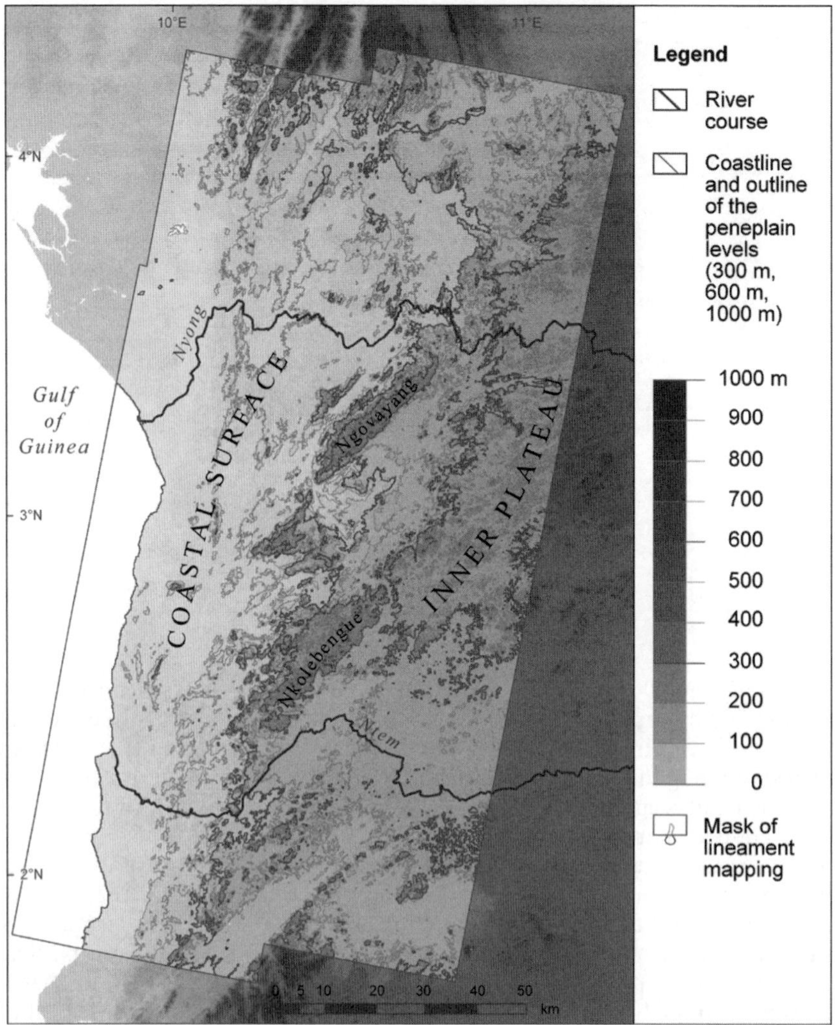

Figure 19. Overview of the peneplain step highlighting the lineament analysis study area (based on SRTM data).

lines, summit alignments, and steps or terraces. A combination of features having a linear appearance were referred to a unique group, however, it was not sub-divided due to several possible combinations. Some of the depth contours were also linear occurring as indentations into ridges. River courses were only matched to this group if a river was marked at the same place on the topographic map. However, there are several depth contours embodying a water course not referenced in the map either as a result of limited scale or missing field studies by the cartographer. All lineaments were divided automatically into units of 10°, each with the same orientation; hence, the number of lineaments was increased to 5534. Density maps were extracted for each angle range.

Later a density map was made that considers all lineaments, highlighting lineament clusters; these are areas showing an accumulation of lineaments. In this way, the

Table 1. Edited lineaments (number and percentage).

ID	name	number	%
1	thalwegs (t)	3808	68.8
11	t_prominent	349	
12	t_obvious	3174	
13	t_visible	285	
2	footslopeline (f)	230	4.2
21	f_obvious	158	
22	f_visible	72	
3	watercourse (w)	188	3.4
31	w_linear	158	
32	w_sub-linear	30	
4	ridgelines (r)	249	4.5
41	r_prominent_ridgeline	132	
42	r_summit_lineation	117	
5	terraces (te)	47	0.8
51	te_prominent	25	
52	te_visible	22	
6	combinations	1012	18.3
		5534	**100**

perimeter of the regional occurrence of linear structures can be determined and these may be interpreted with respect to the whole distribution pattern (Figure 20).

Four regions, characterized by values of up to 1.2, are circumscribed on the peneplain step: the southern outliers of the Cameroonian swell (Dorsale Camerounaise) in the upper North of the study area (R1), the Ngovayang escarpment (R2), as well as the Nkolebengue escarpment (R3), and the ridge dividing the Ntem and Mbini catchments in the South (R4) (Figure 21). West and east of the step, the abundance of lineaments decreases related to the plateau-like character of the landscape, and the identification of linear units along river courses only. Depth contours or linear sequences of summits are less visible. A detailed view of R1 is mainly defined by N-S orientations (N0–10°W, N0–40°E). Moreover, the lineament value "n" is the highest of all for these orientations. The overview of the study region (Figure 19) identifies primarily the tributaries of the Sanaga, which cross the region from North to South and have formed the mostly linear ridges. The Ngovayang escarpment (R2), is mainly characterized by clusters having a NE-SW orientation (N40–50°E, N50–60°E). However, the northeastern border shows E-W oriented lineaments (N70–90°E). This part of R2 is severed from the escarpment by a 10 km wide syncline. Due to the small separation and the different lineament directions, region 2 was divided into R2a (Ngovayang escarpment) and R2b (Inner Plateau). Also the Nkolebengue escarpment (R3) has E-W oriented lineaments (N60–90°W, N80–90°E). A further cluster shows a northeastern direction (N50–60°E). This orientation occurs also south of the Ntem in region R4. Further clusters orient towards N30–40°E and N40–50°E. The high lineament value "n" associated with high mean length highlights the bulging shape in the relief. Under the four main regions, another area delineated is on the eastern Ntem

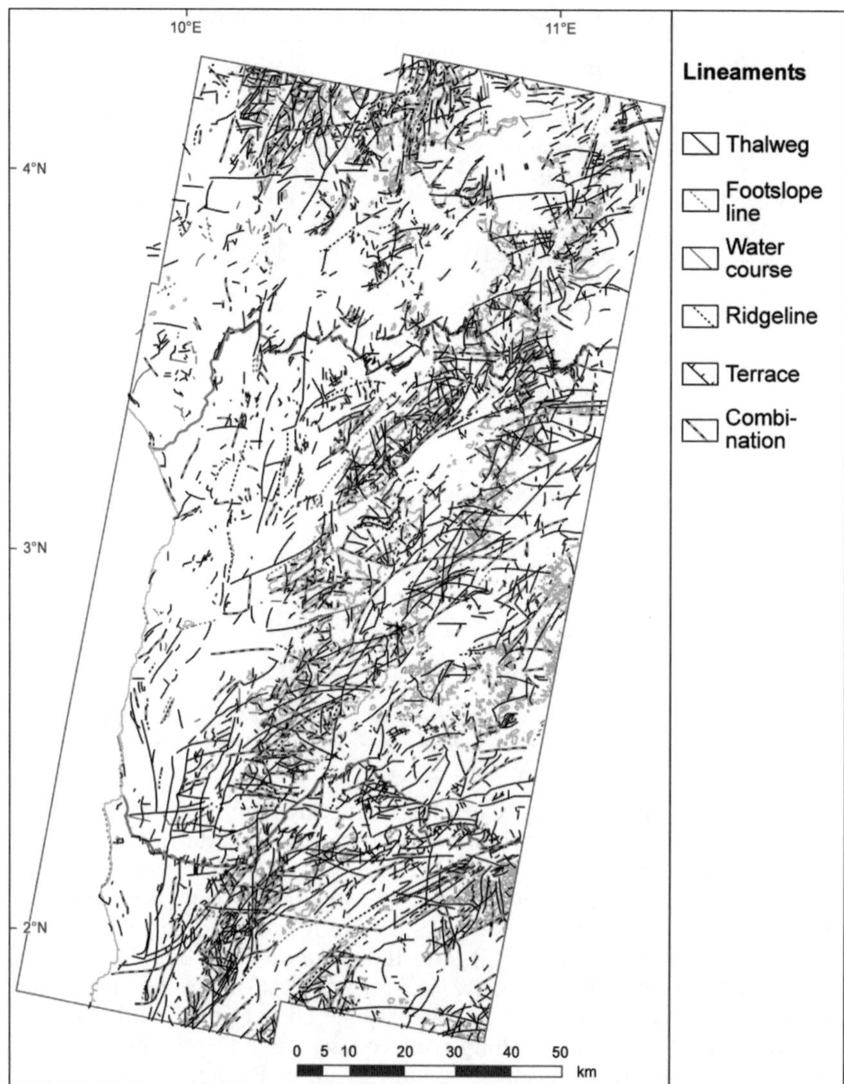

Figure 20. Lineament analysis using InSAR, SAR, and multispectral data.

course at N50–70°E, which has a high "n" value and high mean length (449, 7.1 km; 363, 6.5 km). With respect to the whole study area, there are only minor foci oriented N30–60°W. The low "n" value associated with low mean length (148–259, 2.8–3.0 km) reflects the extensive occurrence of lineaments in contrast to clusters.

Nyong River crossing the peneplain step

Between Makak and Njok, the Nyong follows a fan-shaped geological structure having its origin east of the Ngovayang escarpment and opening to the North. Its orientation varies between N and NE (N0–30°E; c.f. Kuete, 1990:219). Due to the

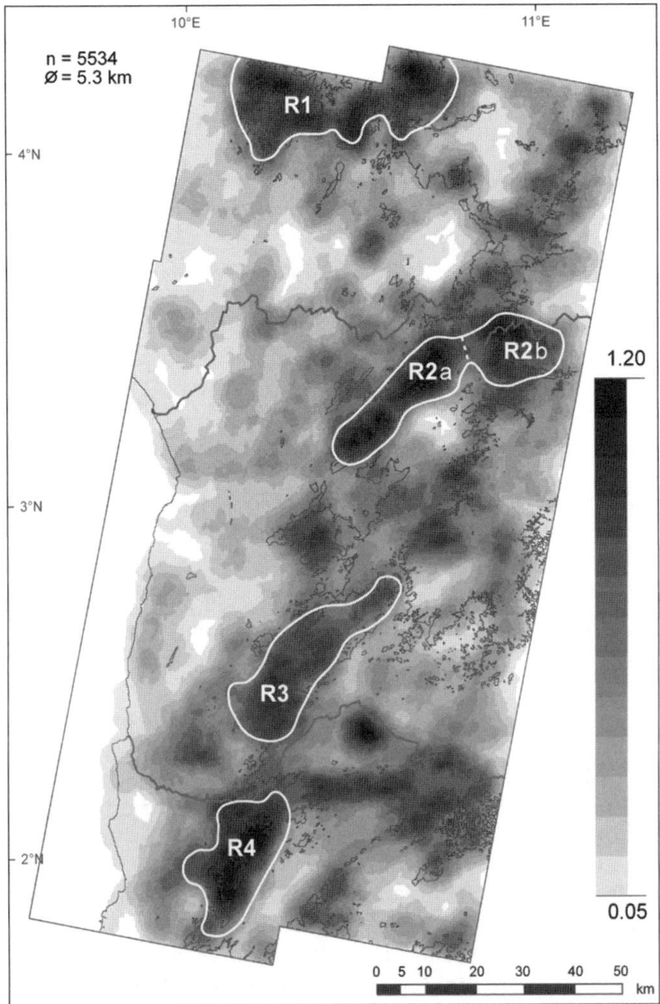

Figure 21. Density map of edited lineaments.

orientation of the structure, which is nearly rectangular to the mean flow direction of the Nyong, the river follows only partially the main structure connecting the different units by linear parts, here, too, probably of structural origin. At Makak, the Nyong still crosses the planation surface with nearly no incision. The riverbed is shallow characterized by outcrops within the river of quarzitic shists of the Yaoundé series, and sandy sediments along the river bank.

Close to Lipombe II, the landscape changes and is marked by a steep slope; drainage enters the fan-shaped structures in a geological environment based on plutonic rocks e.g., tonalite and gneiss (Figure 22). When entering a new structure, the Nyong usually shows distinct changes in flow orientation and also velocity, like at the Chutes de Makai, cataracts situated close to Lipombe II. The pathway from the village to the Nyong leads to a 250 m wide stagnant water zone. The sediments are still sandy, deposited at a bank intermingled with large tonalite rocks. Also in the river, the basement crops out.

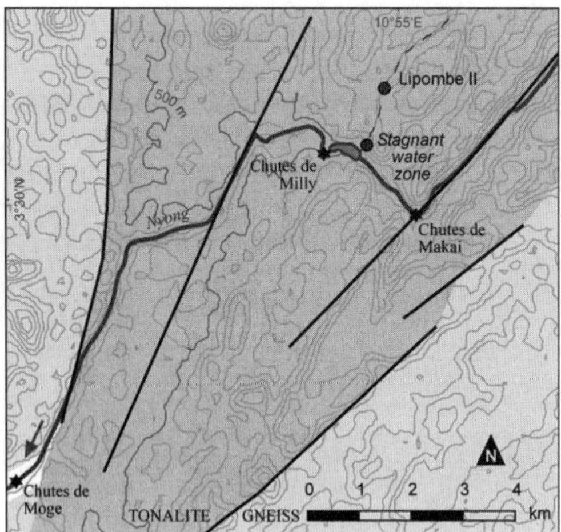

Figure 22. Lipombe II region showing geological units and faults (after Maurizot, 2000; based on GPS editing and SRTM data).

Downstream, the Nyong forms the Chutes de Njok, cascades formed from the basement rocks, crossing three steps of approximately 4 m, 4 m, and 6 m high along 120 m in a WNW direction. Directly below the waterfalls, the river turns to the SSW.

Figure 23 shows a panchromatic aerial photo from 1964/65 of the river in which the high turbulent waters reflect in bright grey. Above the waterfalls, the dark grey reflects a lower turbulent flow of the river. The 25-m isohypses show the steep relief of the surrounding terrain. Above the waterfalls the gneissic basement crops out in the river (top photo of Figure 23) owning a NNE strike direction oblique to the flow direction. Below the waterfalls (bottom photo of Figure 23), huge blocks occur obviously detached from the basement steps. Additionally, unsorted gravels are accumulated at this location fixed in a matrix of iron, aluminium, and manganese and covered by a black polished patina of around 1 mm in thickness of the same content. Unfortunately, it was not possible to loosen a part of the matrix out of the cohesion.

Ntem River crossing the peneplain step

Just before the peneplain step, the Ntem fans out to an interior delta of 210 km^2 in extent. The delta measures about 25 km long by about 10 km wide from SE to NW. At Akom (2°26'N, 10°29'E) the branches rejoin due to an inselberg range, the Collines d'Akom (Kuete, 1990:314), striking into the region from SSW to NNE. The river passes the range in the North, turns to southwest for about 10 km and fans out once more before flowing across the Chutes de Menvé'élé (6–8 m in height) into a linear V-shaped valley, which is oriented towards the SW (Runge *et al.*, 2005; Eisenberg, 2007, 2008; Figure 24).

The whole course of the Ntem delineates a wide westward oriented arch. In the first instance, the possible structural course is quit by an abrupt change of its direction to the NW. At that knick point, the Ntem has formed a small-scale basin (85 km^2), which it exits to the NW by a small passage flowing onto a wide plain slightly tilted towards NW and forming the interior delta. On the plain, the anastomosing Ntem is subdivided into several units along an EW structure. The interior delta has its starting

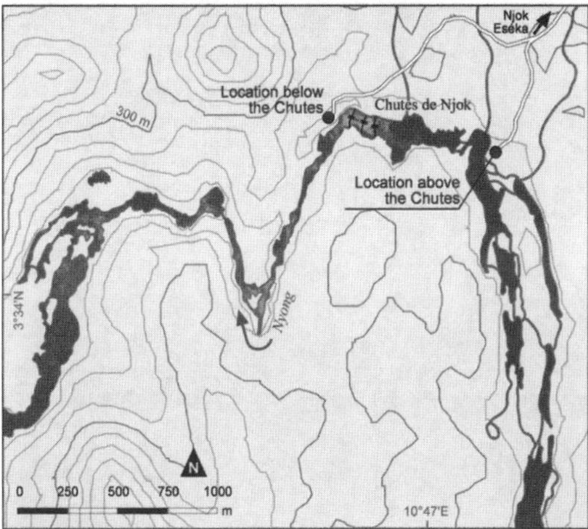

Figure 23. Map of the Chutes de Njok region (center) and photos taken above and below the waterfalls (based on SRTM data, aerial photo of river sections).

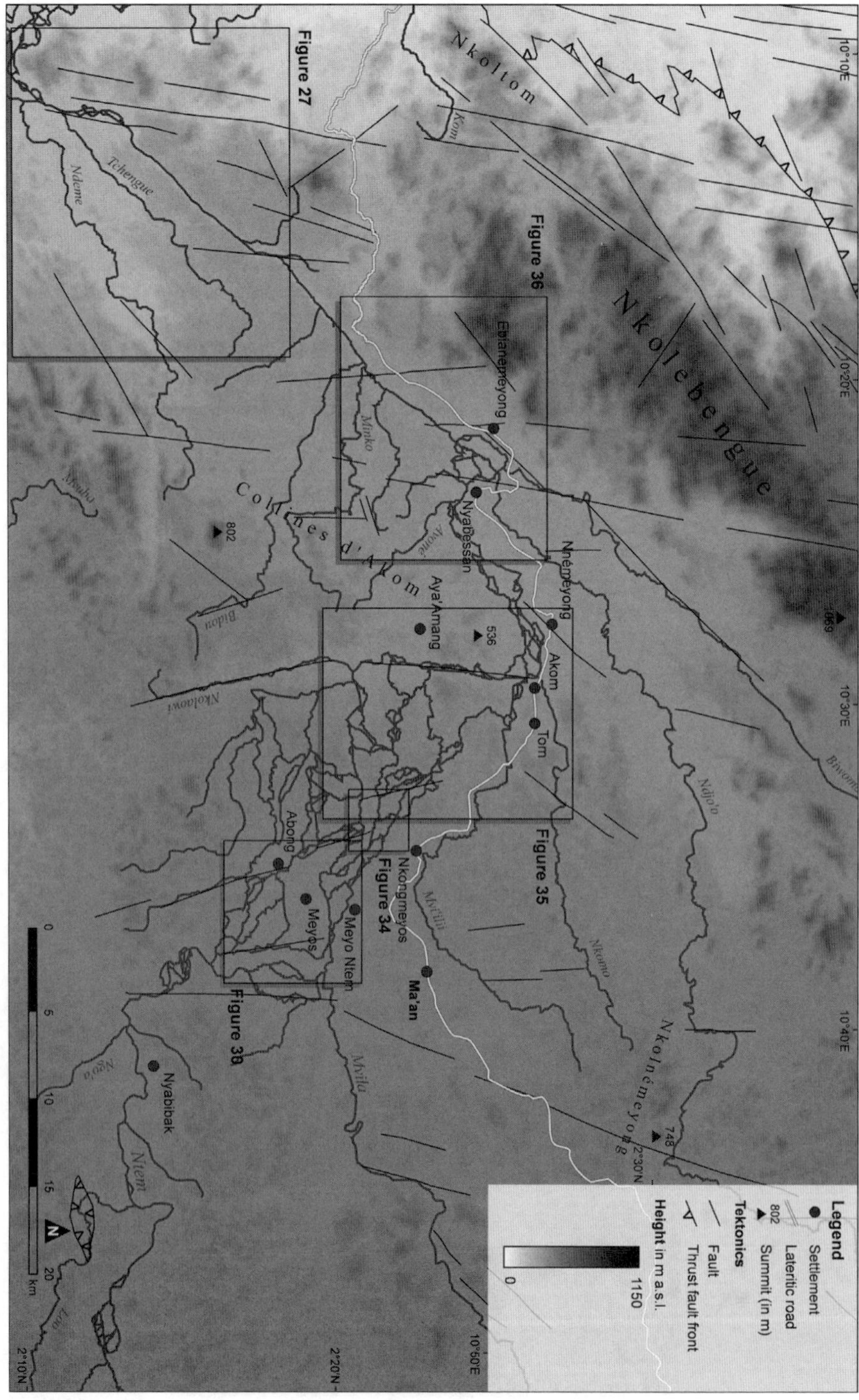

Figure 24. Ntem interior delta and its sourroundings, faults and thrust fault (after Maurizot, 2000; based on SRTM data and topographic map of Kribi, 1976).

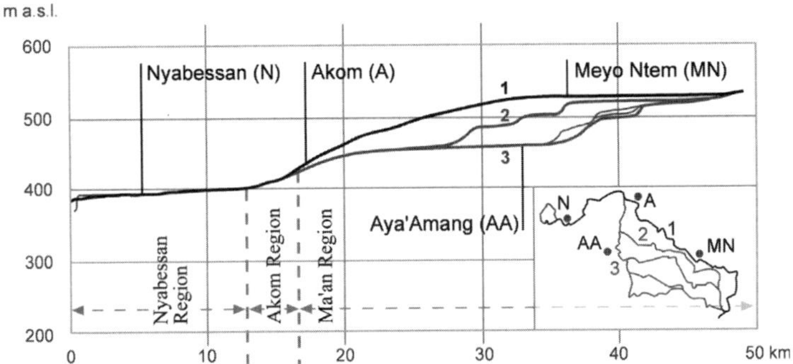

Figure 25. Gradient curves of the Ntem's main branches in its interior delta (based on SRTM data).

point at 530 m a.s.l. and its exit at 390 m a.s.l. at the Chutes de Menvé'élé. Along about 50 km, the Ntem flows down 140 m with a mean gradient of 2.8‰. Drawing a gradient curve for the three main branches in the interior delta highlights that the Ntem passes the first 100 m in altitude until the confluence of the branches at Akom (2.6‰) and a further 30 m over the next 4 km (7.5‰). The last 13 km of the river has a low gradient of 0.07‰ (Figure 25). A division into three units is appropriate following the aforementioned gradient: the anastomosing Ntem (Ma'an region), crossing of the inselberg range (Akom region), and the last section before flowing into the V-shaped valley (Nyabessan region).

Ma'an region: The gradient curve of the various branches is different. Starting with the small passage, all branches show a leveled flow (1.2–2.0‰) before passing the main height difference until the Collines d'Akom, one after the other. The most southern branch (3) shows two distinct steps of 20 m and 40 m in height. Some 5 km downstream, the second branch passes the same height by crossing three steps (20, 10, 30 m). At 450 m a.s.l. both branches have their confluence. After passage of 7 km with a low gradient (0.06‰), it increases to 8.3‰ until crossing the border to the next region. The gradient curve of the main channel in the North (1) has a slightly convex trend starting around 15 km after the small passage, and without any change after the confluence with branches (2) and (3). The Ma'an region is framed by summits in the South surmounting the surface by 140 m (maximum) built of high metamorphic gneiss—this is the same geology as in the southern part of the interior delta, hence, the formation is not based on differential resistance of the basement rocks. However, the small scale of the geological map is not suitable for a more detailed interpretation (1:500,000; c.f. Maurizot, 2000). The summit range has an ENE direction, forming also the Ntem's northern catchment area in its middle reaches on the Inner Plateau. The northern frame of the delta is not characterized by a geological change; nevertheless, the relief energy increases to NNE. The orientation of the whole hydrological network in the region shows a large dispersion of direction and number (Figure 26) with the highest peak in the main flow direction at the NW following the talus of the delta (N40–50°W). The EW orientation is not as obvious when observed on the satellite image of the region, due to having only a few linear units in a much encapsulated network.

Akom region: The Collines d'Akom rise to 130 m above the delta limiting the Ma'an region in the West and the Akom region in the South. Together, the inselberg range in the South and the increasing slope in the North, reduce the passage that the Ntem can use to cross the region. The diagram of the hydrological network shows the

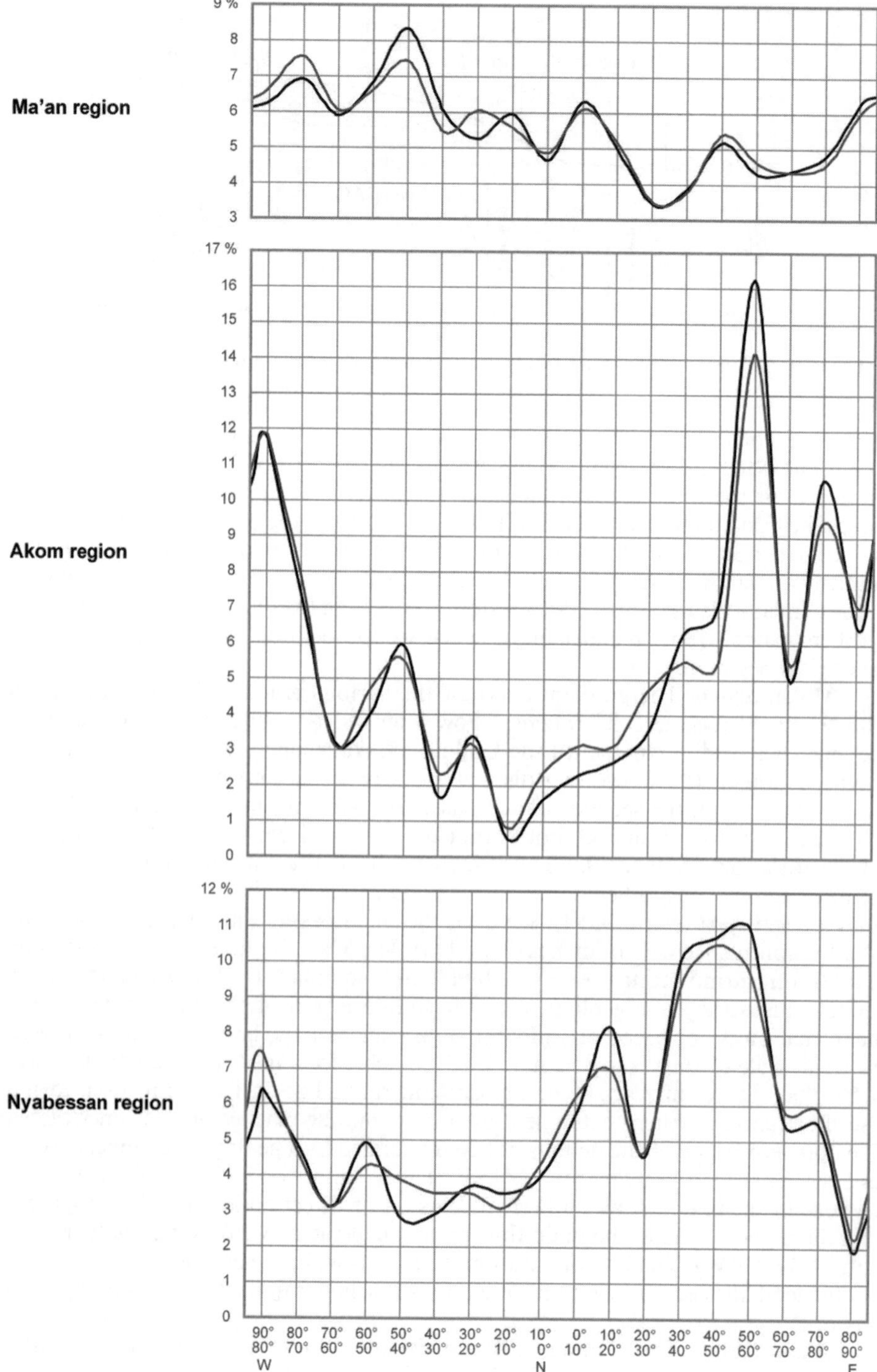

Figure 26. Flow directions of the Ntem in the regions Ma'an, Akom, and Nyabessan (based on Landsat 7 data).

main flow direction towards the SW (N50–60°E, Figure 26). Additionally it shows peaks at N80–90°W and N70–80°E. After passing the Collines d'Akom, the Ntem enters a flat surface.

Nyabessan region: The flat landscape is not only present in the Nyabessan region but also along a strip of 12 km in width on both sides of the V-shaped valley below the Chutes de Menvé'élé. At the level of Nyabessan, the plain is bordered by the Nkolebengue escarpment in the West striking to the SW and rising up to 500 m above the Inner Plateau only divided by a tributary of the Ntem from the plateau. In this region, the escarpment forms the transition between the Inner Plateau and coastal surface and reaches a maximum height of 1069 m a.s.l. at 20 km north of Nyabessan. The escarpment slants down towards the SW and the border between the two surfaces is formed by an accentuated step to the lower level, which becomes eroded by the tributaries of the Ntem's lower course. Also in the South, the plateau becomes eroded by the tributaries of the Ntem. At this location, the transition between the two levels is characterized by two steps. The higher one is frayed but has a distinct border between the two levels, 280 m and 370 m a.s.l., respectively. Above this step the difference between plateau and river amounts to 110 m, below only to 20 m, which probably represents the height of the rain forest of the surrounding area (something also modeled by the DLM); hence, the Ntem flows at the same level as its surroundings. A second step follows in the West, as a southern prolongation of the Nkolebengue escarpment, represented by a few inselbergs.

The Ntem follows a NS oriented fault, leaving it in the direction of the inselbergs, and passes them through a small pathway towards the coastal surface. The surface between the first and second steps is characterized by a slope gradient of 3° in the NE direction (Figure 27). The plateau of the Nyabessan region is bordered by the Collines d'Akom in the East. In the North, the plateau-like character levels off successively along the Biwomé and Ndjo'o Rivers. The primary orientation of the hydrological network of the Nyabessan region is to the SW (N30–60°E; Figure 26).

The mainly structural orientation of the river network in the interior delta seems to be obvious. Also Kuete (1990:312) discusses the hydrological network of the whole Inner Plateau as an indicator of the basement structures. It sketches the already noted tectonic directions of the region. The whole river network of the delta was edited using the topographic map of Kribi (1:200,000) and the DLM. Additionally, each river was assigned to Strahler's (1957) river order to highlight possible influences of neo-tectonic reactivation processes in the region (c.f. Ciccacci *et al.*, 1987; Bellisario *et al.*, 1999). Linear river sections are highlighted in Figure 28, in an effort to clarify the heterogeneous character of the lineaments, e.g., a combination of conspicuous river course changes and linear units as a linear prolongation. Also river segments of different orders complement lineaments. All edited units were drawn in Figure 29: N80–90°E in particular shows a high length peak combined with a distinctly low value of number of considered lineaments due to the long lineaments in this direction in the interior delta.

In the interior delta, alluvions were found in the Ma'an, Akom, and Nyabessan regions, and these were used as indicators in a detailed sketch of landscape and river evolution over the last 50 kyrs:

The main study location in the Ma'an region was Meyo Ntem (2°20'N, 10°36'E; Figure 30). The village is situated 20 m above the northern branch of the river having a gradient of 40°. The Ntem has formed a wide riverbed before it is limited by outcrops to a small and shallow overflow; the presence of some cataracts here highlight the increasingly turbulent flow. During the dry season, the widespread outcrop is particularly visible and it has a diverse fissure morphology. Along several fissures,

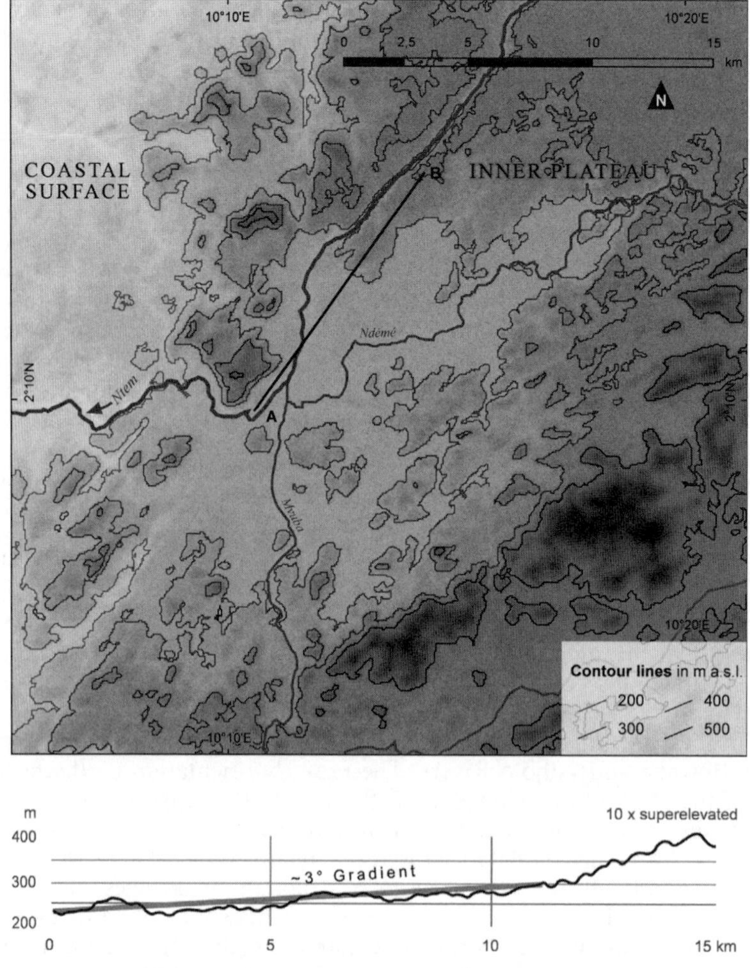

Figure 27. Topography of the peneplain step between the Inner Plateau and coastal surface, transect AB
(based on SRTM data).

large blocks are loosen out of the outcrop. Downstream, the number of rock fragments and gravels increases. They are hardened in a partly porous partly hard matrix
of iron and manganese oxides. At some locations, vegetation islands are formed on
the outcrop or the fixed gravels, representing the water level during the rainy season.
Beneath the steep slope in the North, the branch is framed by shallow steps to a 2 km^2
island in the South. The Ntem crosses the outcrop towards the west and has a depth of
approximately 50 cm. It changes its direction successively towards the NW and crosses
steps formed in an E-W orientation (N80–90°W). The linear fissures and steps of the
outcrop have different characteristics.

In particular, there is an eroded, linear, and angular quartz vein of some 20 cm
high oriented in a N10–20°E direction for about 10 m long, limited by a N80–90°W
fissure filled by river water. North of that fissure, the ridge occurs again with a 20 cm
offset to the East (Figure 31.1). Further lineations occur in the same parent rock.

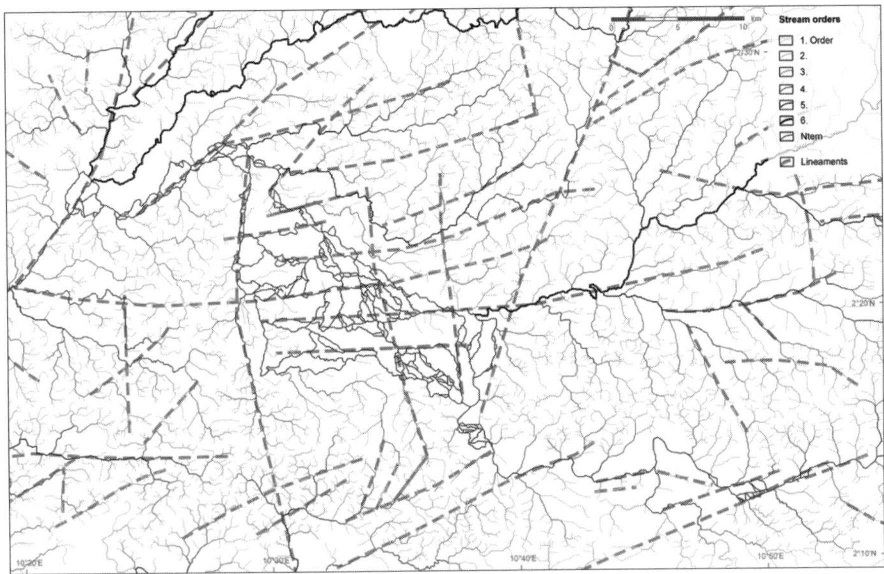

Figure 28. Fluvial network of the Ntem interior delta, linear sections highlighted by dotted lines (based on the topographic map of Kribi, 1976).

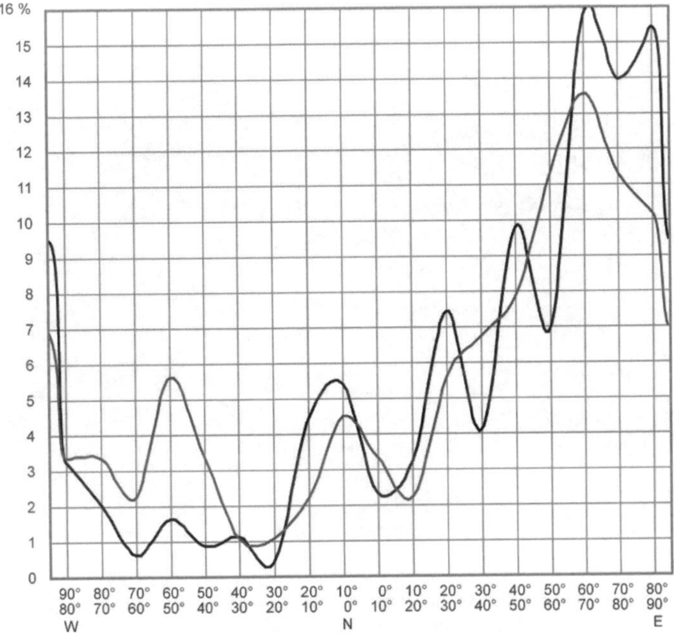

Figure 29. Orientation of linear units in the fluvial network of the delta.

Two meters east of the distinctive vein, another ridge runs parallel to it that is much more eroded and lacks angularity (Fig. 31.2). At an angle of 50° to the described structures, two units oriented to N60–70°E occur, which cut the distinctive vein;

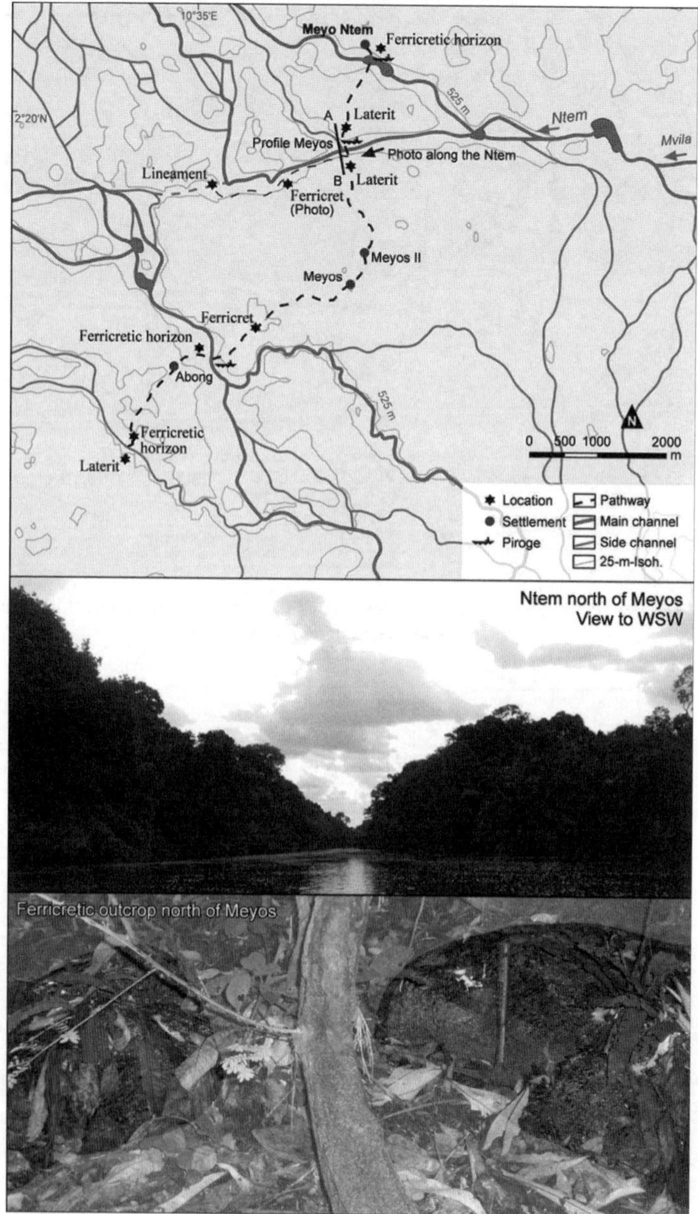

Figure 30. Map of the Meyo Ntem region showing the transect AB, the pathway along which transect Meyos was sketched (see Figure 33), and ferricretic outcrops; photos of the Ntem and of ferricretic outcrops (map based on SRTM data).

however, they are shallower and also more eroded. Parallel to the water-filled fissure, a step of some 5 cm occurs crossing the distinctive vein (black-dotted line, Figure 31.2). Further lineations run parallel to the distinctive vein, which are characterized by a chain of steps of a height of around 1 cm. In the lower west part of the outcrop, two further quartz veins were extracted manually from the surrounding water. The quartz

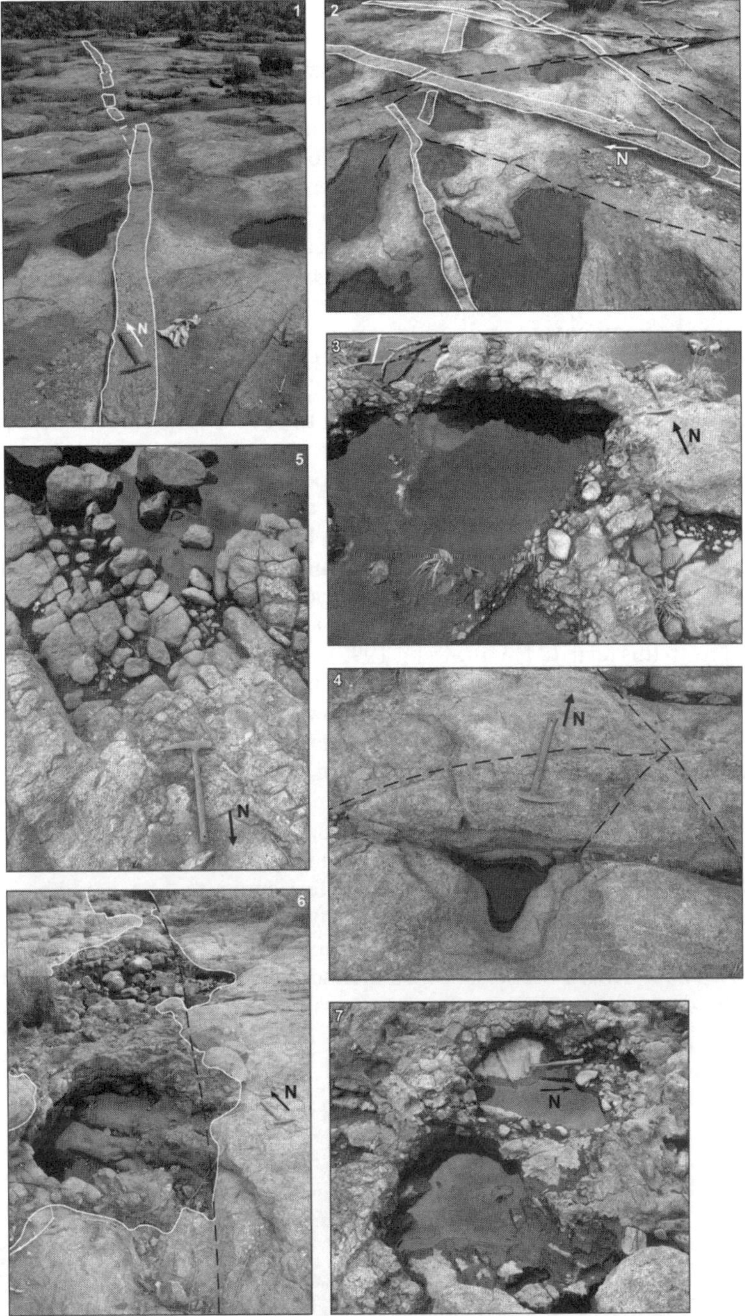

Figure 31.1–7. Significant features of the outcrop near Meyo Ntem: (1) Impregnated quartz vein with an orientation N80–90°W; (2) Quartz veins (white lines) and shallow steps (black dotted lines); (3) Junction of strongly eroded quartz veins; (4) Junction of eroded shallow steps (black dotted line) with impregnated and water-filled joint; (5) Rounded rock fragments in a rectilinear fissure pattern; (6) Pothole in hardened clasts (white line) along a distinct step (black dotted line); (7) Potholes in hardened clasts showing quartz veins of the underlying basement.

was dissected into fragments of 1 cm edge length by weathering; however, it still remains in the original linear character. The one vein orients towards N60–70°E, the other to N10–20°E. Different forms of fissures are visible in the middle of the outcrop: flat incisions are oriented towards N60–70°E, N10–20°E, and N60–70°W; also a more deeply incised water-filled fissure occurs having an N80–90°E orientation (Figure 31.4). Close to this location, small-scale weathering along rectilinear fissures takes place forming palm-sized rocks of the basement. The fissures orient both towards N50–60°E and N30–40°W. Comparable cross-like fissures become apparent at several locations. Downstream, the outcrop areas increase, covered by a pebble matrix fixed by iron and manganese oxide. Sometimes they are exposed; sometimes they occur in wide depressions limited by steps of the outcrop. Figure 31.6 shows a mixed layer of gravels and rock fragments in and out of a matrix covering the 30 cm thick basement along a linear step orienting towards N10–20°E until the slope. Through a pothole, probably formed by the Ntem's water during the rainy season, the underlying rocks appear showing a water-filled crevasse trending to N40–50°W.

Another location having comparable forms exists downstream, two potholes exhibit a 25 cm layer of the matrix. In the pothole, a thin quartz vein trends to N60–70°E (Figure 31.7). A piece of the matrix was separated from the cohesion and cut in the mineralogical laboratory of Frankfurt University. It shows several gneiss and granite rock fragments of differing sizes. Also subangular to rounded fragments in the matrix are found, the likely result of intensive hydrolysis; hereafter, the matrix is called fanglomerate (Figure 32). For now the evolution, normally implied by this designation is excluded from this term (c.f. Murawski, 1992). The areas between the large fragments are filled by small, angular quartz grains, and bright reddish hematite. At protected sites, the fanglomerate is covered by a dark brown to black patina. The color hints to iron and manganese oxide as main components. Several corings taken and exposure sites viewed over the whole interior delta did not reveal a gravel layer under sediments. At the northern bank of the outcrop, several corings were conducted. Close to the outcrop, homogenous sandy sediment has accumulated. With increasing inclination, a hard ferricretic layer was identified approximately 160 cm above water level after a distinct knick point. Below the knick point, no ferricretic material could be excavated. At the southern bank only sandy sediments could be found.

Branches 2 and 3 of the Ntem form an island with the village Meyos in the center (Figure 30). The northern branch shows a distinct linear course (N70°E) having a high flow velocity and a depth of several meters (pers. comm. local guides; Fig. 30, photo).

Figure 32. Cut through a fanglomerate piece.

The confluence of a smaller branch leading off to the NW is characterized by a linear chain of out of the water outcrops parallel to the course of the main branch. It is part of a wide 5 km long, stretched arc starting at the confluence with the Mvila River (2°19'N, 10°37'E) trending to the west and turning slowly to the south (N60°E) before abandoning. At this location, the river turns abruptly to the N and only a small branch continues to the SW.

Starting at Meyo Ntem, a pathway leads through the interior delta from north to south crossing four branches of the Ntem. The islands have maximum heights of 550 m a.s.l. in the undulated landscape; the height of the Ntem River varies between 515 m a.s.l. at Meyo Ntem and 520 m a.s.l. at the fourth branch near Abong. Corings and observations show ferricretic outcrops along the pathway in addition to the ferricretic layer at Meyo Ntem. At the second branch, ferricretic outcrops form a step north and south of the river. Both occur at a height of 50 cm above a terrace and indicate the rainy season level of the Ntem. At the northern bank of the third branch, another ferricretic outcrop forms a step 260 m inland from the river. Below this step, a wide area having mainly clayey sediments and organic layers was measured; it was limited by the Ntem's bank. A quartzite ridge at the near-ground surface divides the area into two parts. Ferricretic residues were not found in the corings. At the southern bank, a ferricretic layer emerges at a depth at 120 cm, however it correlates not obviously with the outcrop on the other bank due to a difference in height of around 300 cm. At the northern bank of the fourth branch, a further outcrop occurs at 350 m from the river, forming a small geomorphologic step. Three hundred meters further, a ferricretic layer was identified in several corings sketching a large extent of the layer. Also at the southern bank of the fourth branch, a ferricretic layer was identified 100 cm above the water level, it continued southwards with a slight slope under a clayish silty soil cover (Figure 33).

At Nkongmeyos (2°21'N, 10°34'E), the Ntem forms three branches for a quite short distance. At the second branch, the Ntem flows into a basin after crossing cataracts (Figure 34). The small step forming the cataracts trends towards N70°E. The whole northern bank is characterized by an accumulation of area-wide palm-sized gneissic rocks, which are barely rounded. The growth of lichens and mosses on the rocks indicates that they have not been moved recently. Some 10 m beneath the river bank, there is an 80 m slope, which is quite steep (15°). Between the rocks, only one coring was possible, it was to 200 cm depth, limited by the basement. The substratum is sandy to clayey having some organic sediment at the base, which was dated to 12.7 cal. kyrs BP (Figure 34). On the southern branch, an over bank deposit occurs made of 80 cm of sandy sediments lying over the basement. No rock fragments occur on this side of the river.

At the northern border of the interior delta, the Mvi'ilii River, a tributary of the Ntem, shows a 2 km linear unit to N80°E crossing 15 m in height before flowing into its receiving stream (Figure 35). The surrounding hills are a prolongation of the Collines d'Akom surmounting the valley up to 100 m. The small valley floor is characterized by large blocks filled-in by sandy sediments. The river bed is flat and exhibits outcropping

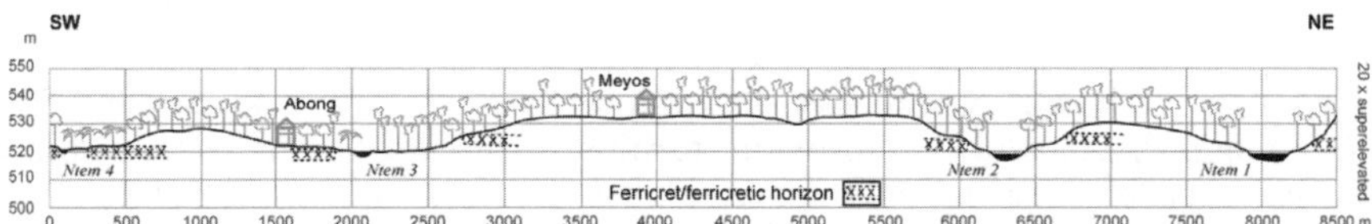

Figure 33. Transect Meyo Ntem-Abong (based on SRTM data and field measurements).

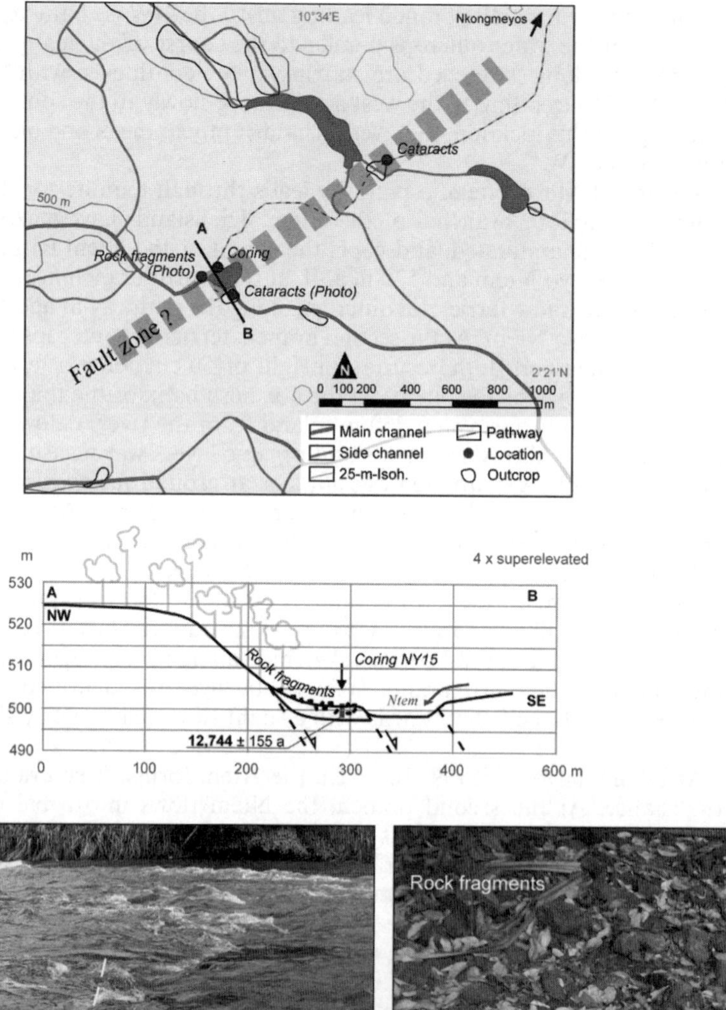

Figure 34. Map of the Nkongmeyos site showing transect AB and postulated fault zone; transect AB; photos of cataracts and rock fragments (based on SRTM data, measurements and GPS editing in the field).

rocks. Above the linear unit, the Mvi'ilii follows a N20°E direction crossing 8 m in height along a distance of 200 m before turning to the West. This section describes several cataracts by crossing rocks rectangular to the flow direction (N80°E).

The interior delta in the Ma'an region is confined in the West by a linear section of the Ntem having a NNE direction (N10°E) and follows a fault sketched in

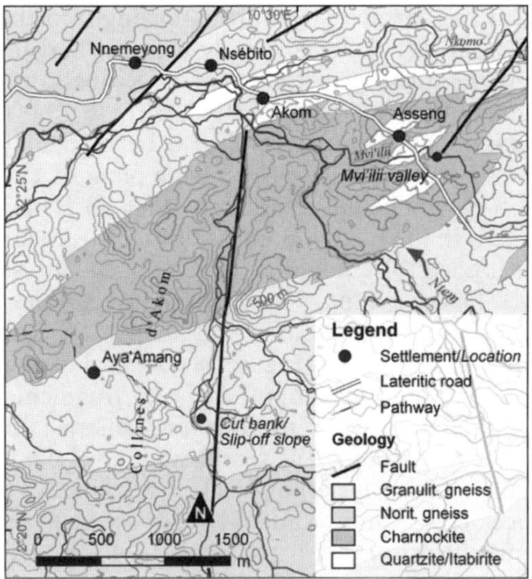

Figure 35. Western part of the Ma'an region showing geology and faults (based on Maurizot, 2000; SRTM data).

the geological map (Maurizot, 2000). The valley of this branch is limited by the steep slope of the Collines d'Akom in the West and inselbergs in the East surmounting the delta some 150 m. Three arms of the anastomosing Ntem flow into this branch at regular intervals before reaching the most northern branch and crossing the inselberg range. At a distance of 8 km, the branch crosses 14 m in height before the confluence (1.75‰). The gradient is distinctly smaller as compared to the mean gradient of the region (2.6‰). The branch functions as an entrainment for the region and has high discharge and low velocity. Close to Aya'Amang (2°22'N, 10°27'E), the branch starts its linear flow. It shows a wide bend coming from the East to NNE and a cut bank is formed at the western riverbank; however, recent processes were not identified. Directly above the bank, the gradient increases towards the inselberg range. On the eastern riverbank, a slip-off bank was formed having a terrace of 50 cm in height at about 100 m inland. On the terrace, a coring revealed organic sediments at a depth of 320 cm having an age of 34.9 cal. kyrs BP (coring L37). A further coring between the terrace and bank revealed a layer having organic sediments at a depth of 140–160 cm with an age of 4.9 cal. kyrs BP (L36). At a comparable level, a layer having organic sediments was revealed on the eastern bank showing an age of 6.0 cal. kyrs BP (L38).

At the level of the Collines d'Akom, the still anastomosing Ntem is reduced to a width of 500 to 1000 m bordered by a step marking the water level during the rainy season. Close to Nnémeyong (2°26'N, 10°27'E), the Ntem has formed cataracts with small-scale potholes of some 20 cm. Sand and coarse sand were identified as the erosion agents. The pisolites sedimented at the location were ruled out as erosion agents because they are too soft.

At Nyabessan the Ntem splits into two branches. The first flows towards the NW, then SW, and finally SSE before forming huge waterfalls of 6–8 m in height, the Chutes de Menvé'élé, which empty into a westward shift of the SW oriented V-shaped valley below the interior delta. The second branch flows towards the SW and drains as a waterfall directly into a basin to which the valley opens at this location.

Additionally, the Ntem has formed small arms flowing into the valley north of the second branch. The Ntem has sketched nearly a circle; and linear units occur, too (Figure 36). Just before the opening of the Ntem, at Nyabessan, there is thick organic sediment of a maximum age of 4.3 cal. kyrs BP (C 27).

Huge blocks of basement and ferricret of up to 2 m in size, lay below the waterfalls. Particularly below the Chutes de Menvé'élé, a wide plain of clasts occurs. The plain is limited by a step of 3–4 m in height trending to N60–70°E in the West across which the waterfalls are formed. In the East, it is limited by a dip-slip fault of the same trend. The clasts consist of rounded to subangular gravels mixed with several rock fragments and covered by a dark brown patina. X-ray diffraction studies indicate a high content of organic material, as well as goethite and manganese hydroxide, which is supported by the dark color of the patina (Figure 38). After destroying the surface with a hammer, it shows a hardened matrix composed of rock fragments and gravels. Directly below the waterfalls, the Ntem has cut into this fanglomerate and formed a cut bank and a slip-off slope. At the cut bank, the recent watercourse and the waterfall reveal the gravel matrix, which forms a layer about 100 cm thick above the water level. Below the layer a hollow space occurs whose content already seems to be eroded. The attempt to dig into the fanglomerate at another place failed due to the resistance of the matrix. For this reason, the material filling the hole could not be determined.

However, two samples of the matrix were taken; [1] at the slip-off slope at a site directly under the influence of the spume close to the waterfall and [2] below a small tributary trending into the V-shaped valley (Figure 37). [1] could be broken out of the matrix. A cut through the fanglomerate shows rounded gravels covered by small subangular pebbles and angular particles of quartz (Figure 39). Below the gravel, a reddish to dark brownish colored crust of up to 2 cm thick was formed. The color suggests that manganese and iron oxides are the principal contents. The coarse grain size of the cover does not occur in the crust. [2] contains a subangular gravel; however, primarily

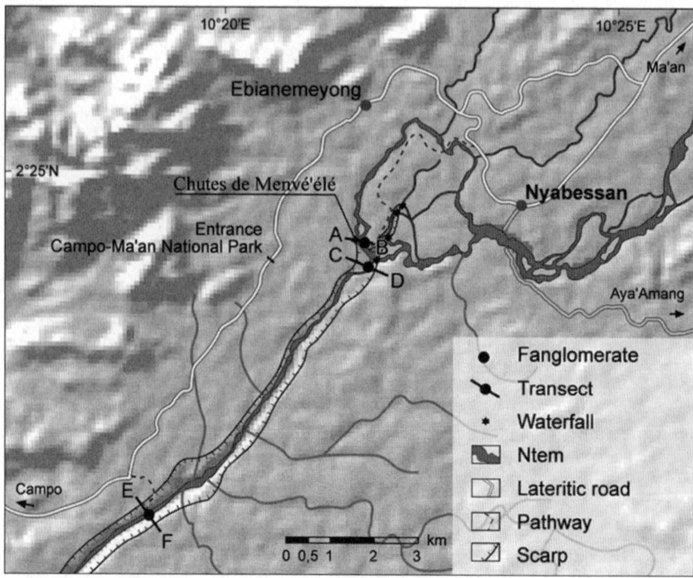

Figure 36. Locations of waterfalls and transects in the Nyabessan region (based on SRTM data, topographic map, and GPS editing in the field; modified from Eisenberg, 2008).

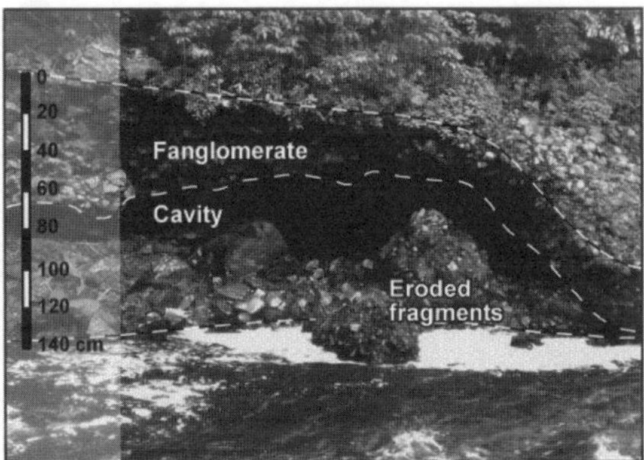

Figure 37. Cut bank and slip-off slope below the Chutes de Menvé'élé.

it shows barely subangular fragments of up to 10 mm in diameter. Additionally the fanglomerate contains two rounded concretions of quartzitic culm in a bright brown, finely grained matrix.

The Chutes de Menvé'élé supply a sub-linear branch of the Ntem oriented towards the NNE-SSW along 350 m. At the SSW border, the branch makes a distinct turn east passing a step of 2 m into a basin, which is the confluence of the second branch of the river. At the up to 30° steep eastern slope of the basin 150 cm above water level, a distinct fringe of the fanglomerate occurs. It is likely that it also occurs on a small island of the Ntem; however, the island could not be studied because it was unreachable. Moreover, it occurs on the opposed bank of the basin, at the location where the

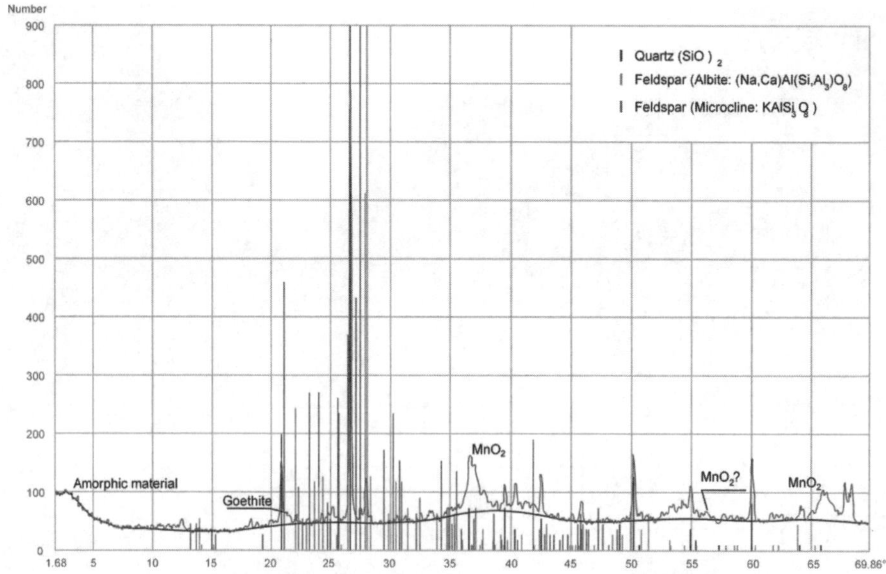

Figure 38. X-ray diffraction of the patina below the Chutes de Menvé'élé (Philips PW1730 40 KV/30 mA, 3410 points (1.68 to 69.86° in 0.02° steps), Tube: 1.541874Å Cu; Automatic Divergence slit, step scan; User: R. Petschick, Frankfurt University).

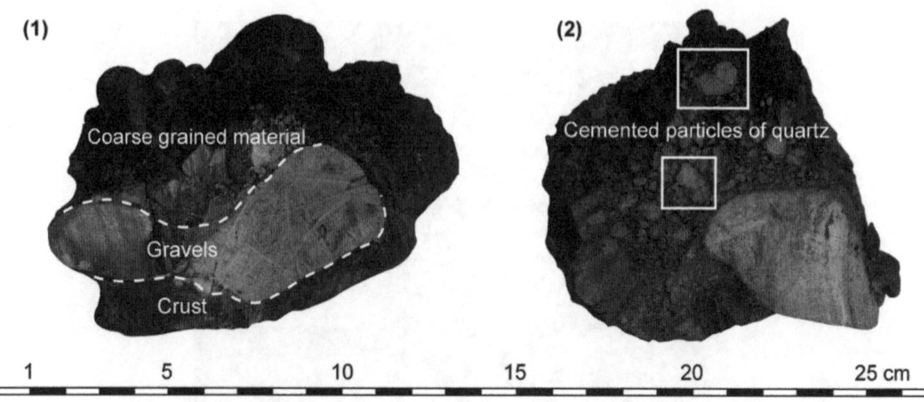

Figure 39. Cut through two fanglomeratic pieces taken from the fanglomerate below the Chutes de Menvé'élé (modified from Eisenberg, 2008).

first branch of the Ntem flows into the valley. The level of the island corresponds to the fringe observed on the eastern slope. Two samples of the fanglomerate were taken out of the fringe: [1] out of a crevice 50 cm wide doubtless containing erosion and colluvial deposits from the slope above; [2] directly out of the fringe (Figure 40). [1] has a breccial character. The components are partially subangular and seem to be weathered out of the same parent rock; the linear gaps are filled with a fine-grained dark brown material. [2] has well rounded gravels and angular rock fragments having a coarse grained material in between.

Below the interior delta, the Ntem crosses the peneplain step along a V-shaped valley. Except for two interruptions, it consists of a more or less linear unit to the

Figure 40. Cut through two fanglomeratic pieces taken from the fanglomerate found in the basin.

SW for about 30 km. Then the river passes the prolongation of the Nkolebengue escarpment with a distinct turn west before entering the coastal surface. The V-shaped valley of the Ntem has its origin north of the confluence of the first and second branch of the Ntem in a prolongation of the Biwoumé River. The river has its source in the Nkolebengue escarpment in which it eroded a wide bend before flowing into the first branch of the Ntem. In a prolongation of the river south of the branch, the V-shaped valley shows several small-scale steps reaching a depth of approximately 10 m. The first interruption along the valley appears as a smooth turn to the WSW (2°20'N, 10°19'E). Further 7 km downstream (2°18'N, 10°16'E), the valley orients at a distinct knick point to the South for a length of 1.5 km before trending again to the SW. In connection with a linear valley of a tributary joining the Ntem from the North, the second interruption forms a lineament sketched as a linear geomorphological feature in the landscape. At 2°24'N 10°21'E, the valley can be reached by foot. The valley floor is characterized by a heterogenous occurrence of fanglomerates and basement outcrops. At some locations the fanglomerate is eroded superficially. Additionally, several loose gravels sedimented in potholes formed behind huge blocks of the basement rocks within both the basement and the fanglomerate.

A fanglomerate was cut to show two different parts (Figure 41): the right portion contains strongly corroded and subangular to well-rounded gravels hardened in a fine-grained matrix. The left portion contains mainly fragments of gravels hardened in a coarse-grained matrix of angular rock grains of about a few millimeters in size and some pisolites. In this section, too, a gravel fragment occurs, which is covered by a patina along the well-rounded part. From the part with the patina, a thin section was made and a photo of it was taken using a binocular microscope (Figure 42). On the left is the whole thin section; there are some white areas, which are holes in the thin section. In the lower third, the patina becomes visible; it probable originally formed around the gravel before fragmentation and hardening within the matrix. The location of the detailed view on the right side is highlighted by the white box in the left photo. The bright beige sections in the upper and lower part represent the adjoining rock fragments of the patina.

The patina is characterized by distinct layers having bright and dark brown colors. The layers are numbered from 1 to 6. In the middle layers (3, 4) the yellow colors are more present as opposed to the adjoining red layers. The difference is probably

Figure 41. Cut through a fanglomeratic piece found in the V-shaped valley.

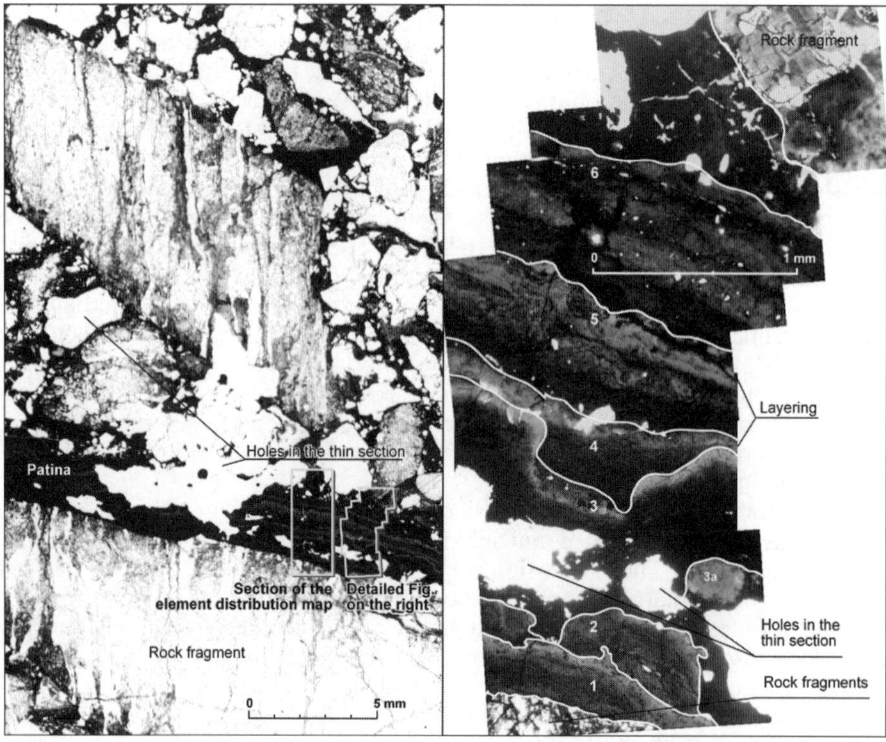

Figure 42. Scanned thin section of the patina and detailed microscope view.

provoked by irregularities of the thin section and therefore by inaccuracies of the lighting during taking the photos. However, all layers show a uniformity of a transition from dark to brighter parts starting at the covered gravel fragment in the lower part of the photo. Particularly layers 3 and 4 show a progressive transition whereas layers 5 and 6 have a distinct difference between dark and bright colors. In layers 1 and 2 the difference is hardly distinguishable. Several holes at the border between layers 2 and 3 are noticeable. Obviously it is less resistant material, which was destroyed during thin section grinding. In detail, layer 3 runs parallel to the holes; hence, layer 4 equates to surface irregularities. Layers 5 and 6 run parallel to layer 4.

An element distribution map was made using an electron probe micro-analyser (Figure 43). It becomes obvious that the main content of the sample is iron. At the border between the layers, iron decreases and aluminium increases slightly. The high aluminium content is the probable reason for the yellow color on these sections.

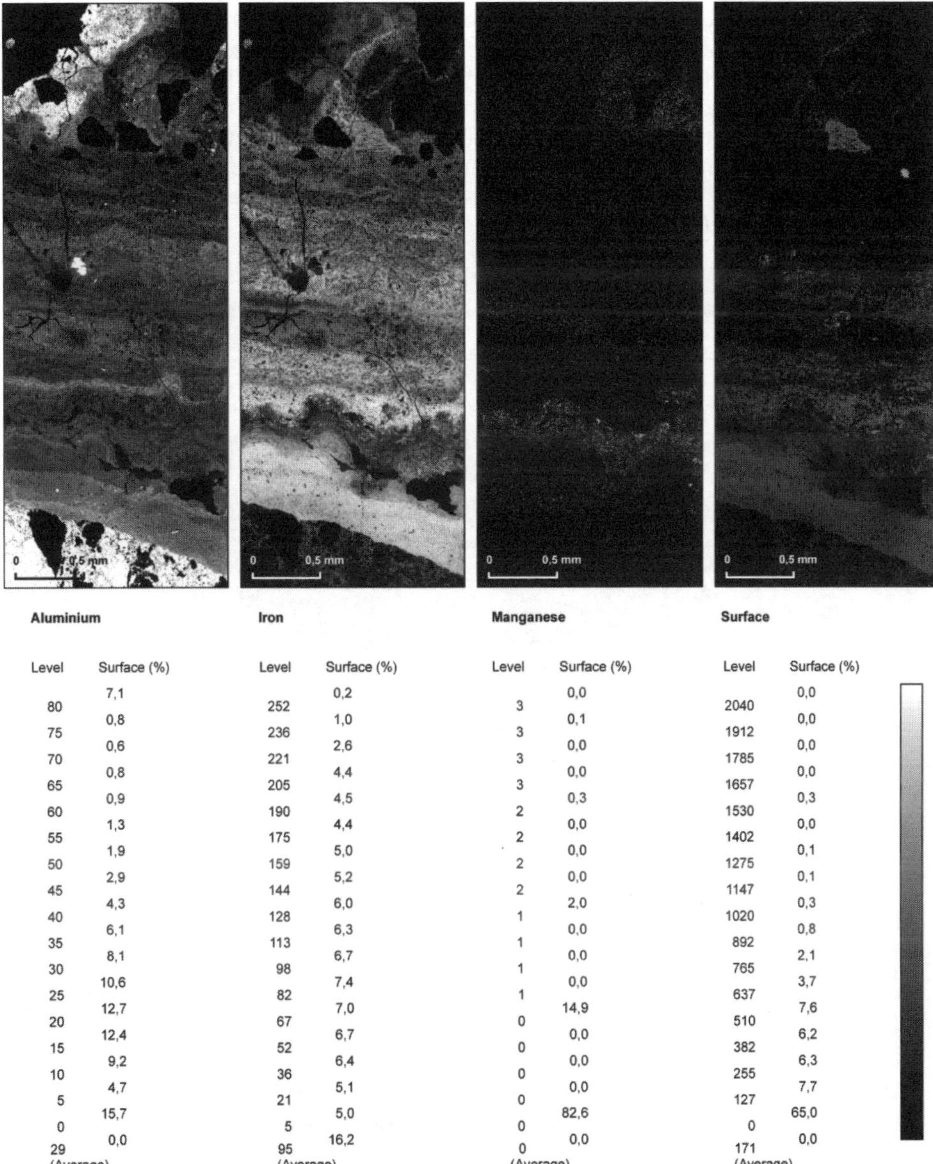

Aluminium		Iron		Manganese		Surface	
Level	Surface (%)	Level	Surface (%)	Level	Surface (%)	Level	Surface (%)
80	7,1	252	0,2	3	0,0	2040	0,0
75	0,8	236	1,0	3	0,1	1912	0,0
70	0,6	221	2,6	3	0,0	1785	0,0
65	0,8	205	4,4	3	0,0	1657	0,0
60	0,9	190	4,5	2	0,3	1530	0,3
55	1,3	175	4,4	2	0,0	1402	0,0
50	1,9	159	5,0	2	0,0	1275	0,1
45	2,9	144	5,2	2	0,0	1147	0,1
40	4,3	128	6,0	1	2,0	1020	0,3
35	6,1	113	6,3	1	0,0	892	0,8
30	8,1	98	6,7	1	0,0	765	2,1
25	10,6	82	7,4	1	0,0	637	3,7
20	12,7	67	7,0	0	14,9	510	7,6
15	12,4	52	6,7	0	0,0	382	6,2
10	9,2	36	6,4	0	0,0	255	6,3
5	4,7	21	5,1	0	0,0	127	7,7
0	15,7	5	5,0	0	82,6	0	65,0
29 (Average)	0,0	95 (Average)	16,2	0 (Average)	0,0	171 (Average)	0,0

Figure 43. Element distribution of iron, manganese, and aluminium from a thin section of fanglomerate patina (see Figure 42 for sample site).

Due to the content of manganese in the covering patina in the whole V-shaped valley (Figure 38), it was also calculated using the micro-analyser; the amount of manganese in the embedded patina was found to be low. The X-ray diffraction results gives only the composition of the sample and not the exact quantity. Beside the change of iron and aluminium content in the layers, the first layer directly on the gravel fragment is more homogenous. Both iron and aluminium are distributed uniformly.

Nyong river on the coastal surface

On the coastal surface, the Nyong's river course is characterized by several outcrops of basement rocks and associated cataracts. The closer the river gets to its mouth, the less turbulent is its flow. The river bank consists mainly of the outcropping basement and mostly sandy sediments mixed with pisolites. The Chutes de Dehané form a distinct transition between Precambrian basement rocks and Tertiary and Quaternary deposits. Downstream from the Chutes, the Nyong is navigable until its mouth into the Gulf of Guinea. Just before the transition, the river course trends towards the geological set up structures in a SSW direction. The river leaves this direction, flowing towards the WNW using three branches. The most southern and the main branch together form the Chutes de Dehané (3°29'N, 10°4'E), which are cataracts crossing two

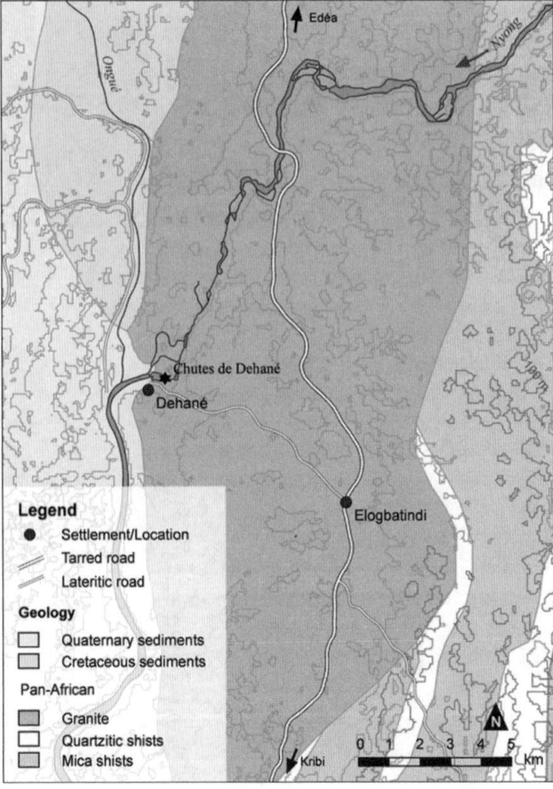

Figure 44. Transition from basement rocks to Kretazic and Neogene sediments in the Dehané region (based on Maurizot, 2000; SRTM data).

steps of approximately 1 m in height rectangular to the flow direction. Directly below the cascade, a fringe of subangular gravels and gravel fragments occur fixed in a poriferous matrix, which could not be sampled for cutting, due to its consistency. Below the cataracts no more outcrops occur until the mouth. The branches of the Nyong River course rejoin accompanied by a distinct reduction in flow velocity. Landsat imagery (Path 186, Row 058, 26.04.2001) and topographic maps show wide meanders as the major geomorphological forms until the Nyong's mouth (Figure 44).

Ntem river on the coastal surface

Starting at the coastal surface, the Ntem River forms the Île de Dipikar, an island 340 km² in size. Just before the confluence of the two branches, the Ntem's flow velocity is low. The sediments are sandy, and cover the gneissic basement cropping out in a few places. Consistent with the topographic map, no more steps occur until the mouth into the Gulf of Guinea (c.f. Figure 6).

3.2.3 Summary

The data shows that there exists a wide range of geomorphological forms in the catchment areas of the Nyong and Ntem Rivers. The watersheds at the level of the Inner Plateau are difficult to delineate. A sub-recent shift in portions of the rivers is apparent and requires an interpretation in terms of morphodynamics and evolution of both catchment areas. The accentuated peneplain step and its structurally controlled river course, and the Ntem interior delta sediments, having young radiocarbon ages, strongly suggest neotectonic reactivation of older structures. Remote sensing and field work have enabled a rough typing of form units by which morphodynamic and morphogenetic processes may be interpreted.

Peneplain and inselbergs

Close to Makak at the Nyong River and to Ngoazik at the Ntem River, the landscape is formed out of an undulated peneplain, which is drained by a uniform river network orienting towards the structural conditions of each respective region. The rivers are characterized by a small floodplain having sandy fluvial facies and outcrops of the basement in and nearby the river. The peneplain is marked by flat watersheds and shallow slopes. Sparse inselbergs surmount the surface by some 10 m only partially covered by a thin soil cover and low vegetation. The rock crops out extensively. The transition between inselberg and surface is abrupt showing debris at the footslope. It is fixed by some fine-grained particles and vegetation. The peneplain is covered by deeply weathered ferralsols, and by either a thick primary, or a secondary forest. The transition between peneplain and floodplain shows a shallow concave knick point with only a little change of the vegetation. The river bank is marked by a distinct step in the river representing the water level of the rainy season.

"Demi-oranges" and "bas-fonds"

The upper catchment areas of the Nyong and Ntem Rivers are characterized by polyconvex summits of the *demi-oranges* and the wide and partially hydromorphic floodplains. The naming of the hills has its source in the French description "*demi-oranges*" meaning "half orange" because of the resemblance these bear to the geomorphological form (Kadomura, 1977a:11). By means of satellite imagery these form units can be easily identified as settlements or plantations with light cyan colors at the top of the hills and red colors representing the forested floodplain between them. Despite the steep convex slopes of the hills, they are covered by a continuous soil and weathered layer (saprolite; c.f. Martin, 1967: *demi-oranges* in the Belabo region of the Sanaga catchment north of the study areas). When anthropogenic influence is low, the hills are mostly covered by secondary forest.

No debris occurs at the footslope. At several locations, pisolitic layers of a few centimeters in thickness could be observed ranging into the floodplain and sometimes into the river bed. Some pisolites were recovered from a depth of 4.5 m out of the Nyong River some 20 m from the bank. They occur mainly along paths leading downslope in a linear manner.

At different locations, residues of ferricrets crop out of the lower slope forming distinct knick points and terraces. In and around the tributaries of the Nyong River, sandy sediments cover caolinitic material of the deeply weathered saprolite. A small dark organic layer limits the sandy sediments at the top. The sediments' thickness increases to around 2 m towards the receiving stream. The aquatic floodplain has prevented sediment recovering; however, due to a comparable appearance of the floodplain receiving stream and tributaries in satellite imagery, a thin sediment layer of a few meters is assumed covering planated rocks or saprolite, respectively. At the level of Ayos and Akonolinga, corings and satellite imagery strongly suggest a continuous shift of the river in its floodplain. Sometimes terraces occur oriented at a structural strike direction to the NNW-SSE suggestive of a structural influence of the river course besides that of fluvial activity. The river gradient is very low. Along 200 km, the river surmounts 20 m in height. The drainage density is also low (1.08 Dd) reflecting the wide floodplains' high moisture content, and this only shows up as a single river course in the analysis.

Anastomosing river

At the transition between the peneplain step and Inner Plateau, the Ntem anastomoses over a length of more than 10 km. The Ntem interior delta (Runge *et al.*, 2006) is formed on a surface slightly inclined to the NW. The orientation of the river to the NW is interrupted by several linear units of an E-W orientation. The interior delta is divided by an inselberg range, the Collines d'Akom, striking into the delta from the SW to NE, slanting down to the NE. In the North the basin is limited by a steep slope. The drainage network north of the delta drains towards the South following the inclination; however, just before the steep slope the rivers turn to the WNW, homogenously flowing into the Ntem at the level of the Collines d'Akom. The most southern branch of the Ntem outlines the sediments of the basin before a smooth rise limits the delta. Within the delta the relief energy is low characterized by only some minor hills covered by thick regolith. At some places ferricretes crop out; these were also identified by corings along a transect in the delta. They do not occur at one level; and they are probably stepped.

Apparently erratic, flat outcrops of the basement occur in the delta. Sometimes large quarzitic linear veins of several centimeters in height are eroded out of the basement. Also linear steps of some centimeters arise out of the rock. The hydrological network shows rapids at the outcrops. Beside sandy to clayey, sometimes organic sediments, at some locations extensive fanglomerate occur, consisting of rounded to angular clasts hardened in a matrix with iron and manganese oxide. Sometimes it forms layers of 50 cm in thickness.

The interior delta ends in a deeply incised V-shaped valley into which the branches of Ntem River flow.

Comparable river sections as found in the Ntem interior delta do not exist any more along the Ntem and Nyong Rivers. Infrequently, the main channel fans out to some branches along short distances. At the Chutes de Dehané, the Nyong River leaves the structural pattern by crossing the transition between basement rocks and sediments using three branches which flow together directly after the transition. The cataracts are defined by sandy sediments and high inclination. A comparable structural pattern occurs in the interior delta; however, the Ntem crosses the structures several times, which acts as a sediment trap in the study area. Hence, the huge extension and step-like formation are important evidence for this form unit and, therefore, for the utilization as a palaeoenvironmental archive.

Peneplain step with inselberg escarpment and deeply incised valleys

The main unit of the peneplain step in the study area, are the Ngovayang and Nkolebengue inselberg escarpments in front of the step, which force the rivers to flow around. The step itself is much accentuated. At a distance of 100 km, the Nyong and Ntem Rivers cross the step. In a related manner, the watershed is also accentuated. The river courses flow mainly in deeply incised V-shaped valleys. They orient distinctly at the tectonic structures and have a partially rectilinear hydrological network. The changes of direction are combined with gradient increase marked by rapids, cataracts, and a turbulent flow. The sandy sediments are deposited in sparsely occuring slack waters. Extensive sedimentation areas do not occur (c.f. Kuete 1990:44ff.).

Sometimes the basement crops directly out of the river bed. Only a few sections have deep water. Combined with the high gradient, it leads to the formation of potholes throughout the region and also in parts of the lower streams along the coastal surface.

Infrequently, clasts are deposited in the Nyong catchment below the Chutes de Dehané and the Chutes de Njok. Extensive areas of fanglomerates hardened in a matrix of iron and manganese oxide occur below the Chutes de Menvé'élé at the Ntem's interior delta and also in the V-shaped valley along which the Ntem crosses the peneplain step. The content of the fanglomerate differs at different locations. Sometimes rock fragments or angular clasts prevail; at others, they contain well-rounded pebbles. The petrography is related to the geology of the respective region.

3.3 DISCUSSION

3.3.1 Peneplain and inselbergs

Manifold theories concerning the evolution of peneplains and inselbergs exist. They were evolved and enhanced by several geomorphologists since the beginning of the last century. An overview is given by Thomas (1994) and Wirthmann (1994). The work of Büdel (1957: etchplanation—"Doppelte Einebnung") and of King (1953: pediplanation) highlight the difference between climatic-based and structural-geomorphologic-based explanatory approaches. There is no explicit theory prevailing in tropical geomorphology.

The various theories are controversial. Citations from two researchers concerning pediplanation illuminate the schools of though: Wirthmann (1994) considers that "zwischen Rumpfflächen auf Gondwana-Untergrund und typischen Pedimenten [...] selbst bei flüchtiger Betrachtung, sowohl was die Neigungsverhältnisse und die allgemeinen Formencharakteristika (Schwemmfächer bei Pedimenten, Mulden bei

Rumpfflächen) betrifft wie auch hinsichtlich der Durchgangssedimentdecke und der morphodynamischen Kopplung mit dem gebirgigen Rückland trotz gewisser Übergangsformen so gravierende Unterschiede [bestehen], dass die von King ausgehende Begriffsauflösung nur als Rückschritt für die Geomorphologie anzusehen ist" ["There are grave differences between Gondwana peneplains and typical pediments concerning both gradient and general forms (alluvial fans on pediments, depressions on peneplains) as well as morphodynamic connectivity with the mountainous hinterland. Therefore, King's explanation is a regression of geomorphology"]. Conversely, Ahnert (2003:297) writes in his introduction to geomorphology: "Diese These des südafrikanischen Geomorphologen L.C. King (1953), dass viele Rumpfflächen in Wahrheit aus durch Pediplanation zusammengewachsenen Pedimenten entstanden sind, hat viel für sich, zumal Pedimentbildung rascher vor sich geht und damit effizienter ist als andere Entstehungsweisen von Rumpfflächen." ["This theory of the South African geomorphologist L.C. King (1953) explaining the peneplain formation by consolidation of pediments has something to be said for it; especially regarding high velocity and, therefore, higher efficiency than other formation processes of peneplains"].

Segalen (1967:165; 1969) also bases his work on King's theory of pediplanation (1953). However, following a single theoretical doctrine is too limiting; a combination of different theories is more appropriate, particularly in view of the diversity of geomorphological forms found in Southern Cameroon.

Pisolites in a regolith layer and planation processes (I)

Bitom and Volkoff (1993) describe different conditions of the process of iron crust decomposition in a tropical humid environment based on an exposure site close to Sangmelima in Southern Cameroon (ca. 3°N, 12°E); here there is a transition from hard ferricret to a soft iron layer having hematitic concretions ('nodules'; c.f. Thomas, 1994:95f). The process starts with goethitisation, the conversion of the cementing material, which originally cemented the nodules to a softer material. The eluviation of iron ('déferruginisation') continues reduction of the hematitic material. Finally the soil is subject to discharge of material and, hence, to a change of its texture ('pédoturbation').

This sequence of processes leads to a clayey soil mainly composed of caolinit and goethite and, in some parts, of hematite (Bitom and Volkoff, 1993:1453). On the summit, they identified the highest amount of pisolites. Like the nodules, the pisolites are composed mainly of hematite; however, they show a goethitic crust (Bitom and Volkoff, 1993:1451). Nahon (1986) delineates a reformation of nodules to pisolites as a final state of ferricret evolution wherein the pisolites are embedded in an earthy matrix. The concentrated appearance of pisolites, therefore, tends to be a long and intensive phase of hardening (c.f. Kuete, 1990:131). Repeated alternation of humid and arid climates can lead to a change in the sequence of the processes (c.f. Martin and Volkoff, 1990:133). If the ferricret was be in a humid phase of dialysis, the pisolites may left as a resistant weathering product of the crust.

If the exposure site at Ambam (Figure 17) was to be interpreted in this way, the pisolitic layer would represent the residuum of a ferricret with the mottled zone in the underlying bed. The clear transition between the covering sediment and the pisolitic layer corresponds to a crusted palaeosurface, which is delicately preserved in contrast to the recent surface. In some places an inversion of the relief becomes obvious: Figure 37 shows the outcropping gneiss in the foreground with its spheroidal weathering. The saprolite and the covering layers are cut. In the background, the pisolitic layer forms a depression in contrast to the smooth summit of the surface.

In the center, the surface cuts the layer. Fölster (1969:5) discusses relief inversion in the context of pedimentation: crusts, which were formed in depressions in the watershed and resist due to their hardness, while the surrounding material erodes. The situation at Ambam differs fundamentally. The resistance of the former crust was reduced by reformation to a pisolitic layer (Bitom and Volkoff, 1993), which was cut. The postulated palaeo-surface is more accentuated than the recent surface. Sometimes small steps occur. Its appearance is reminiscent of ferricretes; however, it is not a continuous plane. Obviously, during the evolution of the crust, the climate was different from recent as evinced by the erosion of the covering layer over the plinthic layer (B_u horizon/accumulation horizon; c.f. McFarlane, 1976), grooves and small gullies were cut in the surface and the dessication of the plinthic horizon to the ferricret occurred. According to the hypothesis respecting formation of the palaeo-surface, the recent sediment cover was deposited allochthonous after the phase of denudation and the phase of dessication of the ferricret. So-called "hillwash" has its origin in fine sediments, which occur during pedimentation (Fölster, 1969). If the ferricret had already obtained its final genetic stage (pisolitic horizon in earthy matrix) a mixing must have occurred in the direct contact zone. This mixing could not be observed.

Fölster (1969:50) describes a sequence of stony layer, pediment gravels and hill-wash as 'pedisediment' at his study area in Nigeria. The stony layer consists of unweathered rocks of eroded saprolite like quartzitic rocks of partially weathered quartz veins close to the back weathering edge. The pediment gravels underlay was transported by water and deposited higher up to the weathering edge over the stony layer. Fölster (1969) distinguishes between pediment gravels and stony layer by the fine and mainly sorted dendritic material over the coarse stony layer. The stony layer and pediment gravels are missing at the Ambam site. The pisolitic layer is covered by up to 5 m thick hillwash sediments. Probably, the origin of the sediments is not only from backward weathering but also the result of denudative processes due to repeated redeposition of fine sediments given the sparse vegetation cover (c.f. Fölster, 1969:50f). The covering of the crust led to the 'déferruginisation' of the cementing material. If the deposition and redeposition of the hillwash sediments occurred, equalizing the relief during pediplanation, this pediplain was reformed afterwards. Therefore, three generations of surfaces exist and this is apparent from the exposure site: the palaeosurface represented by the pisolitic horizon, the pediplain represented by the hillwash sediments, and the recent surface defined by the wide summits and depressions.

Segalen (1969:118ff.; c.f. Hurault, 1967) focused on the evolution of stone lines linked with climate induced peneplain formation in the African tropics. He mentions two theories of peneplain formation, which differ in the climatic preconditions. The "pédiplaine multiconcave" evolves under a semi-humid climate defined by seasonality with an intensive dry season and rainy season. The form units are comparable to the work of Fölster (1969), especially the evolution of the stone lines and pediment gravels. The "pénéplaine multiconvexe" evolves out of the "pédiplaine multiconcave" under an equatorial climate having humid conditions and a short dry season. The evolution of this "almost planated surface" is characterized by chemical weathering processes at structurally favourable locations, sub-soil erosion and, therefore, a lowering of the surface at such locations (Segalen, 1969). The processes intensify the relief energy. However, detailed research on dissolved load of south Cameroonian rivers does not exist. Therefore an implicit causal statement to sub-terranean discharge can not be made. However, Wirthmann (1994:88f) underscores the certainty of surface lowering in the tropics due to a reduction of the basement rocks volume by weathering. The exposure site at Ambam shows considerable transformation of the relief by extensive denudation processes at the surface under a semi-humid tropical climate regime. Thus, the formation of the current surface is based on an increase of seasonality with a distinctive dry season. This said, an indicator of selective weathering with surface lowering could not be identified along the exposure site.

Occurrence and evolution of inselbergs

In the northern upper catchments of the Ntem River several inselbergs surmount the peneplain. Kuete (1990:243) describes the forms as incomplete planation in the "biostasy" phase having active soil formation, which will be replaced by a phase of "rhexistasy" during which mainly planation processes will occur. The terms are derived from Erhart (1967). Tectonic, climatic, and anthropogenic changes, too, can lead to a change of processes.

Based on prominent indentations in the rock, Kuete (1990:243ff) identified three different planation phases. He assumed that the thickness of the regolith increases during the biostasy phase by chemical weathering. Pedogenesis leads to chemical decomposition of the inselberg at the level of the soil. The large karren (pseudokarst) at the inselberg, hint to high water discharge along the rock and therefore to an intensification of hydrolysis and discharge of soluted material in the immediate surroundings (c.f. Bremer, 1971). During rhexistasy, the regolith cover will be partially eroded leaving indentations as indicators of the former level.

This interpretation does not require a really thick regolith cover. During the "large morphogenetically crises" (Kuete, 1990:244), a cover of about 80 m thick was eroded. He calculated 1 m of weathered material for each 77 ka at Ako'Akas for a timespan of 6.16 Ma. The value probably comes from a calculation of Leneuf and Aubert (1960) for Ivory Coast; however, only F.M [?] Thomas is cited without specific reference to the date of publication (Thomas, 1974). For the whole inselberg escarpment in the surroundings of Ako'Akas, he calculates a duration of 14 Ma for the biostasy and a comparable duration of rhexistasy and arrives at the Upper Oligocene or Lower Miocene as the starting age of weathering. This result fits Segalen's (1967) timeline and, therefore, the planation period of the African Surface II in the Miocene. Nevertheless, he does not consider the distinctly higher inselbergs, up to 420 m above the surrounding surface, some 15 km WNW of Ako'Akas. Additionally, a clearly limited amount of weathering is scrutinized without knowledge of the climatic parameters. Leneuf and Aubert (1960; cited after Thomas 1994:48ff) indicate a span of 13–45.5 m each million years. The lowest value corresponds to Kuete's (1990:246) calculations. The indentations in the rock as evidence of former weathering levels seems reasonable. Whereas the approach is based on the pediplanation model of King (1953), the indentations should have corresponding steps in the field, which moved northwards towards the higher surface by headward erosion. However, they could not be identified by elevation data of the study area.

The recent transition between the two planation levels is much accentuated and does not correspond to a lithological shift. It is characterized by a thick hydrological network eroding the peneplain step. The peneplain step does not have a step-like character suggestive of different levels of planation since the Miocene. The theory of Büdel (1957) is pertinent in the context of the indentations; according to him they represent the steps of extensive lowering of the landscape by denudation during different semi-humid phases. At large outcrops of comparable inselbergs in the same regional unit, large-scale desquamation could be observed. This seems to be the most plausible process leading to indentation formation. The recently active process of dissolution of the rock (pseudo-karst) leads to an alteration of the forms and therefore to a complication of the interpretation.

I refrain from rating the indentations as pointers for phases of surface lowering, and assume that the inselbergs were formed by headward erosion starting at the Ntem faults as postulated at the beginning of this chapter (c.f. Penck, 1924:175: "Zone der Inselberge"; inselberg zone). It is probable that inselberg formation is based on structural conditions. Below Ako'Akas, there are more inselbergs, strung to a chain in a NE direction. Erosion along fissures vulnerable to weathering has led to the recent landscape independent of geology (Maurizot, 2000: heterogeneity of granite, gneiss, and charnockite). The forming of the cumulative existing inselbergs probably started simultaneously with planation on the coastal surface. Uplift along the Ntem faults was the trigger for headward erosion and inselberg formation, respectively. Parallel to the Ntem faults, the recent step between the Ntem catchment and the plateau in the North occurs mostly covered by the Nyong catchment.

Evolution of abandoned channels

The river network of the Cameroonian swell is for the most part structurally conditioned; therefore, it can be used as a pointer for geological structures in the study area (c.f. Kuete, 1990; also Bremer, 1971, 1989, and Wirthmann, 1994, for tropical regions and their drainage network in general). The structure of tectonically-formed river courses are discussed extensively in chapter 3.3.3.

In contrast to the structural river courses close to Nkolmvondo, the Nyong abandoned a channel in its transition between the upper and middle catchment. Additionally, there are only abandoned channels along the large floodplain of the Nyong at the level of Akonolinga; these channels are formed in the river's alluvia in contrast to the Nkolmvondo channel, which is formed into the peneplain.

The confluence of the Mvondo with the Nyong is rated as an abandoned channel. The geomorphological habitus shows a bend of 100 m in diameter having two connections with the Nyong. Into this nearly lacustrine form, the small Nyong tributaries Mvondo and Essaa flow, which leads to the prevention of a total cut of the channel, due to the regular water supply. The eastern part of the oxbow orients at a fault having a WSW-ENE direction (Maurizot, 2000). The direction can also be identified in the Nyong's course just before the break through of the channel. Therefore, it seems to represent a further main structure of the region (Figure 11). Due to the structural form and the very linear course also at the eastern abandoned channel, a cut-off by a downstream shift of the meander having different shift velocities (c.f. Ahnert, 2003:225) can be excluded. Additionally, there is no cut bank formed at the level of the breakthrough. The Nyong has an almost rectilinear adjustment of its river course. In the West, the channel is limited by outcropping mica schists, which prevent a shift of the channel, too. Obviously, the river experienced a shift to the East from the downstream side of the breakthrough until the cut-off took place.

A wide meander bow was formed in the Nyong upstreams near Edjom. After a nearly linear course to the SSW, the river course changes to WSW along a fault. At this location, the river has shifted westwards in a few distinct steps, before it cut through its own sediments to the NNW. There is no abandoned channel; hence, the river has formed several small draws filled by water. They do not emerge clearly in the landscape because they are covered by a solid net of grasses. Using satellite imagery, the draws are identified as meanders showing different phases of shifting (c.f. Figure 10). A structural corset of the drainage network, which cannot be left by the river, is supported by this observation. Fluvial geomorphological forms exist but they also orient at the structural presets of the region. Both examples show shifts of the river course at locations partly shaped by basement structures alone. A chronological classification of the cut-off is difficult. Organic material at Edjom shows an age of about 13 cal. kyrs BP reflecting only one aspect of a complex evolutionary history.

Summary of the morphogenesis

At the same time of intrusive activity along the CVL, an extensive uplift occurred triggering planation of African Surface I in the Eocene. The collision of the Eurasian and African Plate triggered uplift and planation in African Surface II (Burke and Whiteman, 1973; Burke, 1996; Segalen, 1967). In Cameroon, the Inner Plateau (African Surface I) and the coastal surface (African Surface II) represent both phases.

The recent form units having peneplain, inselbergs and rivers with a very low gradient over long distances and small hypsometric integrals on the upper catchment area are indicators of the planation. With a regional focus on the transition between the northern Inner Plateau and the limiting plateau in the South, the evolution of the Inner Plateau will be sketched.

It is probable that the plateau originally was formed by a wide undulated syncline. With the start of plateau lowering during the Eocene, inselbergs were shaped; nowadays these are found surmounting the plateau at Yaoundé. The inselberg range of the Ntem's northern catchment area having heights above 1000 m, is also part of it. Triggered by the Miocene uplift, geomorphologic reshaping of the Nyong and Ntem catchments started. A tectonic step slightly slanting down to the North was formed by remobilization of the Ntem faults (Figure 45). The structural preset along the Ntem is highlighted by the gradient curves of the tributaries flowing from the South into the Ntem.

The tectonically originated step was shifted northwards by headward erosion—mostly independent from geology—and became accentuated by fluvial erosion. The gradient curves of these tributaries show much more relief than the southern ones. The view at a relatively small scale also shows the linear character of the recent step along fissures oriented towards the NNE-SSW and WSW-ENE along which erosion could take place. Originating from the thalweg, the valleys proceeded to the plain (pedimentation—pediplanation). The inselbergs evolved by fluvial morphological selection along fissure zones primarily as bornhardts (c.f. Wirthmann, 1994:178f; Thomas, 1994:328ff). The inselbergs already formed before surface lowering rose higher by approximately the value of the lowering.

The hypothesis of the shifted tectonically-formed step explains the distinct border between the northern and southern Ntem watershed and the different surface levels of the Inner Plateau between the Nyong and Ntem watersheds. So far, the surface and inselberg formation is presented in a simplified and somewhat exaggerated manner. It remains to be discussed if the individual considerations on the hillwash cover and the peneplain step, both in a homogeneous geology (see 3.2.2), allow accentuation of pedimentation and pediplanation, the postulated primary processes of surface formation.

Both theories are based on the validity of the described processes characterizing the phases of stability and activity (Rohdenburg, 1970, 1982) as well as the phases of biostasy and rhexistasy (Erhard, 1967). The multifaceted exposure site at Ambam

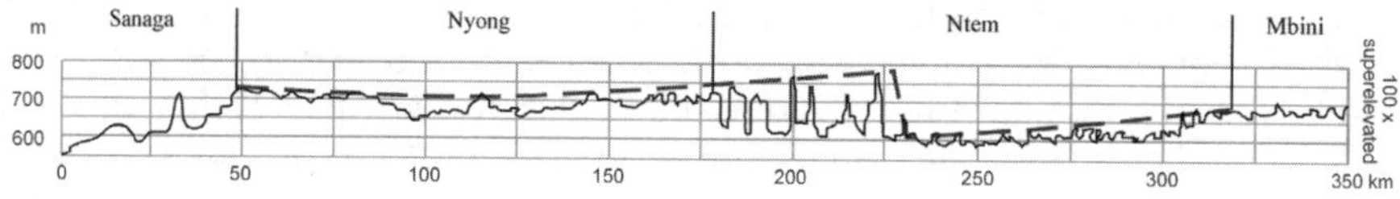

Figure 45. N-S section cutting different levels of the Inner Plateau (based on SRTM data).

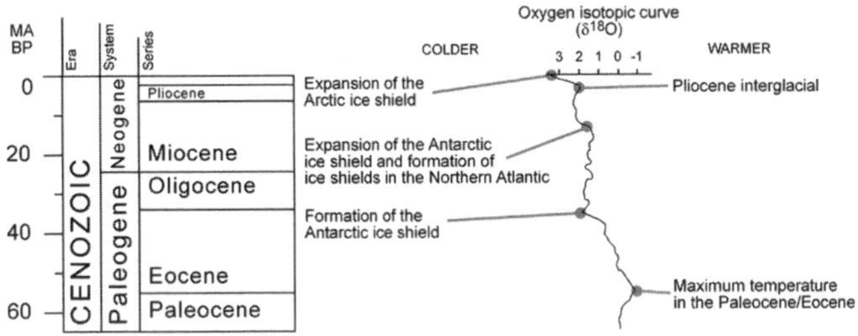

Figure 46. Cenozoic glacial and interglacial periods according to oxygen isotopic stages (after Eyles, 2008).

reflects climatic changes by the following processes: evolution of plinthic horizon (humid), hardening to a crust (arid), overburden of the crust (arid), solution of the crust (humid), and the reformation of the whole profile (arid).

The formation of the deeply weathered Latosol with the plinthic horizon probably had already occurred during the Tertiary given its warm and humid climate; the crust formed during increased aridization in the Pliocene and Lower Pleistocene. The accumulation of the hillwash cover and the transformation of the peneplain to an accentuated surface can be correlated to climatic fluctuations during Pleistocene times (c.f. Eyles, 2008; Thomas, 1994:215ff; Figure 46). In this late phase, the cut-off of the Nyong at Nkolmvondo and the shift of the Nyong's meander at Edjom took place, related to the LGM and the Younger Dryas.

3.3.2 "Demi-oranges" and "bas-fonds"

The disproportion between river and floodplain led to comprehensive considerations of Kuete (1990) concerning the evolution of the Nyong's wide floodplain or the "bas fonds des régions humides" (Raunet, 1985). In common parlance, the Nyong is called "fleuve paresseux" (sluggish river). In what set of circumstances should the valley have been formed by the river?

Rohdenburg (1982:82ff) explains the evolution of so-called "meias laranjas" (*demi-oranges* or half oranges) in his study area in Brazil by lateral and vertical erosion during an arid phase of geomorphodynamic activity. Along newly incised thalwegs, steep slopes (>35°) will be successively shifted backwards by lateral erosion. The wider the valley floor is developed, the more lateral and vertical erosion is placed to the rear in behalf of separate erosional processes of the slope. Therefore, Rohdenburg (1982:84) assumes a relief composition of the valley floor and pediment, which is over formed by flooding processes. Due to thick saprolite—only partly composed of kaolinite—denundation processes can easily commence at the slopes. The summits of the *demi-oranges* represent former surfaces or valley floors. Sometimes, additional generations of landscapes can be identified whose surface residues were already cut to more flat *demi-oranges* overformed during a further phase of geomorphological activity. Gravels, indicators of the former floodplain, were redeposited in the slope area of the younger generations.

Ferricret formation and planation processes (II)

Kuete (1990:206) describes a terrace of the Nyong near Akonolinga above the recent floodplain, which is built by a ferricret of strongly weathered mica schists. He postulates that incision of the valley took place around the Late Tertiary/Early Quaternary.

During my field studies, a residue of a crust was found within the sediments directly below the slope. It probably originates from the ferricretic terrace by scarp retreat during a phase of geomorphodynamic activity; afterwards, it was covered by the alluvial sediments of the Nyong, dated to 32.8 cal. kyrs BP using organic material within the sediments. Hence, the activity must have been prior to this date. Another organic layer likely a palaeosurface was dated to 46.7 cal. kyrs BP, frames a further phase of activity. These datings are not in agreement with Kuete's (1990) hypothesis of a Late Tertiary origin; however, they reflect phases of geomorphodynamic activity after the initial incision, which were probably triggered by climate-induced changes of the landscape. Both dates fall within the Ndjilien (40–30 kyrs BP, OIS 3), a humid phase in Late Pleistocene times. Before this phase, a long arid phase occurred, the Maluékien (70–40 kyrs. BP, OIS 3/4), when scarp retreat probably started (c.f. De Ploey, 1969; Lanfranchi and Schwartz, 1991; Kuete, 1990; Figure 47). Residues of crusts also occur at Abong Mbang directly in front of the slope of the flooplain, which is aquatic in this region. The scarp retreat and, therefore, the separation of the ferricretic blocs seem to be a sub-recent process due to their occurrence directly on the floodplain without any sediment cover. A comparable situation was also identified at a location near Akonolinga where a source inundates the near-slope floodplain. Obviously, the permanent inundation leads to the successive scarp retreat at the source location and at the aquatic floodplain at Abong Mbang.

A further trigger for erosional processes is anthropogenic impact. The Abong Mbang location is characterized by a frequently used footpath to the river, and tillage farming above. The location at Akonolinga shows a cacao field laid out on the thin soil of the ferricretic terrace.

The hypothesis of Rohdenburg (1982) on the evolution of *demi-oranges* could not be proved nor falsified due to missing insights into the below ground layers at

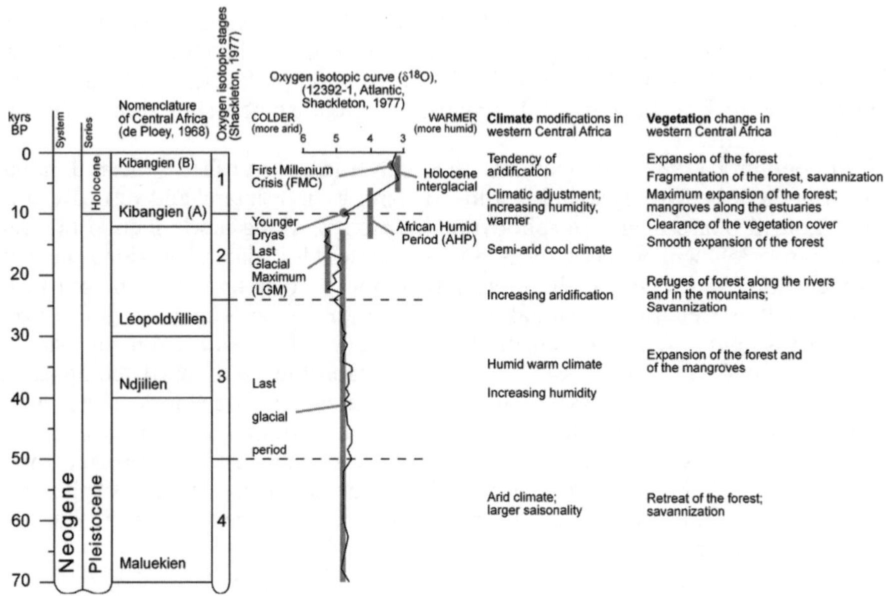

Figure 47. Late Pleistocene glacials and interglacials correlated to Central African nomenclature (de Ploey, 1968), oxygen isotopic stages (Shackleton, 1977), as well as climate and vegetation changes (after de Ploey, 1968; Lanfranchi and Schwartz, 1991, 1993; Oslisly and Mbida, 2001; Runge, 2001).

the transition between floodplain and hill. At road works close to Ayos, a regolith cover of approximately 6 m over the basement could be observed (Figure 48). If the cover continues with the same thickness at the transition zone of the footslope, then Rohdenburg's approach (1982) could explain the phenomenon, e.g., that the weathered saprolite is not so much reduced to kaolinite as found in Brazil. Kaolinitic saprolite was identified at several places along the Nyong's tributaries below a thin layer of organic material and sand, as well in colluvial material directly at the slope. Grain sizes and thickness of the alluvial cover increases from the source (80–100 cm) to the confluence with the receiving stream (250–300 cm). Distinct channels could only be identified at some tributaries close to the confluence. This observation recalls forms and processes of dambos in the semi-humid tropics (c.f. Mäckel, 1975, 1985; Raunet, 1985; see also Büdel, 1965, 'Spülmulde'; Louis, 1968, 'Flachmuldental'). Mäckel (1985:14) describes the dambo soil as a dark and sandy clayey loam of a thickness between 70 to 200 cm. He pictures a central humid area and a lower and upper denundation belt at the border of a dambo limited by a convex slope. During a warm and semi-dry climate (800–1300 mm annual precipitation, Mäckel, 1975:121), an enlargement of the dambo took place at the denundation belt as an active process of surface formation (Mäckel, 1975:34; c.f. Raunet, 1985:43). Aside from the geomorphological form, physiogeographic features such as climate and vegetation are different from that of the East African region where dambos occur. A dating of 1.9 cal. kyrs BP of organic material found directly above the saprolite at 260 cm depth of a Nyong tributary will be discussed in detail as there is no other reference dating (see Figure 8). Given that where dambos occur no drastic changes in the ecosystem over the last 2000 years happened, an enormous accumulation took place of approximately 13 cm/100 years. It is likely that continuous humidity during the rainy season led to landslips at the slope. Additionally, anthropogenic influence by clearing and tillage farming on the summits and also on the slope could cause denudation. However, landslips could not be identified. The ferricretic blocs at the surface of the floodplain hint to scarp retreat but no blocs were found close to the dated material. The similarity of the source

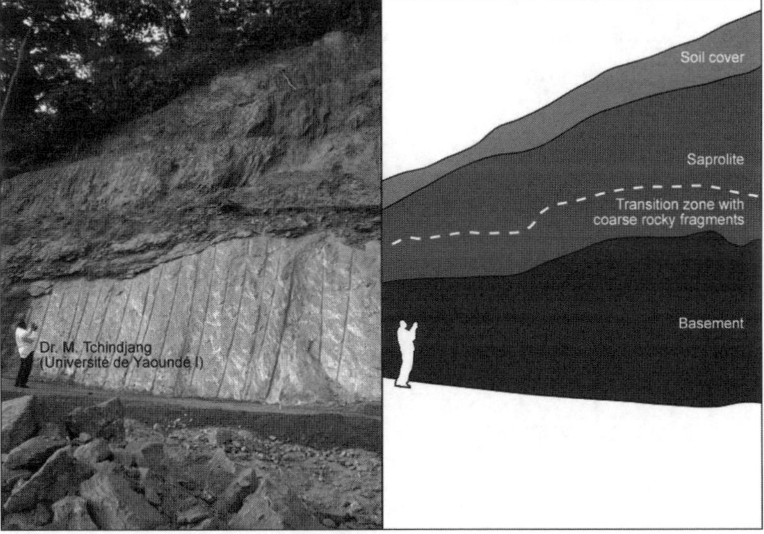

Figure 48. Quarry south of the Nyong.

areas of the tributaries and the dambos lead to the hypothesis of a formation of this geomorphological form under an arid climate—probably by denudation processes— and a reformation under the current humid tropical climate. To test the hypothesis, catenae have to be constructed running through the source areas of the tributaries to compare the corings with those of a dambo. After Webster (1965:36), the central humid area is characterized by a high clay content, which is reduced in the denudation belt.

Denudation at the source areas of the Nyong's tributaries and linear erosion followed by lateral erosion and pedimentation of the slope in the main valley could have occured in combination during an arid phase of geomorphodynamic activity. Due to the oldest date in the Nyong flooplain at Akonolinga, a connection to the Maluékien (70–40 kyrs) can be made. This classification does not support the hypothesis of Kuete (1990; Late Tertiary, Early Quaternary).

Antecedent drainage stream

Kuete (1990:467ff) traces the large Nyong floodplain between Abong Mbang and Akonolinga to tectonic activity. He assumes a N-S oriented dam by a quartzitic ridge impounding the water in a large lake, which is reflected by the recent floodplain. The structural orientation reflects one of the main directions of the lithological relief units of the plateau. Another direction is SW-NE. The Nyong has to cross both directions by its mean westerly flow direction. Indeed, downstream from Akonolinga, an abrupt change of the flow direction of the Nyong occurs with rapids in its course at the point of the former dam (Figure 49).

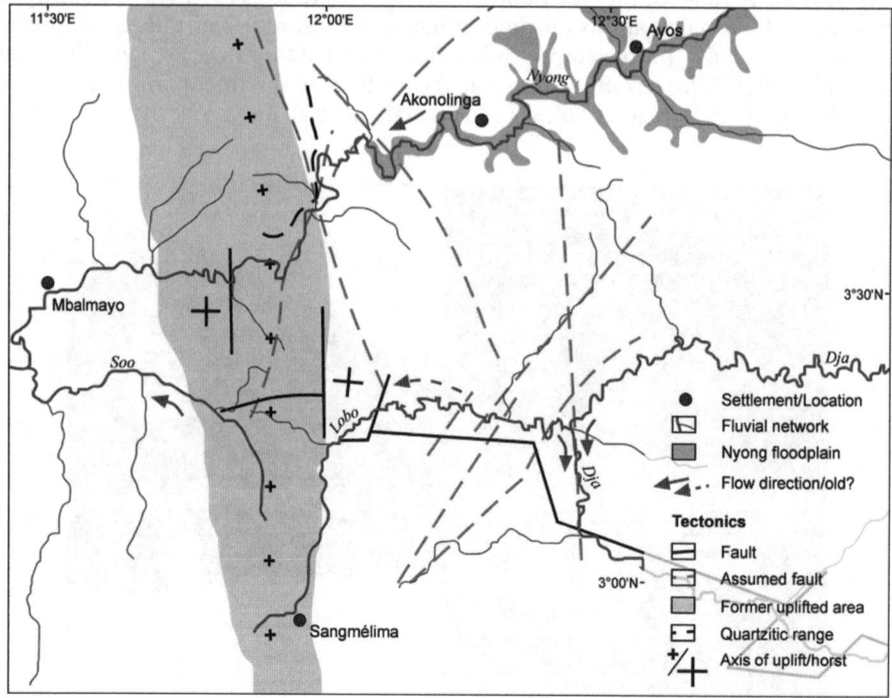

Figure 49. Sketch map explaining the evolution of the *bas-fonds* in the upper catchment of the Nyong (Kuete, 1990, modified).

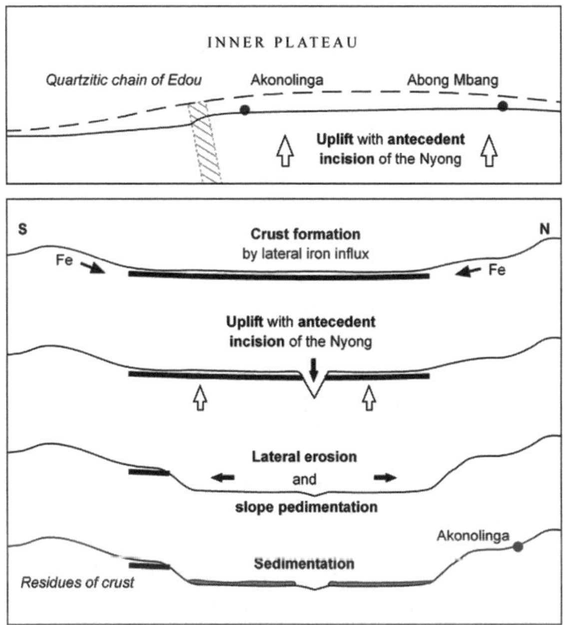

Figure 50. Hypothetical trigger and geomorphic evolution of the *bas-fonds* at Akonolinga.

The theory of Kuete (1990) will be considered at another scale. Given that the Ntem watershed has comparable forms in a region with different geological presets, it becomes obvious that the form unit of *demi-oranges* and large floodplains have to be discussed at a larger scale. The *demi-oranges* seem to be formed over a wider area dominating the landscape—both floodplains are 180 km apart on different surface levels.

A trigger for incision must have occurred as an initial process leading to planation and *demi-orange* formation. Kuete (1990:208) describes a slight rise of the Inner Plateau's summits to the West, which contrasts with the flow direction of the Nyong. This observation would militate for antecedent incursion of the river into an uplifted region; hence, Kuete (1990) assumes a small-scale uplift in the West of the wide floodplain without lateral erosion. His observations concerning a rise of the summits to the West could not be confirmed. Probably an extensive uplift of low magnitude led to the antecedental incision during an arid climate. The magnitude of uplift needs to be 10 m, representing the difference between the ferricretic crust and the recent level of the river; the temporal placement corresponds to the formation of the crust and beginning of sedimentation in the floodplain (Figure 50). The linear erosion kept up with the uplift except at the quartzitic ridge of Edou (c.f. Kuete, 1990:467) representing a lithological limit along which the Nyong formed rapids in recent times (Figure 50).

Terraces

Another hypothesis of Kuete (1990) refers to the formation of the Nyong along a graben filled up successively by river sediments. The Nyong has a wide floodplain, but its meanders orient primarily towards structural lines. Due to missing dendritic material Kuete desists from his hypothesis (1990:208).

At a large scale, the limit of the Nyong floodplain orients towards one direction; however, at a smaller scale, the orientation changes several times. Therefore, it could not be assigned to a graben fault as a whole.

East of Akonolinga, the limit of the Nyong floodplain has linear units suggestive of a descended bloc. The satellite image (Landsat, Path 185, Row 57, 16.01.2002) shows a distinct transition between different textures. Particularly on the left bank, two sublinear terrace levels orienting to the NNW-SSE a distinct transition, which becomes really obvious due to a light texture, representing the herbaceous vegetation above the scarp of the terrace, and a darker texture, representing the higher moisture content close to the river below. Both terraces show a slight rise to their front. Probably this rise originates from a sub-recent remobilization of a underlying graben structure.

The oblique position of the lower terrace might also be strengthen by fluvial processes due to high moisture on the surface during the dry season, leading to to high moisture penetration into the terrace body and to higher linear erosion during the rainy season. The presence of woody vegetation along the terrace fronts indicates flooding of short duration during the rainy season. The vegetation leads to a higher resistance against erosion and therefore to an intensification of the oblique impression.

The substratum of the higher terrace underlay an earlier formed soil having bright red mottles suggestive of oxidation processes. This substratum is eroded below—obviously along the postulated tectonic weakness. The lower terrace is probably formed by sub-recent material showing lots of charcoal residues in the mainly clayey and silty darkish profile. The charcoal suggests that sedimentation started when the land was cleared by burning in the floodplain, something which began around 1000 yrs BP (pers. comm. M. Tchindjang). The clearings engender an increasing erosion of sediments due to missing protection by vegetation mixed with the charcoal formed by fire. A comparable transition of the sediment's texture also occurs south of Akonolinga, close to the Nyong. This area shows neither comparable linear units nor terraces, which probably have been changed by fluvial and weathering activity without any tectonic impact.

The evolution of the terraces has to be classified chronologically before the formation of the oblique position; hence the charcoal fragments and, therefore, the clearings along the floodplain limit the neo-tectonic impetus (Figure 51).

River captures

Due to the dendritic orientation of the Nyong's drainage network, on a plateau between 650–750 m a.s.l. towards the thalweg of the syncline, river captures can be identified. The river capture by the Téré, a tributary of the Sanaga River, cut the largest area off the Nyong watershed. The position of the capture is easily to identify due to a distinct elbow of the capture forming a valley-floor divide in the floodplain.

The dendritic drainage network of the plateau encompasses 22,832 km^2, of which 87.53% are part of the Nyong basin. Against the main drainage orientation of the syncline, 5.31% flow to the Sanaga basin in the North, 6.05% to the Dja basin in the East, 0.68% to the Boumba basin also in the East, and 0.45% to the Ntem basin in the South (Figure 52; c.f. Figure 7; see also Kuete, 1990; Olivry, 1986 concerning the river captures of Dja and Téré). Detached from the syncline parts of the Dja and further small-scale catchments of the Ntem tributaries, and rivers flowing directly into the Atlantic Ocean are also part of the plateau in the East, the South, and the West. Kuete (1990) assumes that the Dja originally was a source river of the Nyong captured by the Congo drainage system (c.f. Olivry, 1986:106) during the damming of the Nyong River. The Nyong could incise antecedently, the Dja was not able to (c.f. Figure 49;

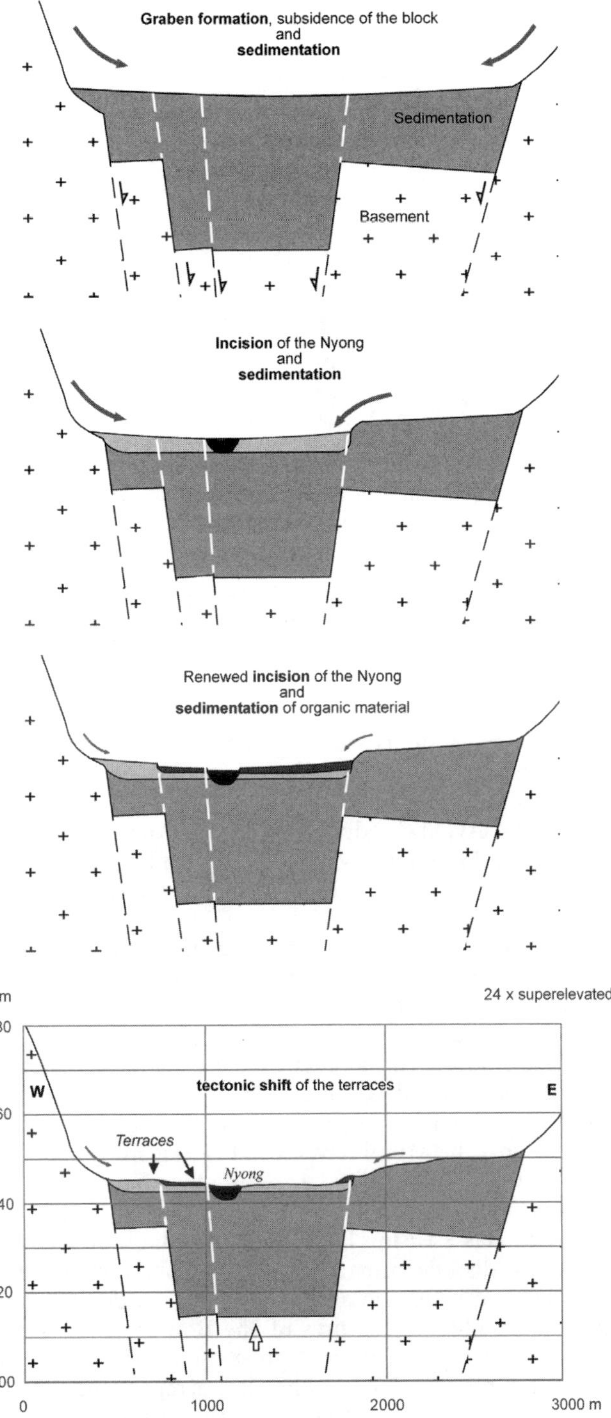

Figure 51. Sketch of a transect cutting the Nyong floodplain west of Akonolinga (Figure 9, transect A'B'; based on SRTM data, GPS editing in the field).

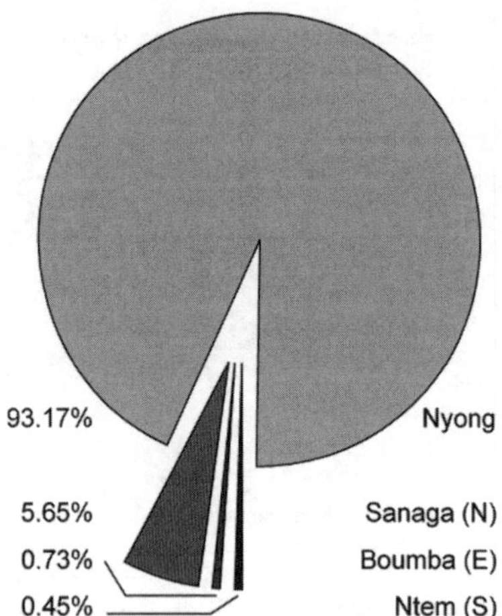

Figure 52. Pie chart showing the percentage of the former Nyong catchment area burden by river captures (based on SRTM data).

former drainage orientation in the region). The river course of the Dja supports this hypothesis: the Lobo, a tributary of the Dja in the West, is almost connected to the Nyong basin. However, the Dja forms a wide bend to the East going around the Dja Fauna Reserve, confluencing with the Boumba and flowing into the Sangha, a large tributary of the Congo.

Comparable drainage systems are also visible in the upper portion of the Ntem catchment. The watershed between the Kom, a source river of the Ntem, and the Ayina, a source river of the Ogooué, is difficult to identify on satellite imagery and the DLM. The Kom and Ayina floodplains are extensive as compared with the upper catchment of the Nyong floodplain as visualized by remote sensing. The upper tributaries of the Ayina orient towards the Kom and flow with an abrupt knick into their receiving stream.

Just below the watershed, the Ayina has incised into the landscape, but not the Kom (c.f. Santoir, 1995). Furthermore, parts of the southern catchment area of the Ntem were captured by other Ogooué tributaries towards the South.

In contrast to the reduction of the upper Nyong catchment, the Ntem catchment extends mainly to the South. An increase of the catchment's surface occurs at the transition between the Nyong and Mbini catchments comparable to the capture of the Nyong catchment by the Téré River. The plateau in the South, which shows a dendritic drainage network towards the Mbini River, will be extensively captured by tributaries of the Ntem (c.f. Figure 13, black dotted line).

Due to the transformation of the hydrological network the river captures can be classified into three classes using DLM data: (i) The first class is represented by river captures having a distinct orientation of the river towards the former receiving stream. The watershed is formed as a valley-floor divide. It is represented by the Téré capture and the small-scale captures of the Boumba and Ntem of the Nyong catchment, and the Ayina capture of the Nyong catchment as well as the Ntem captures of the Mbini catchment. (ii) The second class is represented by the Lobo capture of the Nyong catchment towards the Dja. The river capture is discussed by Kuete (1990) and Olivry (1986); however, its orientation to the Dja catchment is less obvious than that of the Téré to the Sanaga catchment. (iii) The third capture is represented by the Dja catchment, which was captured by the Congo system. The transformation of the capture is well advanced but the elbow of the capture is still obvious.

It is probable that the river captures can be associated to different phases of geomorphological activity; however, no datings could be conducted in the respective locations. By connecting valley formation of the upper Nyong with river captures, a rough temporal classification can be made. Downstream from Abong Mbang, the rest of the captured Téré River flows into the Nyong. This situation clearly indicates that the formation of the wide floodplain was not linked to the discharge of the Téré, as the width of the floodplain upstream from the confluence is the same. However, downstream from the confluence, the aquatic floodplain with its swamp forest vegetation gives way to sub-aquatic floodplain with its grassy vegetation. The vegetation of this ecosystem is directly linked to water content. Water content and vegetation are changed by the river capture; at very least, it promoted land clearance by slash-and-burn in the Nyong's floodplain.

It seems that river captures lead to a successive reduction of the flat plateau in the upper catchment of the Nyong in favour of the capturing lower regions characterized by the Congo system in the East and drainage towards the Atlantic Ocean in the North and the South. Moreover, the asymmetric form of the southern border of the Nyong catchment suggests a much larger former surface.

North of the E-W oriented thalweg of the Ntem, several inselbergs surmount the surface, these can be interpreted as residues of the higher surface in the North (Figure 45). Southwards only a few inselbergs occur on the surface that rises slowly to the adjacent plateau in the South. Figure 45 highlights a likely peneplain step, which was formed along the Ntem faults by tectonic processes. River capture is the main process of the extensive backward erosion and, therefore, of inselberg development after the step's formation. The former step is marked by a dotted line. The more accentuated and steeper gradient curves of the northern tributaries, in contrast to the southern, hint to the young drainage network.

Summary of the morphogenesis

The starting point of the morphogenetic interpretation is an undulated plateau formed during Eocene times (African Surface I). It is characterized by different levels generally occupied by the Nyong and Mbini catchments (650–750 m a.s.l.) and the Ntem catchment in the center (550–650 m a.s.l.). Due to the lower erosion level, the tributaries of the Congo system, the Dja and Boumba, cut backwards into the dendritic drainage pattern of the surface from the East. The same process took place from the South and Southeast by the Ogooué, and from the North by the Sanaga tributaries. Figure 45 highlights the former extension of the plateau.

Since the Miocene, the swell of the Congo Craton rose slightly due to thick sediment layers on the craton (c.f. Lucazeau *et al.*, 2003; Karner *et al.*, 1997). The uplift triggered the incision of the upper Nyong. On the peneplain, ferricretes were formed by a lateral influx of iron (Figure 50). The incision into, and the removal of, the ferricrete was encouraged by a long lasting arid phase. The oldest datings of organic sediments of the floodplain points to Maluékien times (70–40 kyrs BP). During Ndjilien times (40–30 kyrs BP), the accumulation of the sediments started. The uplift probably led to the remobilization of Pan-African and pre-Pan-African structures. In the study area, it is evinced by the shift of the block, the small-scale graben fault close to Akonolinga. The tilt of the terraces point to a (neo-) tectonic transformation.

The backward shift of the slope at the border of the Nyong floodplain seems to have been active until sub-recent times. Large residues of ferricrete occur in and on the sediment layer of the floodplain. Recent processes can not be identified; morphological stability is prevalent. Accumulation of colluvia is induced anthropogenically. Soil material and pisolites will be transported along pathways and cleared areas at the slope, triggered by heavy rainfall.

3.3.3 Anastomosing river network

"An anastomosing river is composed of two or more interconnected channels that enclose floodplains. This definition explicitly excludes the phenomenon of channel splitting by convex-up bar-like forms that characterize braided channels" (Makaske, 1998:149). Makaske (1998) describes avulsions primarily forming an anastomosing section. The river course shifts, divides, and will be connected again by avulsions. The branches downgraded by the avulsions in regard to their water discharge are active only for a short time.

The formation of anastomosing river units can be rated as a reaction of climatic changes, extreme flooding, or a tectonically induced disequilibrium in the river's gradient curve (c.f. Makaske, 1998; see also McCarthy, 1993; McCarthy *et al.*, 1993; Schumm *et al.*, 2000; Wang *et al.*, 2005). Furthermore, any combination of these features can trigger the formation. After a tectonic subsidence or barrier formation, the river reacts by the formation of an anastomosing river bed geometry, trying to reduce the inclination along the gradient curve (Schumm *et al.*, 2000). The attempt of a river to regain its stability leads to the equation of the tectonic impact by accumulation in front of uplifted barriers or in swells and to erosion at the uplifted unit. Therefore the existence of anastomosing river sections should, if triggered by tectonics, provide clues to the neotectonic influence that caused the ongoing disequilibrium. Tectonic influence of an anastomosing river formation is likely when forms occur along structural presets in combination with fault throws of up to several meters and rapids as well as cataracts limiting these sections. This is found at the end of the Ntem interior delta and at Dipikar Island in the Ntem catchment and at Dehané in the Nyong catchment.

Additionally, clasts provide clues to neotectonic activity. Below the Chutes de Dehané clasts were deposited along the bank. Clasts were also deposited in the Ntem interior delta with a thickness of up to 50 cm. Below the Chutes de Menvé'élé, which marks the end of the interior delta, they occur with a thickness of approximately 100 cm above water level. For the tropics, Bremer (1989:192f) thought that cataract evolution occurred during Tertiary times due to missing erosion tools. This proposition can be refuted for the study area.

It is inappropriate to link evolution of anastomosing units with simple faulting activity. Rather, it appears to be a complex mosaic of mainly Precambrian structures, which were remobilized during the youngest geological past. A first view of the Ntem interior delta by satellite imagery led to the idea that they formed along a transcurrent fault with different phases since the Late Pleistocene. However, the geological map (Maurizot, 2000) does not suggest a transcurrent fault. To test this hypothesis, the hydrological network was edited with special focus on first order rivers because of their sensibility to sub-recent faulting (Bellisario *et al.*, 1999; c.f. Strahler, 1957). The Rose diagrams of river orientations did not show a sinistral or dextral shift. Many clusters of first order river orientations at different locations in the delta illustrate the chess board-like pattern of the study area. Independent of the Strahler order, rivers oriented to the ENE-WSW were highlighted. In the eastern region of the interior delta (Ma'an region), a division of the delta into five units becomes obvious (Figure 53). The profile sketch crossing the units shows an accentuated trend due to

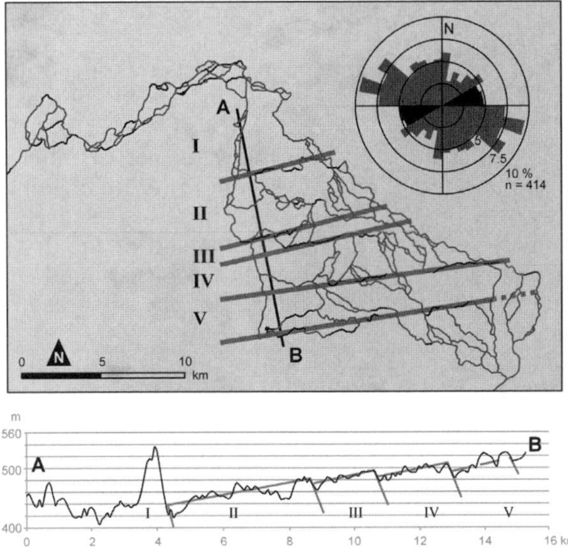

Figure 53. Ntem interior delta with highlighted ENE-WSW orientations of the fluvial network and profile sketch AB cutting the delta rectangular to these orientations (based on SRTM and Landsat 7data).

several branches of the Ntem; however, aside from the inselberg in the northern part, flat sections were identified representing the units divided by ENE-WSW lineaments. The linear geometry and the profile sketch indicate a step fault as the main structure in the Ma'an region of the interior delta. In the South, the alluvial basin is limited by a slope after the southernmost river branch, with a slight rise to an E-W oriented inselberg range. In the North-East it is limited by a steep slope and an adjacent slight rise, too. In the East, and, in particular, in the West, the step fault is limited by sub-linear structures to the NNW marked as a fault on the geological map of Maurizot (2000; Figure 35). At various locations form units and the processes linked to their evolution support the existence of a step fault.

Fanglomerates and fissures

At Meyo Ntem the basement crops out extensively. Probably the outcrop was brought to the exposed position (more exposed to erosion) by tectonic activity (Figure 54). According to the local guides, during the dry season the water depth in front of the outcrop is several meters deep in contrast to about 50 cm depth on the outcrop. The outcrop has a slight incline of 1° from the ENE to WSW. Along release joints in the Archaic socle extensive detachment took place. Vertical fissures are occupied by small plants, which widen the fissures by root pressure and chemical biogenetical weathering. This leads to the separation of rock units as does fluvial activity during the rainy season. This process mainly generates the small-scale steps in the field.

At several locations release joints can be observed arranged in a rectilinear pattern of approximately 10 cm in edge length. Along the fissures weathering took place forming rounded rocks reminiscent of fluvially rounded pebbles (c.f. Figure 31.5). After fluvial transportation of these weathered rocks for a short distance, no difference between pebbels originated by fluvial or weathering activities can be discerned.

For this reason, the index of the degree of roundness to calculate transport distance of pebbles will not be used.

Several rounded, subangular, and angular clasts were deposited on the outcrop and were fixed by iron and manganese oxides in a partly porous, partly hard fanglomerate covered extensively by a dark brown to black patina. The patina occurs particularly in small depressions in which the water persists after the rainy season. Its formation seems to be a recent process (Figure 55). With ongoing evaporation, the concentration of dissolved material in the water increases relatively and forms finally a patina of some micrometers in thickness at the end of the dry season. However, hardening to a hard crust does not occur in the tropical humid environment. Small depressions still show water residues at the end of the dry season because they are refilled by sporadic rainfall.

Alexandre and Lequarré (1975) describe comparable forms along the Shaba River in the southern Democratic Republic of Congo (formerly Zaire). They associate the iron and manganese patina to the current semi-arid climate found in their study area. Low sorted somewhat angular pebbels hardened in a fanglomerate by iron oxides

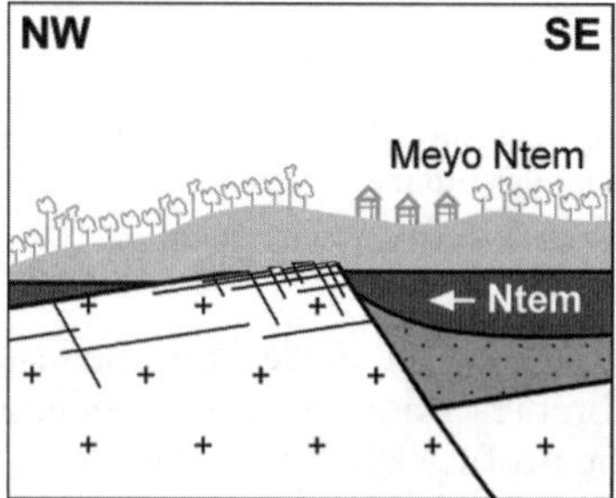

Figure 54. Hypothetical cut along the Ntem at Meyo Ntem.

Figure 55. Formation of a patina at Meyo Ntem.

were interpreted as fluvial deposits fixed under a drier climate regime. No attempt was made to relate the pebbles with morpho-tectonical formation in their study area.

Association of fanglomerate formation with a drier climate can be adapted to the Ntem interior delta; however, dating is difficult. Sangen (2007, 2008) hypothesized the existence of a rainforest refuge for the Ntem interior delta during the LGM. But what was the vegetation in detail? Was it comparable to the present day primary forest? Despite reduced discharge, the Ntem probably provided enough water to support gallery forest. Hardening of the fanglomerate would have been possible in such an environment.

Included under the topic of climate and evolution of the Ntem study area, there must be a discussion of tectonic influences. The formation of the step fault led to increased erosion of the newly exposed sites and to deposition of eroded material in the adjacent depression (Figure 56). Triggered by the tectonic rejuvenation of the region, the Ntem obtained a new river bed according to the structures. The main flow direction of the Ntem to the NW was dammed causing the Ntem to be redirected until it could easily erode what was likely a highly stressed unit and cross the barrage to the NW. This process led to the ENE-WSW oriented river course found in the study area. Erosion and accumulation occurred mainly along the river course, and the extensive outcrops were cleared. The greater the distance to the erosion area, the finer the grain size of sediments. This process led to the current form units: widespread rocky areas having extensively occurring clasts and mostly sandy sediments without clasts within or below them. The hardening of the fanglomerates occurred during a distinctly drier climate.

Currently, the patina is formed and reshapes the older layers of fanglomerate. The fanglomerate will erode in places and potholes will form during the rainy season. Rock fragments will become separated from outcrops by biogene and erosion processes. These can be transported for short distances by the Ntem.

Offset of ferricrets

Along a transect from Meyo Ntem to Abong, several outcrops of ferricrets could be observed, and corings were limited by ferricrets or ferricretic layers. In the small villages along the transect, pisolites occur superficially at places cleared by local people.

Martin and Volkoff (1990:130) relate ferricrets and ferricretic layers that occur consistently in the same region having the same landscape. They classify south Cameroon as having "Plains and hills with thick ferricretic horizons (concentrations

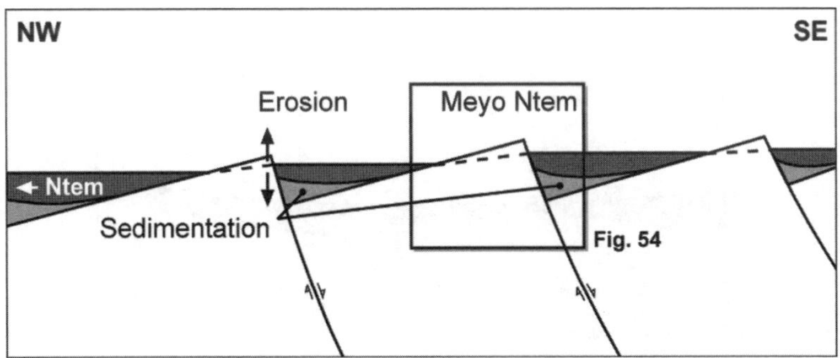

Figure 56. Sketch explaining processes after step fault formation.

and/or crusts); deeply weathered regolith, few outcrops". The observations of Bitom and Volkoff (1993) at an exposure site in south Cameroon support this classification (see 3.3.1: Pisolites in a regolith layer and planation processes (I)).

The start position was a nearly flat formation of ferricretes and ferricretic layers, which must occur at the same height above sealevel. In the interior delta, the heights deviate from each other. Martin and Volkoff (1990) describe a relocation of crustal residues after their evolution to the slope, triggered by an increase in relief energy. However, the study area is characterized by a flat landscape. Martin (1967:213) discussed tectonic faulting as trigger for different levels of ferricrets in his Sanaga River study area (13°E, 5°N; c.f. Martin and Volkoff, 1990:133); however, most of the Ntem study area does not exhibit different levels. Only north of the 4th branch of the Ntem, there is an extensive ferricretic layer as identified by several corings.

Construction of Figure 57 is based in part on a few corings and ferricretic outcrops. Detailed observations were not always possible due to the dense vegetation. For this reason only a rough sketch was drawn. The sketch helps to clarify the hypothesis of a step fault in the interior delta. The transect has a mean direction of NE-SW. If the transect is viewed in relation to the map of Figure 53, it becomes obvious that the offset is represented by Ntem branches 2 and 3 trending towards the ENE-WSW and, therefore, along the main direction of the proposed step fault. Sedimentological research at the northern bank of the 3rd branch (Runge *et al.*, 2006; Sangen 2007, 2008) exhibits no ferricretic material, which was presumably fully eroded, probably induced by the tectonic offset. In contrast, only a thin sediment layer was deposited on a ferricretic layer on the northern bank of the 4th branch. It appears that tecto-genesis at this branch is less straight forward. The bank is formed by the ferricretic layer and during the rainy season water inundates the bank; however, at this branch, fluvial activity is not intense enough to erode the ferricrete. At several locations along the southern bank of the 2nd branch ferricret residues occur. Furthermore, along the flow velocity of the nearly linear section is much faster than at the 4th branch. At the level of an off-branching river section to the North, a linear outcrop occurs in the water, limiting the northern edge of the main branch (Figure 58).

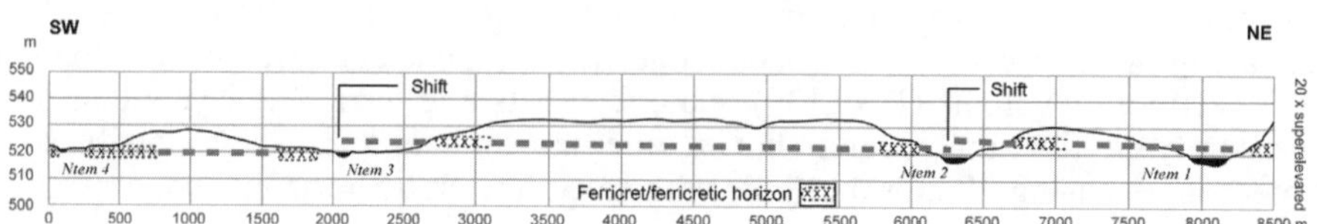

Figure 57. Hypothesized vertical shifts of a ferrictretic layer along the transect Meyo Ntem-Abong (c.f. Figure 33).

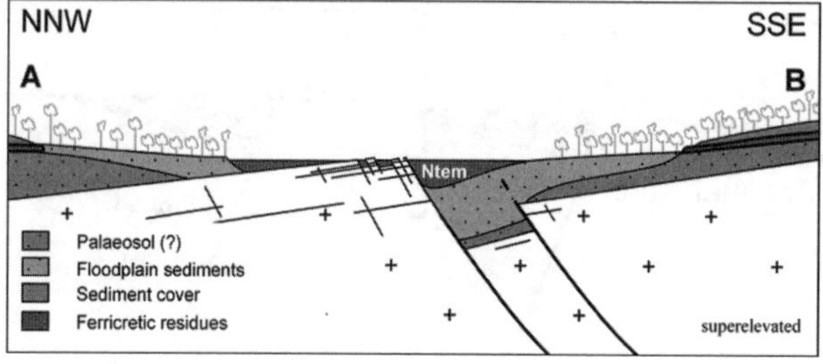

Figure 58. Transect north of Meyos cutting a branch of the Ntem (see Figure 30 for locality).

Sedimentation units

In the interior delta a broad range of sedimentation facies and ages occur, which makes the study area ideal for palaeoecological research (Runge *et al.*, 2006; Sangen, 2007, 2008). Besides having fluvial facies with hiatus and redeposition, some locations are characterized by regular sedimentation over a discrete timeframe. For example, the Meyos site provides detailed insights into the LGM (18–14 kyrs BP, Figure 59).

The northern boundary of the delta has been characterized by several recent datings. Additionally, the most northern branch of the Ntem has several outcrops, and the respective rapids highlight its recent fluvial activity. However, a spatial trend from older to younger sediments in the delta cannot be identified. At three different locations, Late Pleistocene ages were found: two along the Meyos transect; one at the western border of the delta at Aya'Amang. The dated sediments, each of a single sample, are represented by a thin organic layer. Exact delimitation of the sedimentation's timeframe is, hence, not possible. At Nkongmeyos, in the center of the Ma'an region, a palaeosurface of the Younger Dryas could be identified (c.f. Sangen, 2008). Close to Aya'Amang, at the southern limit of the Meyos transect, ages corresponding to the African Humid Period (AHP) were found.

The Late Pleistocene age sites are close to the fault zones of the postulated step fault. After formation of the fault, sediments were first deposited in the newly formed sediment traps, and these should give an idea of the starting time of its evolution. Other Late Pleistocene ages could provide clues to tectonically active locations. Further research along the proposed fault zones could substantiate this hypothesis. By linking datings of the same location, process sequences can be developed. At Aya'Amang, a palaeosurface was formed after step fault formation during the Late Ndjilien (35 kyrs BP, OIS 3) then covered by a thick sediment layer during the Léopoldvillien (30–12 kyrs BP, OIS 2/3). The Ntem incised into its sediments along the structural preset. During the AHP, once more organic sediments were sedimented or *in situ* deposited—the dated materials are layers of leaves (5–6 kyrs BP, OIS 1). At Nkongmeyos, several rock fragments were found covering an area between the footslope of a hill

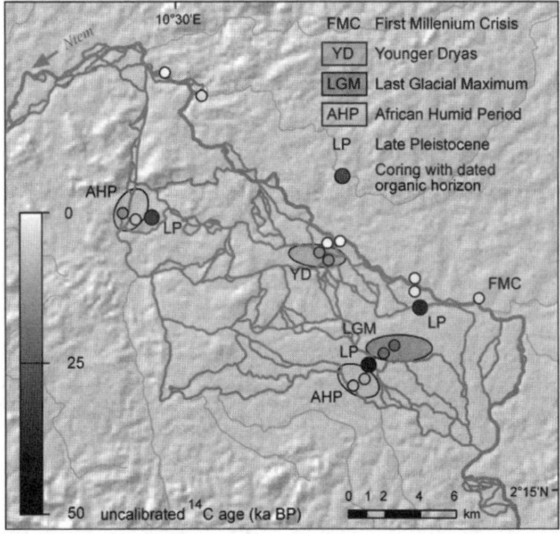

Figure 59. Overview of sediment ages in the Ntem interior delta (based on SRTM data, GPS editing in the field, radiocarbon dating).

and the bank occurring directly below some rapids crossing a small step in the river bed. Organic sediment recovered from a depth of 180–200 cm has an age of around 12.5 kyrs BP (Younger Dryas). The rock fragments covering the sediment were likely deposited after sedimentation. Aridification and a related change of the vegetation, might possibly have been the trigger; however, I consider this unlikely as it would argue against the site having been a vegetation refuge (Sangen, 2007, 2008). It is more probable that downward redeposition of the rock fragments was induced by a tectonic impetus. Below this location, the Ntem has a linear flow in the NNE direction. The prolongation of this line crosses a further river branch showing outcrops and rapids. Along this line the Ntem probably orients at a fault zone highlighted in Figure 34. Deviations from the principal E-W orientation of the step fault, highlights the complex structural situation of the whole region.

Summary: Morphogenesis

The evolutionary starting point of the Ntem interior delta was a flat undulating planation surface of the Inner Plateau. Isostatic induced uplift in Miocene times led to planation processes starting from the coast and forming the coastal surface limited by the peneplain step from the Inner Plateau. The step is characterized by SW-NE trending inselberg ranges. There are several inselbergs on the Inner Plateau and their frequency diminishes towards the East. The hypsometric curve of the Ntem River elucidates these circumstances. The interior delta is situated at the border between the Inner Plateau and the coastal surface defined by a flat landscape having few inselbergs. The complex geology reflects the long geological history of the Precambrian basement eroded by intense weathering and denudation processes. The broad range of fissures and quartz veins in the outcrops provides insights into its complex evolution.

Triggered by the ongoing uplift (Karner *et al.*, 1997; Lucazeau *et al.*, 2003) it is probable that tectonical remobilization took place mainly along E-W and NE-SW structures but also along N-S faults bordering the Ma'an region. Due to the uplift, extensional processes led to the formation of fissures along weak zones, a normal fault in the North-East, East and West of the basin, and the formation of a step fault with normal faults covering the whole study area. Obviously, this activity resulted in an uplift of the northern border of the delta due to the release of pressure in the region. The whole drainage network orienting towards the interior delta from the North was underlain by an abrupt change of its flow direction towards the WNW flowing into the Ntem just at the transition between the Ma'an and Akom regions. The height of the uplift is probably represented by the height of the northern slope. The accentuated course of the slope was mainly caused by different fault throws of faults cutting the slope and to a less extent by erosional activity of the small tributaries having their source on the uplifted area and flowing directly into the delta. Figure 60 sketches the faults and the hypothetical horsts. The drainage network was adapted by the faults. They formed barriers to the different branches of the river. The river reacted, flowed along the faults and crossed them at sites of lower resistance.

In the study area, shifts of the ferricretic layers were induced by the step fault. Therefore its formation took place after hardening of the ferricrete. Martin and Volkoff (1990:131) assume Tertiary times for iron enrichment due to the length of the process. The hardening of the crust is assumed to the global cooling in the Pleistocene.

The Late Pleistocene datings along the main faults suggest that the first sediments were deposited in the newly formed sediment trap during this time.

The eroded quartz veins also point to Late Pleistocene times give their heights out of the surrounding basement. The maximum time is set to 50 kyrs due to

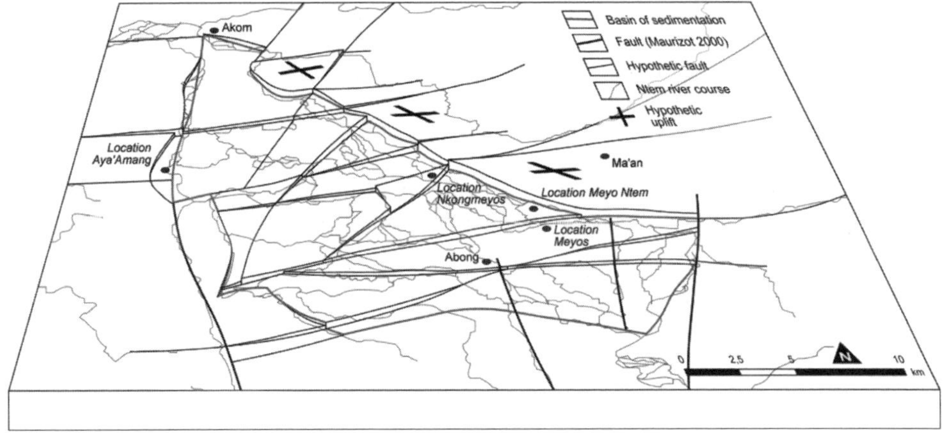

Figure 60. Assumed geotectonic evolution of the Ntem interior delta (based on SRTM data and field work).

the oldest dating for the maximum height of 150 mm. Therefore, the division of both values results in the maximum difference between quartz vein and basement, for each year since exposure date of the basement: 150 mm/50,000 a = 0.003 mm/a.

Given that after the initial exposure the character of the surface was uniform, a mean rate of 0.003 mm/year is plausible. The erosional tools made available by the tectonic event would have strengthened the erosional process.

The tectonic evolution was not the result of a single event. It seems that the formation of the rock fragments' cover at Nkongmeyos was also triggered by another tectonic impetus at around 11 kyrs BP.

3.3.4 Peneplain step, inselberg ranges and V-shaped valleys

In tropical geomorphology, theories concerning peneplain step formation are closely connected to planation processes and inselberg formation (c.f. 3.3.2). Related to this is the formation of the peneplain step between the Inner Plateau to the coastal surface and the Nkolebengue Escarpment. Part of this form unit is also the deep, V-shaped valley of the Ntem with its linear course.

Peneplain step formation

Bremer (1981:76ff) thought that after an initial uplift, inundation increases close to the erosion level forming a thick weathering cover, which will thereafter be denudated successively. Due to differential weathering and erosion, the basement will be revealed in the hinterland. Starting from the exposed rock, an increase of the slope's steepness took place until a peneplain step is formed. The steeper the slope, the less intense is the erosion on the slope. Such evolution occurs also during active uplift of a region. If the upper surface is lowered, then so-called "Aufliegerinselberge" (lie-on-inselbergs) would provide clues to remains of the former step. A comparison of the step at the level of Nyabessan and Ma'an, in the framework of the phases of Bremer (1981) reveals a certain similarity (Figure 61).

Geological differences were not be considered in this theory. The Nkolebengue escarpment is formed out of a resistant charnockitic basement rock in contrast to the gneisses of the Inner Plateau and the highly metamorphic rocks of the coastal surface (Maurizot, 2000). The basement structures are defined by a NE-SW trending thrust fault. The orientation is the same for the inselberg escarpments in front of the peneplain step. Triggered by uplift since the Neogen, Kuete (1990:343ff) proposes a remobilization of Precambrian structures characterizing the step. Geomorphological transformations during arid phases in the Pleistocene led to the recent step (Figure 62). Apart from the influence of faulting Kuete (1990:343) describes erosional incision of the main rivers as the main process leading to surface lowering—in contrast to Bremers (1981) approach.

Penck (1924:175) interprets former valley slopes as peneplain steps between different planation levels in the Fichtelgebirge/Germany. This theory represents the

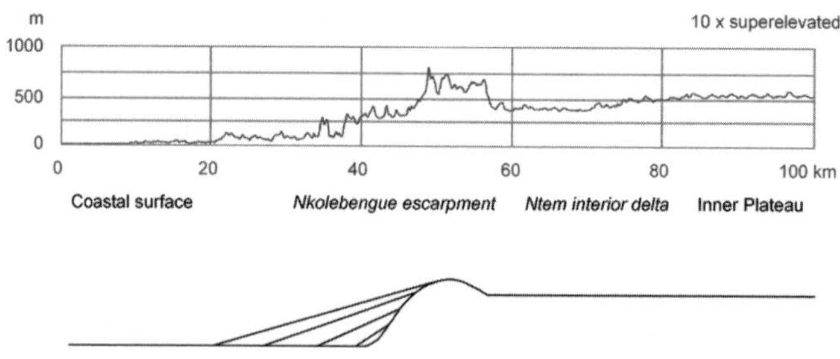

Figure 61. Peneplain step at the level of the Ntem catchment and sketch of Bremer's theory (1981), modified.

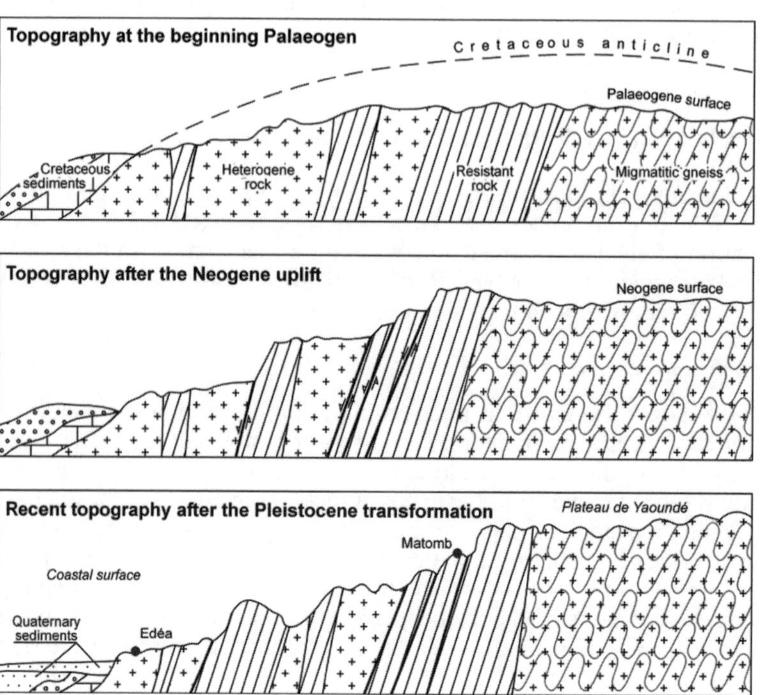

Figure 62. Sketch of peneplain step's evolution (modified from Kuete, 1990:347).

structural geomorphological idea of pedimentation and, therefore, the evolution of peneplain steps (Ahnert, 2003:296ff). Ahnert (2003:298) pictures lateral valley floor pedimentation to be mainly rectilinear pedimentation parallel to the step. In a like manner, the Nkolebengue and Ngovayang escarpments are separated from the upper Inner Plateau by slightly incised valleys. The escarpments are zonal inselbergs, residues of the former step. Bremer (1981) does not assume a regional shift of a peneplain step.

South-west of Nyabessan, lateral valley floor pedimentation is visible in the DLM (Figure 27). The Ntem follows the V-shaped valley before entering a wide flat plain. Reminiscent of the landscape close to Ebolowa, the main river and its tributaries have formed this plain; a petrographical change at the step between the Inner Plateau and the plain is not present (Maurizot, 2000). The plain slants down to the SW at an incli-nation of 3° with a distinct knickpoint directly below the step defining the plain as a pediment (Figure 27).

Another step lower down crosses a NS oriented slip fault along which a pet-rographic change occurs from resistant charnocitic basement, a prolongation of the Nkolebengue escarpment, to gneiss of the Archaic basement, at the coastal surface (Maurizot, 2000). The geological differences do not inevitably lead to other geomor-phological forms. They occur also upstreams along the V-shaped valley, in the form of a flat strip of up to 12 km in width surrounding the linear form. Along the Mbini (Equatorial Guinea), Kienke (draining into the Atlantic Ocean between Ntem and Nyong) and Sanaga Rivers a comparable classification of steps exists. The heights of the steps vary slightly and it is not possible to link the steps precisely; however, the presence of different steps lead to the assumption of several pedimentation cycles since the Miocene.

Additionally, research of Embrechts and de Dapper (1987) reveals different pedimentation cycles at the peneplain step between Yaoundé and Eséka (at the level of the Nyong River) since the Quaternary (Late Neogene). They postulated hillslope pedimentation during arid phases at parts of the slope with only thin vegetation cover as protection against erosion. They connect the last phase to the last glacial period on the basis of other research in tropical Africa (c.f. de Ploey, 1965; Fölster, 1969; Michel, 1973). They preclude recent hillslope pedimentation (Embrechts and de Dapper, 1987:40).

V-shaped valley formation: fanglomerates

The V-shaped valley of the Ntem is characterized by a linear structure having a NNE-SSW orientation. Approximately 15 km downstream from the valley's starting point a further linear valley trends rectangular to the Ntem's valley. Further downstream from this a strike-slip fault forms more disorder in the linear thalweg. The Ntem fol-lows the NS oriented structure before flowing to the West towards the coastal surface. Obviously, the first linear valley represents an eastward shift of the strike-slip fault due to the parallel occurrence and the same sinistral offset to the thalweg in the V-shaped valley.

The physiognomy of the valley suggests a graben fault having two transform faults dividing the peneplain in the Nyabessan region at a length of 50 km in a NNE-SSW direction. The transects AB, CD, and EF highlight different sections of the graben (Figure 63). Blocks of ferricretic residues and unweathered basement in the valley suggest a neo-tectonic remobilization of the structure triggering the break-off of the rocks. Other indicators of tectonic activity are the widespread plains of fanglomerates below the Chutes de Menvé'élé and in the V-shaped valley. Nearly all fanglomerates

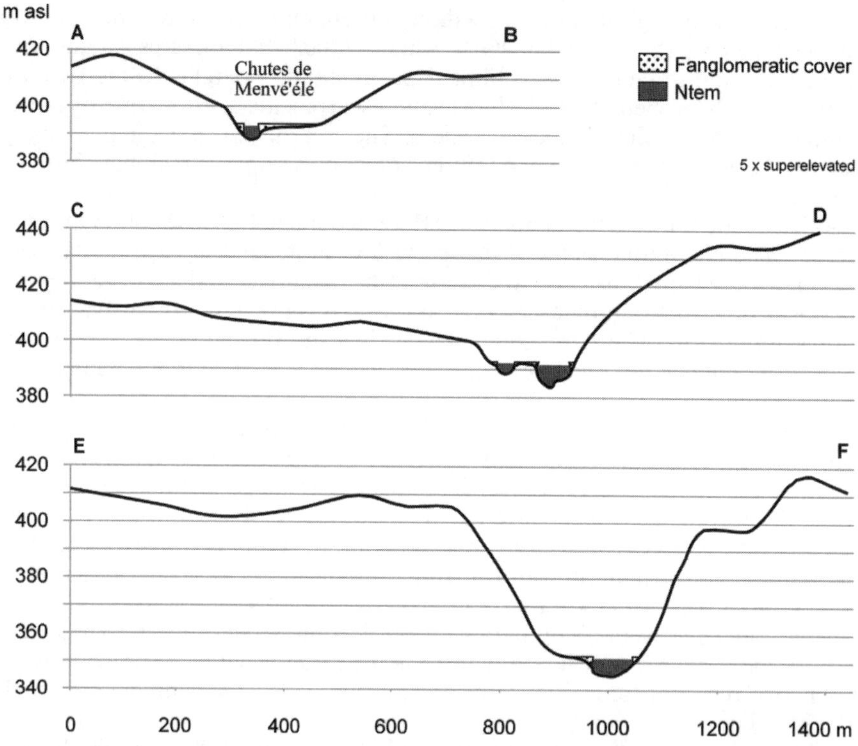

Figure 63. Transects cutting the V-shaped valley of the Ntem (see Figure 36 for locality).

consist of rounded and sub-angular gravels, rock fragments, as well as quarzitic grus, and are fixed in a matrix with iron and manganese oxides.

The gravels testify to fluvial transport and intense weathering processes, which form rounded rocks as described above, the rock fragments and quartz grains suggest colluvial input into the valley. The rock fragments consist of gravel fragments and fragments of the basement. The gravel fragments are sometimes rounded and sometimes broken without any rounding. At Ebianemeyong, a part of the fanglomerate was broken-off out of the whole rock, revealing two sections highlighting two phases (Figure 41): The right part shows rounded gravels having traces of corrosion and linear fissures (Section I), the left part, several gravel fragments (Section II). Probably the gravel fragments were formed out of corroded gravels like those of Section I.

Section I was formed by a fluvial process. Section II also suggests fluvial processes and colluvial transformation afterwards. The gravels were deposited, broken, and mixed with the rocky grus. Additionally, intact gravels were deposited on the colluvial sediment. The whole body hardened to the fanglomerate. Probably the hardening of the fanglomerate occurred by repeated inundation and drying of the whole gravel body (pers. comm. Prof. Brey, Mineralogical Institute, Frankfurt University; c.f. Heim, 1990:197). These circumstances can be provoked by a change of seasonality towards longer dry seasons and shorter rainy seasons.

Furthermore, the layered patina in Section II partially covering a gravel fragment is suggestive of a climate unlike recent having an alternation of arid and wet phases. The bright layers hint to phreatic phases, the darker ones, to vadose phases (pers. comm. Dr. Ries, Mineralogical Institute, Hamburg University). However, the element distribution map of the patina (Fig. 43) presumes a layer formation in similar phases with sorting during deposition. Iron at the base led to the dark red colors, aluminium to the light yellow colors. Probably, the formation of the patina took place in a depression of stagnant water, which filled during the rainy season. During the dry season, the soluted elements in the water increased by evapotranspiration. Only after total dessication the patina formed around the gravel and hardened. Arid climatic phases occurred during the middle of the last glacial period (OIS 4) and during the LGM (OIS 2). The formation of the patina occurred before the fragmentation of the gravel and the colluvial redeposition.

Also recently a patina (or crust) formation was observed. Under constant inundation by the Chutes de Menvé'élé, a crust of 2 cm in thickness was formed in a small cavity. However, its habitus is uniform and differs, therefore, from the layered patina of Ebianemeyong. Only the layer at the bottom had a comparable habitus with a uniform distribution of iron and aluminium. The formation of that layer can be linked to a more humid climate.

A comparable crust could not be identified in the study area. It seems likely that the continual alteration between dry season (and a possible formation under the direct influence of the cataracts) and wet season (higher water level and increasing erosion at the bank) prevents superficial crust formation.

Below the cataracts, cut and slip-off banks were modeled into the fanglomerate. This geometry was formed after the deposition of the clasts and was fixed during the cementation of the clasts. The original extent was sketched by the grey polygon in Figure 37. Currently, the slip-off bank was cut by the cataract probably induced by an increasing discharge under a more humid climate. By the cut, a cavity was exposed below the fanglomerate. It is assumed that easily erodible sediment was deposited in the underlying bed, which has already been eroded by the cataract. Residues could not be recovered. A coring through the fanglomerate to excavate the sediment at a comparable depth was not possible due to the resistance of the material. The small fringe of the fanglomerate around the large basin representing the starting point of the V-shaped valley, gives an idea of the former extent of the deposited clasts. The high discharge at that location led to the erosion of a huge part of the clasts.

The geomorphic evolution of the valley and of the cataracts is subdivided into different stages:

i. The gravel fragments in the fanglomerate point to a former gravel layer deposited at the level of the later V-shaped valley.
i.i. The formation of the patina covering a gravel occurred after its rounding. Probably the patina was formed at another place and the gravel was transported over a short distance. Therefore, stage (i) and (i.i) are not basically different. The origin of the gravels could not be determined. The tectonic formation of the interior delta upstreams could have led to an increased entry of rock material into the drainage network. However, the gravels could also be part of a pedisediment formed during extensive pedimentation into the Nkolebengue escarpment, and then deposited on the plain (c.f. Fölster, 1969). The cemented particles of quartz present in the cut of the fanglomerate are assumed to be micro-aggregates. Embrechts and de Dapper (1987:40) describe a hardening of clay into micro-aggregates induced by pedimentation. The occurrence of such aggregates in the fanglomerate, suggests that there was a pedisediment covering the plain before the tectonic valley formation. During the Maluékien (OIS 3/4), an arid climate having a more pronounced seasonality prevailed

such that active pedimentation into the escarpment was possible. The patina covered gravel further suggests a semi-humid climate.

ii. Iron concentrations at the border of the gravel fragments and along fissures into the rock hint to a long phase with corrosional processes persisting after the break-off of the gravels. Probably abrupt redeposition of the material led to the break and the mixing with the regolithic material. The displacement of the material was triggered by a remobilization of the pan-African strike-slip fault that terminated in graben formation.

iii. Besides colluvial, other gravels were deposited in the valley. Obviously, tectonic activity in the region led to a durable supply of rock material, which was fluvially treated before deposition.

iv. The clasts were hardened in a fanglomerate.

v. Recently, the fanglomerate will be eroded below the Chutes de Menvé'élé and in the V-shaped valley. At the Ebianemeyong location several gravels

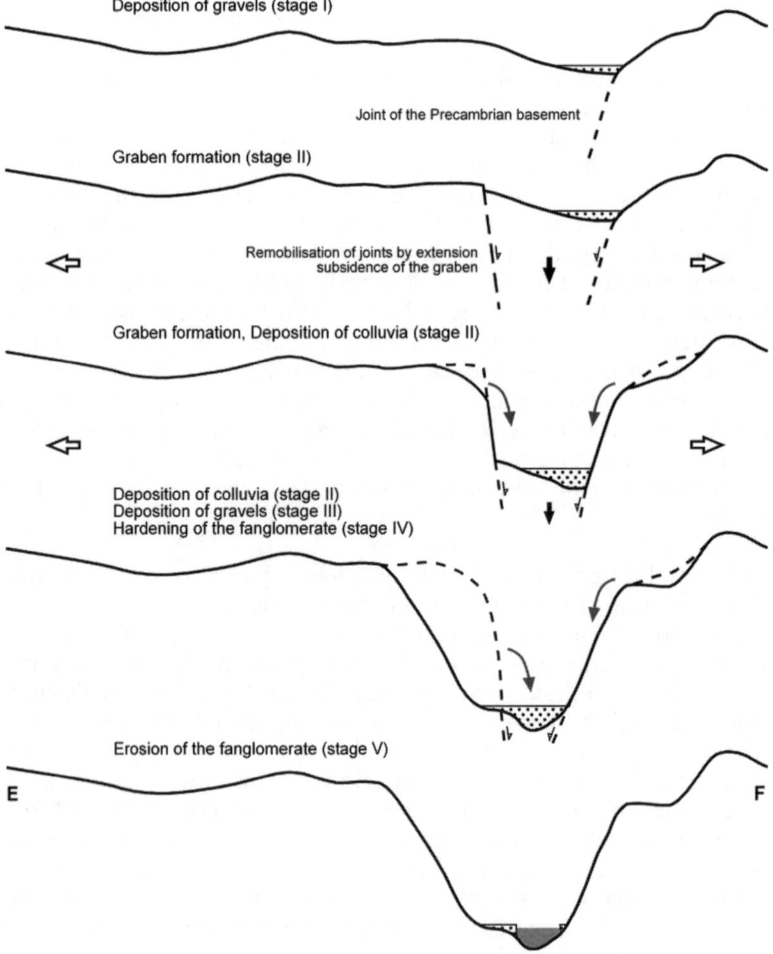

Figure 64. Hypothesized geomorphic evolution of the V-shaped valley of the Ntem at the level of transect EF (Figure 63; based on field observations and interpretation of fanglomerates).

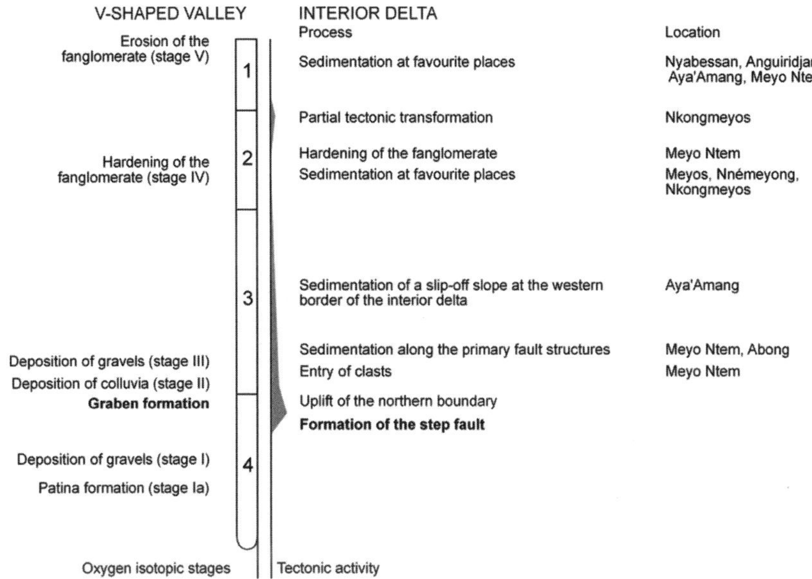

Figure 65. Correlation of the geomorphic evolution of the V-shaped valley and the interior delta (oxygen isotopic stages after Shackleton, 1977).

are deposited in pot-holes acting as erosional tools during the rainy season. Residues of the patina on the gravels hint to their former postion within the fanglomerate. Only at particular waterfall-near sites, under the influence of spray, is the formation of a crust now possible. Based on the transect EF, the evolution of the valley will be elucidated (Figure 64).

In contrast to the Ntem valley, the deep valley of the Nyong close to Njok shows several orientations with abrupt changes in the direction due to the fan-like geological pre-settings at this location. At Njok, deposited clasts were partially fixed in a fanglomerate below the Chutes de Njok suggestive of neo-tectonic activity. Additionally large weathered blocks of the tonalit were broken-off out of the outcropping basement rocks and deposited below the cataracts. A detailed insight into the fanglomerate is missing; therefore, a statement concerning the morphogenesis is not appropriate.

Summary: Morphogenesis

The formation of a new peneplain level in the hinterland since the Miocene occurred in several phases composed of steps of up to 100 m in height. The intermediate levels were formed by lateral valley floor erosion. During the Late Pleistocene, a graben was formed in the highest level by a remobilization of Precambrian structures having a NE-SW orientation. A temporal classification of graben evolution was arrived indirectly at correlating the forms having evidence of severe climatic change with the results from the interior delta upstreams.

Figure 65 correlates the Late Pleistocene and Holocene climatic changes with different stages of geomorphic evolution. It attempts to link results of sedimentation units and datings of the interior delta. However, the timespan is somewhat relative, hence, the timescale is given as oxygen isotopic stages (OIS) of Shackleton (1977).

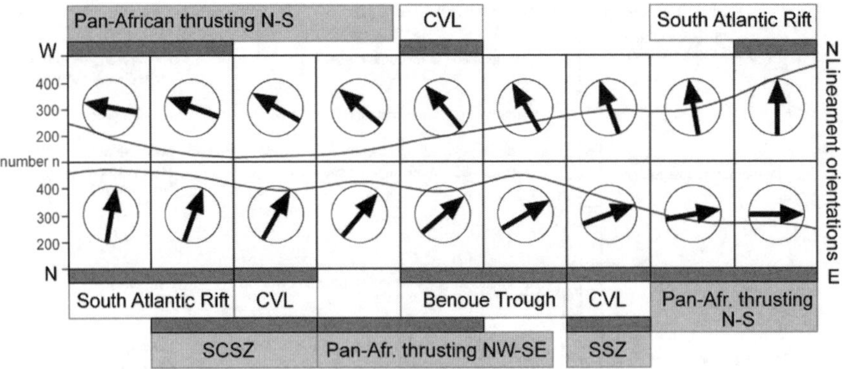

Figure 66. Presentation of the main tectonic events of the study area since the Pan-African orogeny according to the lineament analysis (after Ateba et al., 1992; Déruelle et al., 1987; Dumont, 1986; Feybesse et al., 1998; Fitton and Dunlop, 1985; Kankeu and Runge, 2008; Ngako, 1999; Ngako et al., 2003; Toteu et al., 1990, 2004); n denotes the respective number of lineaments.

3.3.5 Lineaments

Several structural load directions of the Pan-African orogeny and the break-up of Gondwana are highlighted by the lineament clusters (c.f. Figure 21). The main tectonic stages are sketched in Figure 66. The titles having a grey background relate to the Pan-African orogeny, the other ones to younger rift formation and volcanic activity along the CVL. Dark bars accentuate the lineament directions, which are correlated to the stages. Additionally, the value "n" of all edited lineaments is drawn as a curve indicating the number of lineaments of each respective direction. Lineaments of the Congo Craton are assumed to be Archaic; however, it is difficult to differenciate between Archaic and Pan-African structures (c.f. Feybesse *et al.*, 1998). If there is no correlation between a high value of lineaments to a tectonic stage, then it may be an indication of an Archaic structure. It is probable that the break-up of the Atlantic triggered remobilization of several Archaic structures beneath the graben formation because of initial extension. Its frequency is the highest at N0–10°E.

During the Pan-African orogeny, the Congo Craton was affected by two thrust faults. South of Yaoundé runs the E-W oriented Nappe de Yaoundé and in the hinterland the Nyong series trends NE-SW. Addtionally, the SCSZ (SSW-NNE) and the SSZ (WSW-ENE) are linked to this stage. The formation of the Benoué aulacogen (SW-NE) and the South Atlantic rift (N-S) was triggered by the break-up of the Atlantic Ocean. The active volcanic chain, the CVL, has a comparable orientation to that of the Benoué Trough to N30°E (NNE). Furthermore the CVL is flanked by several horsts and grabens limited by fault directions N70°E (ENE) and N130–140°E (NW) (Déruelle *et al.*, 1987).

The number of lineaments having a NW orientation is distinctly lower than those having a NE orientation. In particular, those directions having an overlay of different stages show a high number of lineaments. The number decreases with respect to the lineations of the Nappe de Yaoundé in the East, as a result of the editing mask of the study. Apart from the orientations rectangular to the CVL, mostly the NW orientation has low values due to the absence of associated tectonic stages.

Lineament clusters and tectonic stages

After highlighting the different lineament orientations, the respective tectonic phases can be associated to the different lineament clusters of the study area (Figure 67). The first region (R1) is characterized by the rift of the South Atlantic Ocean. However, the N-S structures correlate also to the Pan-African collision of the West and Central African palaeo-craton (c.f. Toteu *et al.*, 1990, 2004). In the South the influence of

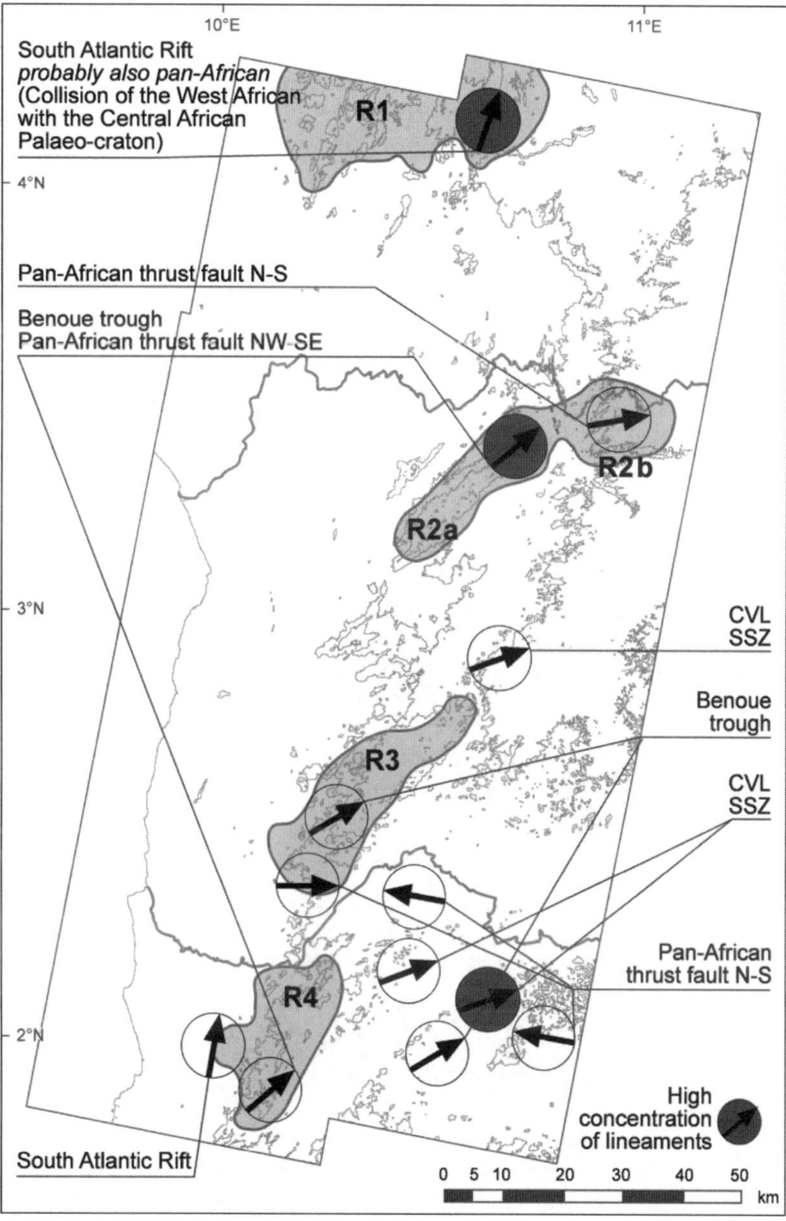

Figure 67. Study area of the lineament analysis highlighting the main lineament clusters.

the collision decreases due to the distance of the former collision zone. The western area of the second region (R2a) is characterized by a cluster of lineaments trending N40–50°E pointing to the Nyong Series and to the aulacogen of the Benoué Trough. This overlay leads to the high n value within the cluster. The western part of the region (R2b) is characterized by the influence of the Nappe de Yaoundé and its E-W thrust fault (N80–90°E).

The Nkolebengue Escarpment (R3) is characterized by orientations of the Benoué Trough (N40–50°E) and in the South, of the Pan-African thrust fault of the Nappe de Yaoundé. Comparable to regions 2a and 3, the inselberg range of the Ngovayang escarpment (R4) shows orientations of the Benoué Trough. Orientations of the South Atlantic rift are present, too. Besides the regions having high densities of lineaments further lineament clusters occur in the South-East of the analysed region referring to the Sanaga fault and the CVL and additionally to the thrust fault of the Nappe de Yaoundé.

All sketched orientations are mainly affected by the Pan-African orogeny. The zonal inselbergs of the Nkolebengue and Ngovayang escarpments highlight the structures of the Nyong series. Towards the East, the influence of the Nappe de Yaoundé distinctly increases. The Pan-African tectonic impacts also led to remobilization activities of the Archaic Congo Craton. Direct contact during thrusting led to the formation of faults trending parallel to the thrust front of the Nappe de Yaoundé (Figure 68; c.f. Toteu *et al.*, 2004).

Obviously, a remobilization induced by younger tectonic events having the same stress orientation led to an increased appearance of lineaments. Only a few lineaments suggest younger tectonic events whose orientation was independent of the Pan-African orogeny.

Linear units in the field

Linear units are summarized in Table 2, based on observations in the field. Units were subdivided into large steps in the landscape (terraces, linear cataracts), fissures and joints as well as quartz veins, eroded out of the surrounding outcrop. According to their size the forms were classified into three different levels (e.g., small step of a few millimeters = 1; step of a few centimeters = 2; step of a few meters = 3).

Compilation of the field-based data shows lineaments are mostly NNE-SSW trending suggestive of the SCSZ and the South Atlantic rift (Figure 69). Mainly the

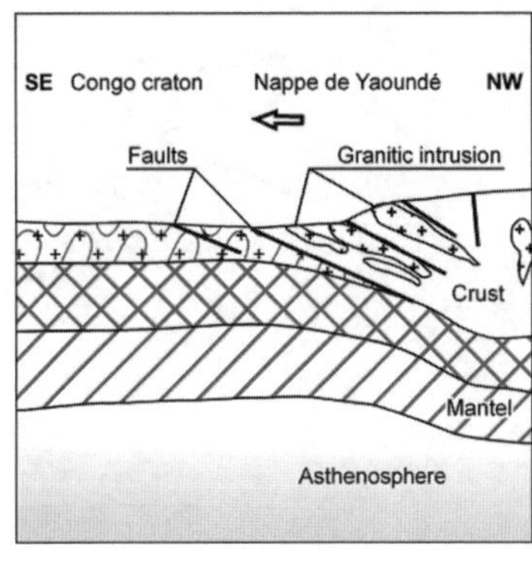

Figure 68. Transect cutting the Nappe de Yaoundé (modified from Toteu *et al.*, 2004:82).

Table 2. Linear form units edited in the field.

Direction	Form	Detail	River, location	Class
Upper catchments				
NNW-SSE	River terraces		Nyong, Akonolinga	–
NNW-SSE	Outcrop of gneiss	Outcrop bordering the river	Nyong, Nkolmvondo	–
Middle catchments				
NNE-SSW	Distinct step of 3–4 m	Waterfall	Nyong, Eséka	3
NNE-SSW	Fissure	Jointing in outcrop	Nyong, Eséka	1
E-W (N80–90°W)	Small step of 0.5 m	Rapids	Ntem, outcrop Meyon Ntem	2
NNE-SSW (N10–20°E)	Quartz vein	Massive angular rise	Ntem, outcrop Meyo Ntem	3
E-W (N80–90°W)	Joint	Joint in the rock filled by water	Ntem, outcrop Meyo Ntem	2
N-S	Quartz vein	Massive angular rise	Ntem, outcrop Meyo Ntem	3
NNE-SSW (N10–20°E)	Quartz vein	Strongly eroded rise	Ntem, outcrop Meyo Ntem	1
NE-SW (N60–70°E)	Quartz vein	Strongly eroded rise	Ntem, outcrop Meyo Ntem	1
E-W (N80–90°E)	Quartz vein	Strongly eroded rise	Ntem, outcrop Meyo Ntem	1
NNE-SSW (N10–20°E)	Small step	Linear step a few millimeters high	Ntem, outcrop Meyo Ntem	1
NNW-SSE (N10–20°W)	Small step	Linear step a few millimeters high	Ntem, outcrop Meyo Ntem	1
NE-SW (N60–70°E)	Quartz vein	Strongly eroded rise	Ntem, outcrop Meyo Ntem	1
NNE-SSW (N10–20°E)	Quartz vein	Strongly eroded rise	Ntem, outcrop Meyo Ntem	1
ENE-WSW (N60–70°E)	Fissure	Low indentation	Ntem, outcrop Meyo Ntem	1
NNE-SSW (N10–20°E)	Fissure	Low indentation	Ntem, outcrop Meyo Ntem	1
WNW-ESE (N60–70°W)	Fissure	Low indentation	Ntem, outcrop Meyo Ntem	1
Upper catchments				
E-W (N80–90°E)	Joint	Joint in the rock filled with water	Ntem, outcrop Meyo Ntem	2
NE-SW (N50–60°E)	Fissure	Rectangular arrangement	Ntem, outcrop Meyo Ntem	1

(Continued)

Table 2. (*Continued*).

Direction	Form	Detail	River, location	Class
Upper catchments				
NW-SE (N40–50°W)	Fissure	Rectangular arrangement	Ntem, outcrop Meyo Ntem	1
N-S (N10–20°E)	Small step of 0.5 m	Limit of clasts	Ntem, outcrop Meyon Ntem	2
NW-SE (N40–50°W)	Joint	Joint in the rock filled with water and rock fragments	Ntem, outcrop Meyo Ntem	2
ENE-WSW (N60–70°E)	Quartz vein	Eroded rise	Ntem, outcrop Meyo Ntem	2
E-W (N80°E)	Small step of 0.5 m	Rapids	Ntem, Mvi'ilii	2
NE-SW (N40–50°E)	River course	Junction with waterfalls	Ntem, Nyabessan	3
ENE-WSW (N60–70°E)	Distinct step of 3–4 m	Waterfalls, slope of a valley	Ntem, Nyabessan	3
NNE-SSW	Distinct step of 3 m	Waterfalls, slope of a valley	Ntem, Nyabessan	3
Lower catchments				
WSW-ENE (N60–70°E)	River course	Occurence of some rapids	Nyong, Ngogtos	3
SW-NE (N50–60°E)	River course	Slope out of basement	Nyong, Son Mbong	3
NNE-SSW	Distinct step	Waterfalls crossing several steps	Nyong, Dehané	3
NNE-SSW	Fissure	Low indentation along the schistosity of the gneiss	Ntem, Mabingo	1

large steps and the quartz veins have their highest peak in this orientation. The fissures and joints have two similar peaks at a NNE-SSW and a NE-SW orientation. The resultant Rose diagrams highlight the NE-SW trend. However, the large steps in the field do not show the orientation of the Benoué Trough nor the Archaic thrust fault of the Nyong series. The geomorphic evolution of the steps is probably based on different processes.

Due to the desquamation along release joints, small steps will be formed (level 1–2). Huge steps are formed by large scale processes like the formation of the V-shaped valley of the Ntem River represented by the Chutes de Menvé'élé. The Rose diagrams emphasize the NNE-SSW trend pointing to the main orientation of the huge steps in the field. Probably, the quartz veins represent Archaic and Pan-African structures revealed during planation processes in the study area. A hint to the age was identified at the Meyo Ntem outcrop: a NNE-SSW trending quartz vein underwent a horizontal

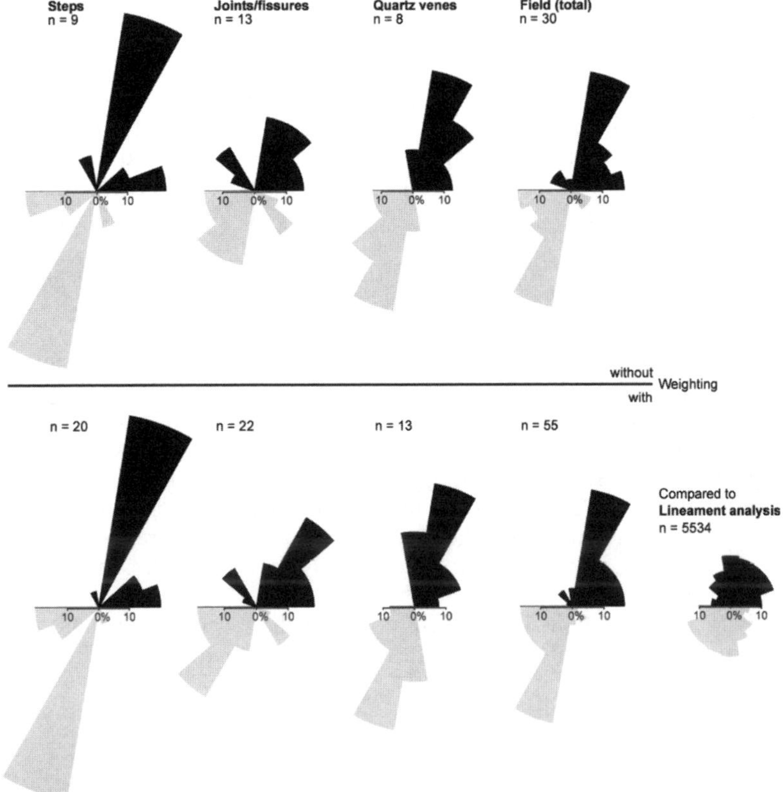

Figure 69. Rose diagrams of the linear form units edited in the field (Table 2) with and without weighting.

shift of some 20 cm at a water-filled joint (E-W). Therefore, the vein was formed before the horizontal shift. Beside the SCSZ trending quartz veins, they show the NE-SW trend of the Nyong series and N-S trend previously discussed as the Archaic orientation. The fissures and joints reflect also the trends of the Nyong series and of the Benoué Trough, which assumes a remobilization along these weak zones triggered by the opening of the Benoué aulacogen and extension of the crust due to ongoing uplift in the region. With respect to the compilation of linear trends, the NW-SE orientation seems to be of minor interest.

Discussion: Remote sensing and field work

Thalwegs and rivers represent the main class (72.2%) of linear features edited by remote sensing lineament analysis due to their real occurrence and also to their visual conciseness in the landscape. For instance, a sub-linear chain of hill tops is less prominent in the landscape.

In comparison to the small-scale remote sensing analysis, fissures and joints correspond best to thalwegs and fissures in the large-scale observations in the field. Some are filled with water. Rivers are the only form unit present in large- and small-scale editing. In the field, rivers were associated with the highest recognition level due to their distinct occurrence. Additionally, rivers orient often to distinctive scarps if they are defined as valley slopes (Chutes de Njok, Chutes de Menvé'élé). Scarps and rivers represent 76% of all lineaments observed in the field; a value close to that estimated by remote sensing analysis.

Comparing both scales highlights the N-S orientation, which is much more evident by remote sensing than by field observation. A huge part of this orientation is located in the northern analysis area, which was not considered during field work. The focus of lineament observation in the field was the Ntem interior delta, and this shows clusters that refer to the Nappe de Yaoundé (E-W). This orientation has the second highest peak considering scarps as well as fissures and joints, in contrast to quartz veins, which lacked any peak in this direction. This observation illustrates the difference between remote sensing and field data. Therefore, the structural discussion always has to consider both large- and small-scale data.

Several authors have hypothesized the remobilization of Precambrian structures in Africa (Summerfield, 1985; Burke, 1996; Burke and Gunnell, 2008; Daly *et al.*, 1989) and especially in Cameroon (Nfomou *et al.*, 2004; Ngako *et al.*, 2003; Skobelev *et al.*, 2004; Toteu *et al.*, 2004) starting in the Late Mesozoic and Early Cenozoic triggered by the opening of the South Atlantic and also by horst formation and correlating fault activity along the CVL. The respective lineament orientations show the highest peaks. The Archaic N-S lineaments of the analysis region are only represented by quartz veins in the field. Steps in the field trend mostly to NNE representing the sinistral SCSZ, which was probably remobilized by rift activity of the South Atlantic. Neotectonic activity along the Pan-African SSZ points to the sensitivity of such structures to remobilization. The formation of a joint or a step highlights their liability to vertical or horizontal shifts.

The Ntem interior delta was formed along E-W, NE-SW, and also NNE-SSW structures until it reached the current landscape habitus. Probably, the E-W orientation is based on the Ntem faults, also represented by the Ntem's source rivers. Along these structures the step fault was formed. The detailed editing of the river network in the study area and its structural interpretation supports this proposition. With a comparable conciseness the interior delta is outlined by NNE-SSW structures in the East and West. It is the same direction as the transform faults of the graben along which the V-shaped valley of the Ntem was formed west of the delta suggestive of the SCSZ. Sometimes, the fluvial network follows NE-SW orientations parallel to the thrust fault front of the Nyong series running 35 km north-west of the delta. Under the hypothesis of the step fault, the Meyo Ntem site is characterized by an E-W trending horizontal fault. In the field, this direction is represented by fissures and joints in the outcrops. The orientations NNE-SSW and NE-SW are characterized by quartz veins having no remobilization along these structures. Fluvial network editing shows several first order rivers flowing into a river of a distinct, higher order. This untypical flow behavior represents a hierarchic anomaly probably related to neotectonic activity in the study area (c.f. Guarnieri and Pirrotta, 2008:267f.).

The geomorphological unit of the peneplain in the Nyabessan region including the V-shaped valley is mainly characterized by structures of the Nyong series. The orientations of the fluvial network show respective peaks in the Akom and Nyabessan regions, and mostly along the V-shaped valley. The main orientation of the thalweg points to the front of the Nyong series, the transform fault to the SCSZ.

At the northern outliers of the Ngovayang escarpment, the Nyong orients towards structures of the Nyong series. In contrast to the Ntem, the Nyong crosses this geological unit; the Ntem flows south-east of the Nyong series. At the level of the peneplain step, the Nyong is characterized by abrupt changes of its orientation to the SSW with several waterfalls (i.e., Chutes de Njok) and cataracts. The gneissic rocks of the region show schistosity and fissures parallel to this flow direction. This linear micro-form reflects the structural properties of the rock and also of the large-scale fluvial network.

Apart from the analysed region of lineament editing, graphs of flow directions of the source rivers of the Ntem (Ntem upper catchment; Kom) and Nyong (Nyong upper catchment) were compiled (Figures 12, 15, 16). Each figure shows two graphs sketching both the number of units (grey) and their length (black). The water courses are composed of single straight lines. For instance, a meander is composed of several short lines; the more linear the river course, the less is the number of straight lines. Therefore, a linear water course having a structural influence is supposed for the relation length > number. The diagrams show accentuated graphs wherein the highest peaks are in an E-W orientation (N80–90°E, N80–90°W). The second highest peak is found in a N-S (N0–10°E) and NE-SW (N30–40°E orientation at the Nyong and Kom, and an N40–50°E at Ntem, respectively). The N80–90°E orientation is the most interesting for the lineament interpretation due to the difference between line number and length in all flow direction diagrams. The mostly linear courses of this orientation are highlighted. At the Nyong's source river this direction is quite obvious in the structurally shaped meanders within the wide floodplain added by N80–90°W orientation. Probably, these structures are influenced by the Nappe de Yaoundé. High unit lengths for the N50–60°E orientation reflect the linear course to the SW downstreams the wide floodplain along the quarzitic chain of Edou (c.f. Kuete, 1990). In contrast to the source river of the Nyong, the Kom has a high value of unit length at N50–60°E. It is likely that this orientation represents an Archaic structure in the upper catchment of the river before it turns towards west (N80–90°E). In contrast to the Nyong in the North, the structures of the Archaic Ntem faults form mainly the river course. Below the main peaks of the Ntem's source river, and suggestive of the Ntem faults, another direction to N0–10°W highlights a structural shaping due to a difference of 2% between unit length and number. However, due to missing information about the tectonic history of the Archean, it is not possible to assign this to a tectonic structure.

3.4 CONCLUSIONS

3.4.1 Geomorphologic reconfiguration of the drainage network by the opening of the South Atlantic

The rift of the South Atlantic was formed since 130 Myrs BP along the triple junction of the Niger delta. It was the center of uplift triggered by a plume and led to a reconfiguration of the whole drainage network (Figure 70a). The rivers firstly oriented towards the Congo basin incising the Gondwana peneplain (Surface Gondwanienne). The Mandara Mountains and parts of the Adamawa Plateau in the North of Cameroon are the last residues of this planation phase.

During ongoing rift activity with accompanying graben formation, a lowering of the continental margin proceeded (Figure 70b). Probably, the deposition of sediments on the continental shelf brought up by juvenile catchment areas along the aulacogen of the Benoué rift, led to isostatic equilibrium by syn-tectonic uplift in the hinterland (c.f. Summerfield, 1985:291). The rejuvenation of the relief was equated by planation at around 90 Myrs BP to the post-Gondwanian Surface (Surface Post-Gondwanienne; Figure 70c). At 65 Myrs BP, further uplift occurred combined with intrusive rocks along the CVL. The post-Gondwanian Surface was dissected. Parts of the Adamawa Plateau are the respective residues. Based on the incision, another planation phase formed the African Surface I (Surface Africaine I) represented by the Inner Plateau stretching over a large area in the South of Cameroon (Figure 71a).

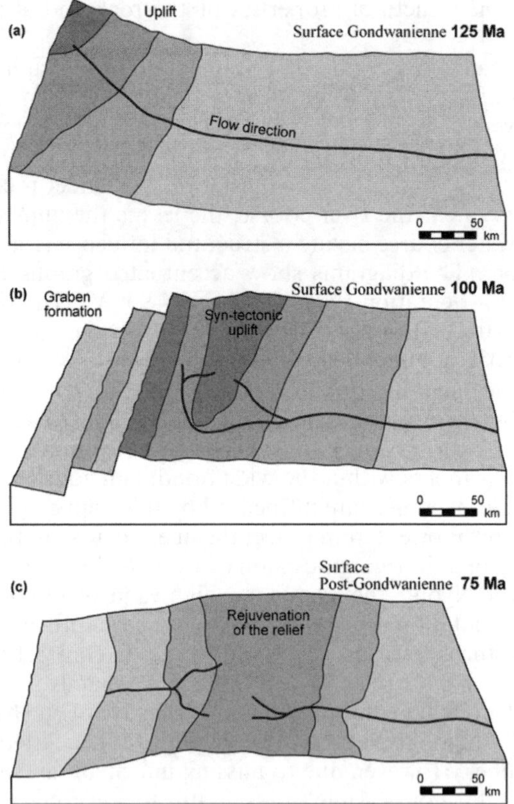

Figure 70. Block diagrams of the studied region at 125 (a), 100 (b), 75 (c) (after Burke, 1996; Burke and Whiteman, 1973; Segalen, 1967; Summerfield, 1985, 1991; own draft).

The planation processes had cut a region of geological heterogeneity shaped by Archaic and Pan-African (Neoproterozoic) deformation. The surface cuts the Nappe de Yaoundé in central Cameroon as well as the Ntem complex in the South with the Ntem and Nyong series. High temperatures (Figure 46) during the Early Cenozoic led to the formation of thick regolith layers supporting the denudation processes during the planation phase.

The West of the Inner Plateau is mainly characterized by the catchment areas of the Nyong and Ntem Rivers. Probably, they received their recent flow direction by the formation of African Surface II (Surface Africaine II) since Miocene times (Figure 71b). The trigger for the renewed planation was an intensive uplift at the transition between the Oligocene to Miocene. Due to the collision between the African and Eurasian plates, the velocity of plate drift decreased resulting in increasing impact of fixed plumes on the lithosphere. Currents of magma in the asthenosphere led to the activation of the East African rift. Along the CVL volcanism started. Additionally, the collision triggered the formation of the basin and swell relief (Summerfield, 1996).

Déruelle *et al.* (1987) described the form units along the CVL as an alternation of graben and horsts. This can be observed in detail at Mount Cameroon, here a prominent horst formed on which the volcanic layers of the stratovolcano were deposited. The formation of the horsts affected the sourrounding form units. The flow

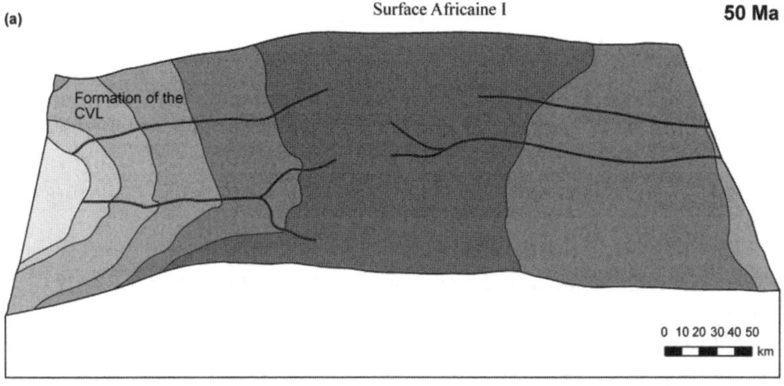

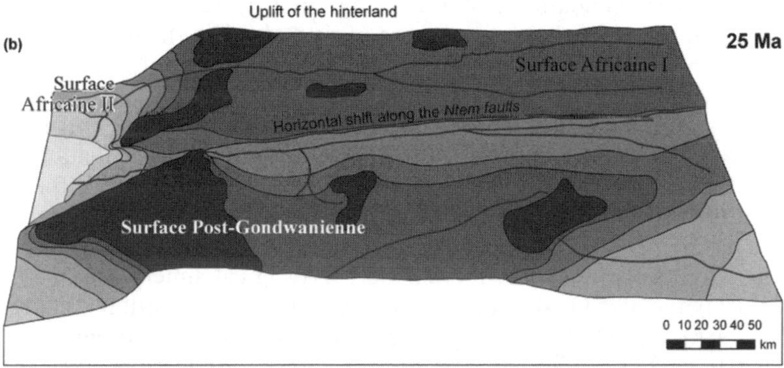

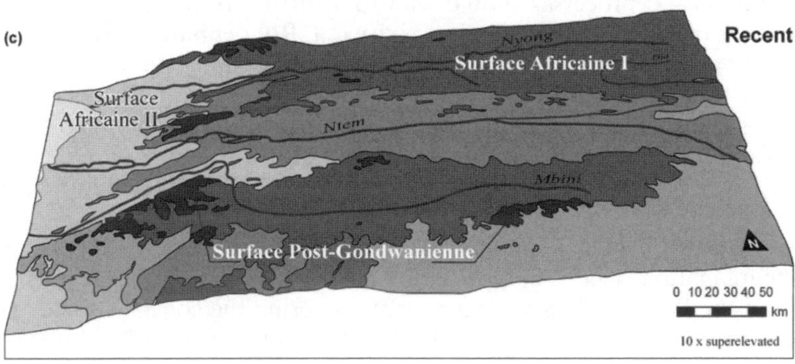

Figure 71. Block diagrams of the studied region at 50 Myrs BP, 25 Myrs BP, and recent times (after Kuete, 1990; Segalen, 1967; own draft based on SRTM data).

direction of the Sanaga River leads away from the so-called Dorsale Camerounaise to the South-East before turning to West at the level of the Sanaga Shear Zone towards the Gulf of Guinea. Therefore, its catchment area is formed asymmetrically. On the left side of its middle and lower catchment the Nyong catchment adjoins to it directly, on the right side it is shaped by the successive increase of inclination to the CVL.

Probably, the Miocene uplift did not occur as a uniform epirogenetic process without faulting. Archaic structures were remobilized and led to the morpho-tectonic evolution of a WSW-ENE trending step along the Ntem faults comparable to peneplain step formation between the Inner Plateau and coastal surface as postulated by Kuete (1990:347). Along Precambrian structures uplift occurred, which formed steps rising to the East. The planation, which formed the African Surface II, reshaped the relief into large escarpments due to resistant geological units surmounting the surface up to 200 m.

The gradient curves of the Nyong and Ntem basically support the evolution of the peneplain step. They seem to be composed of the flat, uniform upper catchment flowing cross the older peneplain, and the heterogenous lower catchment, starting at the prominent peneplain step with several waterfalls and cataracts before flowing into the Gulf of Guinea (c.f. Figure 62).

The order of uplift and planation as described by King (1949), which supposedly led to the peneplain steps in Africa has to be modified in the study area. Phases during which uplift exceeds denudation in contrast to phases during which denudation compensates for uplift or exceeds it seems more likely to lead to the form units observed in the study area. The Miocene uplift happened in several steps. Along the Ntem, two small planated areas occur at the level of the peneplain step between the Inner Plateau (> 500 m a.s.l.) and coastal surface (< 200 m a.s.l.) reflecting different uplift impetuses. The upper level was formed directly below the Ntem interior delta between the Nkolebengue escarpment and the Collines d'Akom (400–430 m a.s.l.), the lower one adjoins it in the South-East (250–280 m a.s.l.). The steps follow the rivers; therefore, the process of lateral valley pedimentation seems most probable (Figure 71c).

In the Mont Fébé region in the northern Nyong catchment area, scarp retreat occurred during the arid Quaternary climate by slope pedimentation (Embrechts and de Dapper, 1986). It might be the main process of peneplain step formation of the step between the Inner Plateau and coastal surface with its inselberg escarpments. Probably, the uplift continues until recent times (c.f. Karner *et al.*, 1997; Lucazeau *et al.*, 2003). The discussed processes should lead to a further transformation of the relief during a phase of geomorphodynamic activity (c.f. Rohdenburg, 1982).

Lateral valley pedimentation also shifted the step along the Ntem faults. Inselbergs bear witness to the former level. The summits of some inselbergs represent the palaeogene surface. The heights of around 1000 m a.s.l. correspond to the residues north of Yaoundé (Kuete, 1990). The mostly monolithic inselbergs are quite resistant to denudation in contrast to the surrounding surfaces. Inselberg evolution is based on selective erosion along joints in the rock susceptible to weathering.

The African Surface I might have been denudated extensively according to Büdel's etchplanation (1957, "Doppelte Einebnung"). The inselberg escarpments were eroded from the West by the river catchments covering the coastal surface. However, they also surmount the Inner Plateau. Probably the difference in height was induced by a tectonic event. For this reason, a vertical shift of 500 m is unlikely at the swell of the Congo Craton. A small shift at the level of the step to the disadvantage of the plateau in the East could lead to surface lowering by denudation, detached from the coastal surface, depending on the tectonic barrier. The barrier corresponds to the post-Gondwanian Surface (c.f. Figure 71c).

On the adjoining surface in the East, the upper Nyong catchment was formed in a slight syncline. The fluvial network was oriented dendritically to the receiving stream, characterized by smooth valley slopes. The Ntem oriented at the E-W trending step along the Ntem faults forming a comparable form unit in its southern catchment area (Figure 71 b). The step was shifted by scarp retreat towards the North. Recently, the upper Ntem catchment has a symmetrically physiognomy. An exposure site at Ambam in the northern catchment area shows several phases of erosion and accumulation of a hillwash cover as well as the hardening and solution of a ferricret to uniformly distributed pisolites. Moreover, the accentuated recent step (independent from geology) is visible in the DLM as a shift by pedimentation.

In the East of both catchment areas, the landscape changes from the flat and undulated plain having some inselbergs, to several *demi-oranges* with widespread valleys in between ('bas-fonds des régions humides', Raunet, 1985). The formation was probably triggered by a slight uplift leading to incision of the fluvial network into the thick regolith cover. Starting at the thalwegs, the large valley floor was formed by lateral erosion and denudation on the slope (c.f. Rohdenburg, 1982). The *demi-oranges* characterized by a poly-convex slope and a flat summit, are the result of these processes.

Not only from the Atlantic and the Ntem side the Nyong catchment was eroded. Also the Sanaga catchment in the North and tributaries of the Congo catchment in the East reduce the area of the upper catchment. The capture by the Dja, a former tributary of the Nyong, devided the upper catchment nearly in half. The course of the Dja to the West, after an abrupt turn to the South, and then to the South-East hint to the capture; however, it is strongly transformed.

More distinct is the capture of the Nyong catchment in the North by the Sanaga catchment. The Téré has a course adjusted towards the Nyong's thalweg. It orients towards the Sanaga River in the North-West by a distinct elbow of the capture. At the level of the confluence of the cut relic of the Téré and the Nyong, the aquatic floodplain abandons to the advantage of a sub-aquatic floodplain. Landsat imagery (Path 184, Row 57, 17.11.1999) shows grass vegetation in contrast to the previous swamp forest. During the dry season, there is no longer inundation, and the vegetation can be burnt by the local fishermen. The distinct contrast between grass and swamp forest highlights the hydrological change of the ecosystem by the river capture.

Moreover, the Ntem catchment will be captured by tributaries of the Ogooué. For instance, the Ayina orients in its upper catchment at the Ntem faults as does the Ntem; hence, it drains towards the East. The catchment area of the Ogooué covers mainly a wide peneplain some 100 m below the Ntem level. By incision of its tributaries into the higher level, the form unit of *demi-oranges* and *bas-fonds* changes. The aquatic floodplain is drained, and the ecosystem adapts.

3.4.2 Geomorphologic modifications by neotectonic and palaeoclimatic influence

The considerable height above sea level of Africa, which contrasts to the other continents, is attributed to a hot plume in the asthenosphere below the thick continental plate leading to epirogenetic uplift. The plume had already formed below Gondwana. Due to the drift activity of the continental plates, a cooling took place below the thinner oceanic plate. The former components of Gondwana drifted apart with a partial high velocity. Only the African plate stays more or less in place leading to conservation of the hot plume, which was supported by the collision of the African and the Eurasian plate (Ayele, 2002). After Eisbacher (1996) the thickness of the continental plate led to an uplift *en-bloc* without faulting activity. The sometimes fast drift of the other plates led to a cooling below them (Bond, 1978; Summerfield, 1996; Burke and Gunnell, 2008; c.f. Eisbacher 1996:204).

The study area in Southern Cameroon was transformed by neotectonic action. It took place by the remobilization of Precambrian structures triggered by extensional stress at the swell of the Congo Craton due to uplift since the Miocene. The uplift was induced by epirogenetic activity (see above) and isostatic equilibrium according to thick sediment deposits in the Congo basin and at the continental shelf at the mouth of the Sanaga. Seismic measurements at Mount Cameroon indicate earthquakes having epicenters along the hypothetical border of the Congo Craton. More over several guides from Ma'an region told me that the earthquake of Kribi in 2002 (Nfomou *et al.*, 2004) was also feltin the interior delta above the peneplain step. A reactivation of faults supported by increasing stress due tothe ongoing uplift and decompression in combination with denudation of the regolith cover is possible. Also, earthquakes connected with a reactivation of the SSZ north of the Congo Craton (Kouankap, 2006) or eruptive activity along the CVL could trigger single faulting in the study area. Ayele (2002:47ff) describes high compression in the Congo basin due to ongoing rift activity of the East African Rift, the mid-ocean ridge of the South Atlantic and the Indian Ocean south-east of Africa. Additionally, the collision of the African and the Eurasian plate is still taking place with a rift velocity of 10 mm/year. Ayele (2002) interpreted five earthquakes in 1998 having epicenters at a depth of 12–15 km. The region has its maximum compression in an E-W orientation and along the fault line trending NE-SW parallel to the East African Rift and to the CVL.

Indications of neotectonic activity increase from East to West. At the level of Akonolinga, a descended bloc could be identified by remote sensing (Landsat imagery, Path 185, Row 57, 16.1.2002). Linear slopes and terraces trending to the NNW-SSE distinctly limit the area. A temporal classification of the faulting is difficult; however the terraces show a slight tilt having a rise towards the front hinting to neotectonic remobilization with a small vertical shift (c.f. Figure 51). Olivry (1986:99) assumes a recent tilting towards the south-east of the Inner Plateau due to an asymmetrically formed fluvial network, and he links the forms to the geographical cycle of Davis (1899): He describes the right-sided valleys of the Nyong as part of the old stage having wide valley floors and a calm flow of its rivers. In contrast, the left-sided valleys represent a juvenile stage having turbulent flowing rivers and high relief energy. However, the interpretation of Landsat imagery and DLM shows a uniform landscape. Also, other rivers do not show such an asymmetry. Only the Sanaga can be discussed this way due to the uplift of the Dorsale Camerounaise (see above, 3.4.1).

Before the formation of the interior delta, the Ntem was probably subject to a shift of its course to the north-west. Figure 71(b) highlights the hypothesis of a former flow orientation in prolongation of the Ntem faults along the Benito Rift, a basin of 20 km in width located in the Equatorial Guinean part of the coastal surface. Currently, only a summit of 20 m in height prevents drainage towards the west. At the level of the assumed shift, the Ntem has formed rapids in its course and shows a distinct turn to the north-west.

Burke and Gunnell (2008:25) mention the Mayombe escarpment between the Congo basin and the South Atlantic, which underlies a steady uplift since the Miocene. If this uplift was higher than the postulated uplift of the study area (c.f. Lucazeau *et al.*, 2003; Karner *et al.*, 1997), the disequilibrium could have led to the proposed shift of the Ntem's course. At the same longitude, the Mbini, south of the Ntem, turns abruptly to the north-west before flowing into the Gulf of Guinea along the Benito Rift (instead of the Ntem). Just before the Benito Rift, the Mbini and its tributaries have lowered a large plain to the level of the basin (Figure 71c).

Since the formation of the vertical shift along the Ntem faults due to increasing uplift since the Miocene, the respective step was shifted by backward erosion towards the North. Pedimentation resulted in a nearly planated surface having some inselbergs in a heterogenous Precambrian basement. A part of this surface was tectonically transformed during the Pleistocene induced by a remobilization of Precambrian structures forming a step fault along faults trending E-W, limited by the NNE-SSW, and cut by NE-SWtrending faults. The orientation of the structures highlights the influence of the Ntem faults, but also the SCSZ and the CVL or the thrust fault of the Nyong series, respectively. Crossing these structures, the Ntem has filled the sediment traps (Late Pleistocene). The Ntem was forced to leave its main north-western flow direction at several barriers until finding a suitable passage. The northeastern border of the interior delta was uplifted slightly; therefore, the respective tributaries could not flow directly into the Ntem. They have adjusted their flow direction to the West before flowing into their receiving stream (Figure 60). Along the northern border, only rivers of a distinctly lower Strahler order (1957) flow into the delta representing a hierarchic anomaly of the fluvial network, which points to a neotectonic influence (c.f. Guarnieri and Pirotta, 2008:267f.). Within the interior delta, layers of fanglomerates could be observed at several places. Cuttings through them revealed nearly no rounded gravels and, if so, the rounding can be linked to both rounding by weathering processes and by fluvial transport. Additionally, ferricretic outcrops were extrapolated to a former extensive ferricretic layer showing distinct vertical shifts along the main faults. At the level of such shifts, the oldest sediments were found (c.f. Figure 59).

Below the interior delta, the V-shaped valley of the Ntem adjoins. It trends somewhat to the NE-SW but mainly to the NNE-SSW. There are two sections along the linear thalweg with abrupt turns to the South. A Precambrian graben structure having two transform faults is presumed. It is interpreted as an eastward shift of the NS trending SCSZ. Thick layers of rock fragments and gravels in the valley suggest neotectonic remobilization. The accumulation of the rock fragments was induced by the graben formation and accompanying break out of the rocks. The gravels were deposited in the valley partly during graben formation, partly by fluvial accumulation. The deposits are fixed by iron, manganese and aluminium oxides in a resistant fanglomerate. Cuttings through the matrix give an idea of five different formation processes (Figure 65): The first phase was characterized by accumulation of gravels before the graben opened. The gravels are either part of a pedisediment (Fölster, 1969) or a reaction of increasing rock input into the fluvial system induced by tectonic faulting in the interior delta above. In the second phase, gravels and additional colluvial material was deposited in the framework of the graben formation. The third phase is characterized by the additional fluvial deposition of gravels before hardening of the whole clast complex to the fanglomerate occurred in the fourth phase. The recent fifth phase is defined by erosion processes, which currently affect the fanglomerate. The hardening of the fanglomerate is induced by repeated moisture penetration of the clasts and a long period of dehydration suggestive of a more arid climate than recent and, therefore, to distinct climatic modifications in the region since graben formation. It is not possible to date the deposition of the gravels in the first phase due to the lack of datable material. The corrosion of the rocks suggests a long period between accumulation and formation of the V-shaped valley. It is probable that graben formation fits to the faulting in the interior delta (c.f. Figure 64). The repeated climate changes as apparent by fanglomerate interpretation fit well with the palaeo-climatic history of the study area.

3.4.3 Recent processes

In the framework of field work pisolites were observed, which were eroded along a pathway running rectilinear to the slope and deposited in the Nyong floodplain west of Akonolinga. In the study area, pathway construction and tillage farming triggers erosional processes. Additionally, this was observed at the Nyong's source region at Abong Mbang and also in the interior delta close to Meyo Ntem. The anthropogenic influence is quite substantial. The expansion of road infrastructure cuts the tributaries of the Nyong close to Abong Mbang. Dams will be raised in the floodplain with small pipes for maintaining discharge. The impact on the ecosystem and its processes is unforeseeable. Close to townssand will be exploited as construction material by artisanal mining at nearly all rivers.

South of Akonolinga, large ferricretic blocs were observed on the floodplain directly below the slope. At this location the floodplain was inundated by a source. Probably the ferricretic residues point to a sub-recent scarp retreat triggered by year-round inundation. Also at Abong Mbang, several large ferricretic blocs were deposited on the floodplain ranging out of the water.

In and below the Ntem interior delta, the thick fanglomerate layers are currently being eroded. In the V-shaped valley close to Ebiannemeyong gravels were deposited in a pothole showing residues of a dark patina identifying them as a former component of the fanglomerate. The patina will also be directly eroded from the as yet still fixed gravels of the fanglomerate.

3.4.4 Outlook

Due to the ongoing compression and uplift, tectonic activity is very likely in the future. Faulting could be triggered by earthquakes along the hypothetical border of the Congo Craton or along the SSZ. Due to the discussion (c.f. Chapter 3.3), mainly the Precambrian E-W trend of the Ntem faults, the NNE-SSW trend of the SCSZ, and the NE-SW trend of the thrust fault of the Nyong series and of the Tertiary CVL, respectively, are linear units along which a remobilization may occur. Due to this fact, the construction of a dam above the V-shaped valley should be requested. In the outcrops, the mentioned orientations appear as fissures and joints of differing degrees.

Additionally, the reduction of the upper Nyong catchment will proceed to the advantage of the Congo, Sanaga, and Ntem catchments. The captured tributaries will shrink the upper surface by lateral valley erosion down to the level of the receiving stream. In this manner the erosion equates the active uplift of the Congo Craton's north-western swell. The lowering occurred mainly under sub-humid conditions, more arid than the recent. The active planation proceeds slowly under the recent climate; however, it exists.

Further research

Besides documenting and sampling of alluvial deposits, roadsides and pathways were also observed in the search for clues of palaeoclimate and evolution of the river catchments in the framework of the ReSaKo project. The interpretation of several locations visited and documented only once during field work turned out to be more important than originally thought. Therefore, several statements are based on short observations, which have to be studied in further research. Some hypotheses were made on the basis of DLM interpretation; they have to be verified in the field, too.

The hypothetic course of the Ntem along the Benito Rift can be proved by further locations at the watershed between Ntem and Mbini catchments at the level of the abrupt knickpoint of the water course. Probably, fossil deposits highlight the former course. High resolution elevation models could identify a fossil course.

The river captures postulated by Olivry (1986) and Kuete (1990) are not yet dated. Triggered by a capture, the river incises into its own sediments. The river capture of the Nyong by the Sanaga's tributary Téré occurred in the upper Nyong catchment in a region with wide floodplains of the tributaries, too. Therefore, working in the Téré catchment at the level of the elbow of the capture should be promising. A detailed classification of the capture linked with the form units of the location could lead to an improved interpretation of its formation.

The *bas-fonds* and *demi-orange* form units were compared to the dambos of a drier semi-humid climate. Using different transects cutting these form units it might be possible to identify the upper and lower denudation belt. Detailed research should contribute to understanding of recent processes in this ecosystem.

Pedologic research on the exposure site at Ambam in the upper Nyong catchment could help distinguish the different layers and support the genetic discussion. Probably the postulated palaeosurface could be dated by luminiscence to prove and date the hillwash process. The digging of ancillary exposure sites would help to extrapolate the large-scale documentation at Ambam.

The results of the geomorphic evolution of the Ntem interior delta could be proven by geophysical methods. An improved dating method adapted to fanglomerates would give an idea of the age of interior delta and V-shaped valley.

REFERENCES

Ahnert, F. 2003, *Einführung in die Geomorphologie*, (Stuttgart: UTB), p. 477.

Alexandre, J. and Lequarré, A. 1975, Essai de datation des formes d'erosion dans les chutes et les rapides du Shaba. In *Geomorphologie dynamique dans les regions intertropicales. Colloque de Géomorphologie de l'Environnement dans les régions intertropicales, Lumbumbashi,* edited by Alexandre, J. (Presses universitaires du Zaire), pp. 279–286.

Ateba, B. and Ntepe, N. 1997, Post-eruptive seismic activity of Mount Cameroon (Cameroon), West Africa: a statistical analysis. *Journal of Volcanology and Geothermal Research,* **79**, pp. 25–45.

Ateba, B., Ntepe, G., Ekodeck, G.E., Soba, D. and Fairhead, J.D. 1992, The recent earthquakes of South Cameroon and their possibile relationship with main geological features of Central Africa. *Journal of African Earth Sciences,* **14**(3), pp. 365–369.

Ayele, A. 2002, Active compressional tectonics in central Africa and implications for plate tectonic models: evidence from fault mechanism studies of the 1998 earthquakes in the Congo Basin. *Journal of African Earth Sciences,* **35**, pp. 45–50.

Bellisario, F., Del Monte, M., Fredi, P., Funiciello, R., Lupia, E., Salvini, P. and Salvini, F. 1999, Azimuthal Analysis of stream orientations to define regional tectonic lines. *Zeitschrift für Geomorphologie, N.F.,* Supplement-Band **118**, pp. 41–63.

Bitom, D. and Volkoff, B. 1993, Altération déferruginisante des cuirasses massives et formation des horizons gravillonaires ferrugineux dans les sols de l'Afrique Centrale humide. *Académie des Sciences Paris,* **316** (Série II), pp. 1447–1454.

Bond, G. 1978, Evidence for late Tertiary uplift of Africa relative to North America, South America, Australia and Europe. *The Journal of Geology,* **86**, pp. 47–65.

Bremer, H. 1971, *Flüsse, Flächen- und Stufenbildung in den feuchten Tropen,* (Würzburg: Würzburger Geographische Arbeiten), p. 194.

Bremer, H. 1981, *Zur Morphogenese in den feuchten Tropen: Verwitterung und Reliefbildung am Beispiel von Sri Lanka,* (Berlin: Borntraeger), p. 296.

Bremer, H. 1989, *Allgemeine Geomorphologie*, (Berlin: Borntraeger), p. 450.

Büdel, J. 1957, Die "Doppelten Einebnungsflächen" in den feuchten Tropen. *Zeitschrift für Geomorphologie, N.F.*, **1**, pp. 201–228.

Büdel, J. 1965, *Die Relieftypen der Flächenspülzone Süd-Indiens am Ostabfall Dekans gegen Madras.* (Bonn: Colloquium Geographicum 8).

Bumby, A.J. and Guiraud, R. 2005, The geodynamic setting of the Phanerozoic basins of Africa. *Journal of African Earth Sciences*, **43**, pp. 1–12.

Burke, K. 1996, The African Plate. *South African Journal of Geology*, **99**(4), pp. 341–409.

Burke, K. and Gunnell, Y. 2008, The African erosion surface: A continental-scale synthesis of geomorphology, tectonics, and environmental change over the past 180 Million years. *The Geological Society of America Memoirs*, **201** (Boulder), p. 67.

Burke, K. and Whiteman, A.J. 1973, Uplift, Rifting and the Break-up of Africa. In *Implications of continental rift to the Earth Sciences,* edited by Tarling, D.H. and S.K. Runcorn (London and New York: Academic Press), pp. 735–755.

Ciccacci, S., Fredi, P., Lupia Palmieri, E. and Salvini, F. 1987, An approach to the quantitative analysis of the relations between drainage pattern and fracture trend. *International geomorphology*, Part **II**, pp. 49–68.

Daly, M.C., Chorowicz, J. and Fairhead, J.D. 1989, Rift basin evolution in Africa: the influence of reactivated steep basement shear zones. In *Inversion tectonics,* edited by Cooper, M.A. and Williams, G.D. (London: Special Publication of the Geological Society of London), **44**, pp. 309–334.

Davis, W.M. 1899, The geographical cycle. *The Geographical Journal*, **14**, pp. 481–504.

de Ploey, J. 1965, Quelques aspects de la recherché Quaternaire et géomorphologique en Afrique Equatoriale. *Bulletin de la Société Belge d'Etudes Géographiques,* **34**, pp. 159–169.

de Ploey, J. 1968, Quaternary phenomena in the Western Congo. In *Means of correlation of Quaternary successions,* edited by Morrison, R.B. and Wright, H.E. (Salt Lake City: University of Utah Press), pp. 501–515.

de Ploey, J. 1969, Report on the Quaternary of Western Congo. *Palaeoecology of Africa,* **4**, pp. 65–68.

Déruelle, B., N'ni, J. and Kambou, R. 1987, Mount Cameroon: an active volcano of the Cameroon line. *Journal of African Earth Sciences,* **6**(2), pp. 197–214.

Dumont, J.F. 1986, Identification par télédétection de l'accident de la Sanaga (Cameroun). Sa position dans le contexte des grands accidents d'Afrique centrale et de la limite nord du craton congolais. *Géodynamique,* **1**(1), pp. 13–19.

Eisbacher, G.H. 1996, *Einführung in die Tektonik* (Stuttgart: Enke), p. 374.

Eisenberg, J. 2007, Neotektonische Prozesse und geomorphologische Entwicklung des Ntem-Binnendeltas, SW-Kamerun. *Zentralblatt für Geologie und Paläontologie,* Teil **1**, pp. 37–45.

Eisenberg, J. 2008, A palaeoecological approach to neotectonics: the geomorphic evolution of the Ntem River in and below its interior delta, SW Cameroon. In *Palaeoecology of Africa 28: Dynamics of forest ecosystems in Central Africa during the Holocene. Past–Present–Future,* edited by Runge, J. (London, Leiden, New York: Taylor and Francis), pp. 259–271.

Embleton, C. 1987, Neotectonic and morphotectonic research. *Zeitschrift für Geomorphologie, N.F.,* Supplement-Band **63**, pp. 1–7.

Embrechts, J. and de Dapper, M. 1987, Morphology and genesis of hillslope pediments in the Febé area (South-Cameroon). *Catena,* **14**, pp. 31–43.

Erhart, H. 1967, *La genèse des sols en tant que phénomène géologique. Esquisse d'une théorie géologique et géochimique, Biostasie et Rhexistasie*, (Paris: Masson et Cie, Éditeurs), pp. 1–177.

Eyles, N. 2008, Glacio-epochs and the supercontinent cycle after ~ 3.0 Ga: Tectonic boundary conditions for glaciation. *Palaeogeography, Palaeoclimatology, Palaeoecology*, **258**, pp. 89–129.

Fairhead, J.D. and Girdler, R.W. 1969, How far does the rift system extend through Africa? *Nature*, **221**, pp. 1018–1020.

Fairhead, J.D. 1985, Preliminary study of the seismicity associated with the Cameroon Volcanic Province during the volcanic eruption of Mt. Cameroon in 1982. *Journal of African Earth Sciences*, **3**(3), pp. 297–301.

Feybesse, J.L., Johan, V., Triboulet, C., Guerrot, C., Mayaga-Mikolo, F., Bouchot, V. and Eko N'dong, J. 1998, The West Central African belt: a model of 2.5–2.0 Ga accretion and two-phase orogenic evolution. *Precambrian Research*, **87**, pp. 161–216.

Fitton, J.G. and Dunlop, H.M. 1985, The Cameroon line, West Africa, and its bearing on the origin of oceanic and continental alkali basalt. *Earth and Planetary Science Letters*, **72**, pp. 23–38.

Fölster, H. 1969, Slope development in SW-Nigeria during Late Pleistocene and Holocene. In *Göttinger Bodenkundliche Berichte 10: Beiträge zur Geomorphologie der wechselfeuchten Tropen*, edited by Fölster, H. and Rohdenburg, H. (Göttingen), pp. 3–56.

Gemerden, B.S. and Hazeu, G.W. 1999, *Landscape ecological survey of the Bipindi-Akom II-Lolodorf-Region Southwest Cameroon*, (Tropenbos Cameroon Programme).

Goudie, A.S. 2005, The drainage of Africa since the Cretaceous. *Geomorphology*, **67**, pp. 437–456.

Guarnieri, P. and Pirrotta, C. 2008, The response of drainage basins to the late Quaternary tectonics in the Sicilian side of the Messina Strait (NE Sicily). *Geomorphology*, **95**, pp. 260–273.

Haman, P.J. 1976, Angular and spatial relationships of Landsat lineaments of the united states. In *Second International Conference on Basement Tectonics*, edited by Podwysocki, M.H. and Earle, J.L. (IBTA Publishing), pp. 353–360.

Heim, D. 1990, *Tone und Tonminerale: Grundlagen der Sedimentologie und Mineralogie*, (Stuttgart: Enke), p. 157.

Hurault, J. 1967, L'érosion régressive dans les régions tropicales humides et la genèse des inselbergs granitiques, (Paris: Institut Geographique National), p. 68.

Kadomura, H. 1977a, Some aspects of geomorphology in the forest and savanna areas of Cameroon, with special reference to South-North variation. In *Geomorphological studies in the forest and savanna areas of Cameroon. An interim report of the Tropical African Geomorphology Research Project 1975/76*, edited by Kadamoura, H. (Hokkaido), pp. 7–36.

Kadomura, H. (Eds.) 1977b, *Geomorphological studies in the forest and savanna areas of Cameroon. An interim report of the Tropical African Geomorphology Research Project 1975/76*, (Hokkaido).

Kadomura, H. (Eds.) 1986, *Geomorphology and environmental changes in Tropical Africa. Case studies in Cameroon and Kenya. A Preliminary Report of the Tropical African Geomorphology and Late-Quaternary Palaeoenvironments Research Project 1984/85*, (Sapporo: Laboratory of Fundamental Research, Hokkaido University).

Kadomura, H. 1989, *A preliminary report of the tropical African Geomorphology and Late Quaternary Palaeoenvironments Research Project (TAGELAQP) 1987/88*, (Tokyo: Department of Geography, Tokyo Metropolitan University).

Kankeu, B. and Runge, J. 2008, Late Neoproterozoic palaeogeography of Central Africa: Relations with Holocene geological and geomorphological setting. In *Palaeoecology of Africa 28: Dynamics of forest ecosystems in Central Africa during the Holocene. Past—Present—Future,* Runge, J. (London, Leiden, New York: Taylor and Francis), pp. 245–257.

Karner, G.D., Driscoll, N.W., McGinnis, J.P., Brumbaugh, W.D. and Cameron, N.R. 1997, Tectonic significance of syn-rift sediment packages across the Gabon-Cabinda continental margin. *Marine and Petroleum Geology,* **14**(7/8) pp. 973–1000.

Keller, G.R., Wendlandt, R.F. and Bott, M.H.P. 1995, West and Central African rift system.Chapter 13. In *Continanetal Rifts: Evolution, structure, tectonics. Development in Geotectonics 25,* edited by Olsen, K.H. (Amsterdam: Elsevier), pp. 437–452.

King, L.C. 1949, The pediment landform: some current problems. *Geological Magazine,* **86**, pp. 245–250.

King, L.C. 1953, Canons of landscape evolution. *Geological Society of America Bulletin,* **64**, pp. 721–752.

Kouankap, N.G.D. 2006, *Neotectonic markers in the Panafrican belt formations of Cameroon: elements of interpretation and their environmental impacts,* (Göttingen: Abstract zur Postersession der TSK 11).

Krenkel, E. 1939, *Geologie der deutschen Kolonien in Afrika,* (Berlin: Borntraeger), p. 272.

Kribi NA-32-XVII—*Carte du Cameroun 1/200.000,* 1976), (Yaoundé: Centre Géographique National).

Kuete, M. 1986, Palaeoforms and superficial deposits: evolution of footslopes and river valleys in the West-Central part of the Cameroon Plateau. In *Geomorphology and environmental changes in tropical Africa. Case studies in Cameroon and Kenya. A Preliminary Report of the Tropical African Geomorphology and Late-Quaternary Palaeoenvironments Research Project 1984/85,* edited by Kadomura, H. (Sapporo: Hokkaido University), pp. 11–30.

Kuete, M. 1990, *Géomorphologie du plateau sud-Camerounais à l'ouest du 13°E,* (Thèse de Doctorat d'Etat, Université de Yaoundé 1).

Lanfranchi, R. and Schwartz, D. 1991, Les remainements de sols pendant le Quaternaire supérieure au Congo. Évolution des paysages dans la région de la Sangha. *Cahiers ORSTOM, série Pédologie,* **26**(1) pp. 11–24.

Lanfranchi, R. and Schwartz, D., 1993, Les cadres paléoenvironnement de l'évolution humaine en Afrique Centrale Atlantique. *L'Anthropologie,* **97**(1), pp. 17–50.

Lauer, W. and Frankenberg, P. 1987, Erde—Klima. In *Diercke Weltatlas,* edited by Diercke, C. (Braunschweig: Westermann).

Lavier, L.L., Steckler, M.S. and Brigaud, F. 2001, Climatic and tectonic control on the Cenozoic evolution of the West African margin. *Marine Geology,* **178,** pp. 63–80.

Lee, D.-C., Halliday, A.N., Fitton, J.G. and Poli, G. 1994, Isotopic variations with distance and time in the volcanic islands of the Cameroon line: evidence for a mantle plume origin. *Earth and Planetary Science Letters,* **123**, pp. 119–138.

Letouzey, R. 1968, *Etude phytogéographique du Cameroun,* (Paris: Lechevalier).

Letouzey, R. 1985, *Notice de la carte phytogéographique du Cameroun au 1/500.000,* (Yaoundé: Institut de la Carte Internationale de la Végétation, Toulouse and Institut Recherches Agronomique).

Leturmy, P., Lucazeau, F. and Brigaud, F. 2003, Dynamic interactions between the Gulf of Guinea passive margin and the Congo River drainage basin: 1. Morphology and mass balance. *Journal of Geophysical Research*, **108** (B8; 2383): pp. 8/1–13.

Louis, H. 1968, *Allgemeine Geomorphologie*, (Berlin: De Gruyter), pp. 1–522.

Lucazeau, F., Brigaud, F. and Leturmy, P. 2003, Dynamic interactions between the Gulf of Guinea passive margin and the Congo River drainage basin: 2. Isostasy and uplift. *Journal of Geophysical Research*, **108**(B8; 2384), pp. 1–19.

Mäckel, R. 1975, Untersuchungen zur Reliefentwicklung des Sambesi-Eskarpmentlandes und des Zentralplateaus von Sambia. *Gießener Geographische Schriften,* **36** (Sonderheft 3).

Mäckel, R. 1985, Dambos and related landforms in Africa—an example for the ecological approach to tropical geomorphology. *Zeitschrift für Geomorphologie, N.F.,* **52**, pp. 1–23.

MacKenzie, K. 2001, Evidence for a plate tectonics debate. *Earth-Science Reviews,* **55**, pp. 235–336.

Makaske, B. 1998, Anastomosing rivers: forms, processes and sediments. *Nederlandse geografische studies,* **249**, (Utrecht: Universität Utrecht).

Martin, D. 1967, Géomorphologie et sols ferralitiques dans le Centre-Cameroun. *Cahiers ORSTOM, série Pédologie*, **4**(2), pp.189–218.

Martin, D. 1970, Géomorphologie et sols ferralitiques dans le Centre-Cameroun. *Cahiers ORSTOM, série Pédologie*, **8**(2), pp. 189–218.

Martin, D. 1973, Les horizons supérieurs des sols ferralitiques sous forêt et sous savane du Centre-Cameroun. *Cahiers ORSTOM, série Pédologie*, **11**(2), pp. 155–179.

Martin, D. and Volkoff, B. 1990, Signification paléoclimatique des cuirasses et des nappes de nodules ferrugineux dans les sols d'Afrique Centrale (rive droite du Zaïre). In*Paysages Quaternaires de l'Afrique Centrale Atlantique*, edited by Lanfranchi, R. and Schwartz, D. (Paris: Editions de l'ORSTOM), pp. 129–135.

Maurizot, P. 2000, *Carte Géologique du sud-ouest Cameroun. Geological map of southwest Cameroon. Échelle 1:500.000*, (BRGM, Orléans, France).

McCarthy, T.S. 1993, The great inland deltas of Africa. *Journal of African Earth Sciences*, **17**(3), pp. 275–291.

McCarthy, T.S., Green, R.W. and Franey, N.J. 1993, The influence of neo-tectonics on water dispersal in the northeastern regions of the Okavango swamps, Botswana. *Journal of African Earth Sciences*, **17**(1), pp. 23–32.

McFarlane, M.J. 1976, *Laterite and Landscape*, (London, New York, San Francisco: Academic Press), pp. 1–151.

Michel, P. 1973, Les basins des fleuves Sénégal et Gambie. Étude géomorphologique. Mémoire, *ORSTOM 63*, pp. 1–752.

Milesi, J.P., Toteu, S.F., Deschamps, Y., Feybesse, J.L., Lerouge, C., Cocherie, A., Penaye, J., Tchameni, R., Moloto-A-Kenguemba, G., Kampunzu, H.A.B., Nicol, N., Duguey, E., Leistel, J.M., Saint-Martin, M., Ralay, F., Heinry, C., Bouchot, V., Doumnang Mbaigane, J.C., Kanda Kula, V., CHene, F., Monthel, J., Boutin, P. and Cailteux, J. 2006, An overview of the geology and major ore deposits of Central Africa: Explanatory note for the 1:4,000,000 map "Geology and major ore deposits of Central Africa". *Journal of African Earth Sciences*, **44**, pp. 571–595.

Murawski, H. 1980, *Afrika-Kartenwerk, Ser. W, Westafrika (Nigeria, Kamerun), Bl. 3, Geologie,* edited by Freitag, U. and Kayser, K. (Deutsche Forschungsgemeinschaft), p. 70.

Murawski, H. 1992, *Geologisches Wörterbuch*, (München: Enke).

Nahon, D. 1986, Evolution of iron crusts in tropical landscapes. In *Rates of chemical weathering of rocks and minerals*, edited by Coleman, S. and Dethier, D.P. (New York: Academic Press), pp. 169–192.

Nfomou, N., Tongwa, A.F., Ubangoh, R., Bekoa, A., Metuk, N.J. and Victor, H.J. 2004, The July 2002 earthquake in the Kribi region: geological context and a preliminary evaluation of seismic risk in southwestern Cameroon. *Journal of African Earth Sciences,* **40,** pp. 163–172.

Ngako, V. 1999, *Les deformations continentals panafricaines en Afrique Centrale,* (These de Doctorat d'Etat des Sciences Naturelles), p. 301.

Ngako, V., Affaton, P., Nnange, J.M. and Njanko, Th. 2003, Pan-African tectonic evolution in central and southern Cameroon: transpression and transtension during sinistral shear movements. *Journal of African Earth Sciences,* **36,** pp. 207–214.

Nwajide, C.S. 1987, Provenance and palaeographic history of the Turonian Makurdi Sandstones in the Benue Trough of Nigeria. *Palaeogeography, Palaeoclimatology, Palaeoecology,* **58,** pp. 109–119.

O'Leary, D.W., Friedman, J.D., and Pohn, H.A. 1976. Lineament and linear, a terminological reappraisal. Paper presented at *Second International Conference on Basement Tectonics* in Newark, Delaware, USA. July 13–17.

Olivry, J.-C. 1986, *Fleuves et rivières de Cameroun,* (Paris: Éditions de l'ORSTOM), p. 733.

Oslisly, R. and Mbida, C. 2001, *Surveillance archéologique de l'axe routier Lolodorf-Kribi-Campo,* (Rapport final. IRD, Yaoundé), p.186.

Owona, S., Mvondo Ondoa, J., Ratschbacher, L., Mbola Ndzana, S.P., Tchouba, F.M. and Ekodeck, G.E. 2011a, The geometry of the Archean, Paleo- and Neoproterozoic tectonics in the Southwest Cameroon. *Comptes Rendus Geoscience,* **343,** pp. 312–322.

Owona, S., Schulz, B., Ratschbacher, L., Mvondo Ondoa, J., Ekodeck, G.E., Tchoua, F.M. and Affaton, P. 2011b, Pan-African metamorphic evolution in the southern Yaoundé Group (Oubanguide Complex, Cameroon) as revealed by EMP-monazite dating and thermobarometry of garnet metapelites. *Journal of African Earth Sciences,* **59,** pp. 125–139.

Panizza, M. and Castaldini, D. 1987, Neotectonic research in applied geomorphological studies.*Zeitschrift für Geomorphologie, N.F.,* **63,** pp. 173–211.

Penck, W. 1924, *Die morphologische Analyse. Ein Kapitel der physikalischen Analyse,* (Stuttgart: Engelhorn Nachf.), 1–283.

Raunet, M. 1985, Les bas-fonds en Afrique et à Madagascar. *Zeitschrift für Geomorphologie, N.F.,* **52,** pp. 25–62.

Reyment, R.A. and Tait, E.A. 1972, Faunal evidence for the origin of the South Atlantic. *Proc. 24th Int. Geol. Cong. Montreal,* **7,** pp. 316–323.

Robert, C. 1987, Clay mineral associations and structural evolution of the South Atlantic: Jurassic to Eocene. *Palaeogeography, Palaeoclmatology, Palaeoecology,* **58,** pp. 87–108.

Rohdenburg, H. 1970, Hangpedimentation und Klimawechsel als wichtigste Faktoren der Flächen- und Stufenbildung in den wechselfeuchten Tropen. *Zeitschrift für Geomorphologie, N.F.,* **14,** pp. 58–78.

Rohdenburg, H. 1982, Geomorphologisch-bodenstratigraphischer Vergleich zwischen dem nordostbrasilianischen Trockengebiet und immerfeucht-tropischen Gebieten Südbrasiliens. Mit Ausführungen zum Problemkreis der Pediplain-Pediment-Terrassentreppen. In *Beiträge zur Geomorphologie der Tropen (Ostafrika, Brasilien, Zentral- und Westafrika),* edited by Ahnert, F., Rohdenburg, H. and Semel, A. (Braunschweig: Cateny Suppl. 2).

Rosendahl, B.R. and Groschel-Becker, H. 1999, Deep seismic structure of the continental margin in the Gulf of Guinea: a summary report. In The oil and gas habitats of the South Atlantic. *Geological Society*, **153**, edited by Cameron N.R., Bate, R.H. and Clure, V.S. (London, Special Publications), pp. 75–83.

Rosendahl, B.R. and Groschel-Becker, H. 2000, Architecture of the continental margin in the Gulf of Guinea as revealed by reprocessed deep-imaging seismic data. In Atlantic rifts and continental margins. *Geophysical Monograph,* **115**, edited by Mohriak, W. (American Geophysical Union), pp. 85–103.

Ruhe, R.V. 1952, Topographic discontinuities of the Des Moines lobe. *American Journal of Science,* **250**, pp. 46–56.

Runge, J. 2001, Landschaftsgenese und Paläoklima in Zentralafrika. *Relief Boden Paläoklima,* **17**, (Berlin, Stuttgart: Gebr. Borntraeger), pp. 1–294.

Runge, J., Eisenberg, J. and Sangen, M. 2005, Ökologischer Wandel und kulturelle Umbrüche in West- und Zentralafrika—Prospektionsreise nach Südwestkamerun vom 05.03.-03.04.2004 im Rahmen der DFG-Forschergruppe 510: Teilprojekt "Regenwald-Savannen-Kontakt (ReSaKo)". *Geoökodynamik,* **26**, pp. 135–154.

Runge, J., Eisenberg, J. and Sangen, M. 2006, Geomorphic evolution of the Ntem alluvial basin and physiogeographic evidence for Holocene environmental changes in the rain forest of SW Cameroon (Central Africa)—preliminary results. *Zeitschrift für Geomorphologie, N.F.,* Supplement, **145**, pp. 63–79.

Sangen, M. 2007, Physiogeographische Untersuchungen zur holozänen Umweltgeschichte an Alluvionen des Ntem-Binnendeltas im tropischen Regenwald SW-Kameruns. *Zentralblatt für Geologie und Paläontologie, Teil I,* Heft **1/4**, pp. 113–128.

Sangen, M. 2008, New evidence on palaeoenvironmental conditions in SW Cameroon since the Late Pleistocene derived from alluvial sediments of the Ntem River. In: Runge, J. (Hrsg.): Dynamics of forest ecosystems in Central Africa during the Holocene. Past—Present—Future. *Palaeoecology of Africa,* **28**, pp. 79–102.

Santoir, C. 1995, L'oro-hydrographie. In Atlas régional Sud-Cameroun, edited by Santoir, C. and Bopta, A. (Paris: Éditions ORSTOM), pp. 4–5.

Schumm, S.A., Dumont, J.F., Holbrook, J.M. 2000, *Active tectonics and alluvial rivers,* (Cambridge: Cambridge University Press), p. 276.

Segalen, P. 1967, Les sols et la géomorphologie du Cameroun. *Cahiers ORSTOM, série Pédologie,* **5**(2), pp. 137–188.

Segalen, P. 1969, Le remaniement des sols et la mise en place de la stone-line en Afrique. *Cahiers ORSTOM, série Pédologie,* **7**(1), pp. 113–127.

Shackleton, N.J. 1977, The oxygen isotope stratigraphic record of the Late Pleistocene. *Philosophical Transactions of the Royal Society.* London B, **280**, pp. 169–182.

Skobelev, S.F., Hanon, M., Klerkx, J., Govorova, N.N., Lukina, N.V. and Kazmin, V.G. 2004, Active faults in Africa: a review. *Tectonophysics,* **380**, pp. 131–137.

Stewart, I.S. and Hancock, P.L. 1994, Neotectonics. In *Continental deformation,* edited by Hancock, P.L. (Oxford, New York: Pergamon Press), pp. 370–409.

Strahler, A.N. 1952, Hypsometric (area-altitude) analysis of erosional topography. *Bulletin of the geological society of America,* **63**, pp. 1117–1142.

Strahler, A.N. 1957, Quantitative analysis of watershed geomorphology. *Transactions, American Geophysical Union,* **38**(6), pp. 913–920.

Suh, C.E., Ayonghe, S.N. and Njumbe, E.S. 2001, Neotectonic earth movements related to the 1999 eruption of Cameroon Mountain, West Africa. *Episodes,* **24**(1), pp. 9–12.

Summerfield, M.A. 1985, Tectonic background to long-term landform development in tropical Africa.In *Environmental Change and Tropical Geomorphology,* edited by Douglas, I. and Spencer, T., pp. 281–294.

Summerfield, M.A. 1991, *Global geomorphology: an introduction to the study of landforms,* (Essex: Longman), p. 537.

Summerfield, M.A. 1996, Tectonics, geology, and long-term landscape development. In: *The Physical Geography of Africa,* edited by Adams, W.W., Goudie, A.S. and Orme, A.R. (Oxford), pp. 1–17.

Thomas, M.F. 1974, *Tropical Geomorphology: a study of weathering and landform development in warm climates,* (London: McMillan).

Thomas, M.F. 1994, *Geomorphology in the tropics. A study of weathering and denudation in low latitudes,* (New York, Toronto, Singapore: Wiley), p. 460.

Thomas, M.F. 2004, Landscape sensitivity to rapid environmental change—a Quaternary perspective with examples from tropical areas. *Catena,* **55**, pp. 107–124.

Toteu, S.F., Bertrand, J.M., Penaye, J., Macaudiere, J., Angoua, S. and Barbey, P. 1990, Cameroon: a tectonic keystone in the Pan-African network. In: Lewry, J.F. and Stauffer, M.R. (Hrsg.): The Early Proterozoic Trans-Hudson Orogen of North-America. *Geological Association of Canada* (Special Paper), **37**, pp. 483–496.

Toteu, S. F., Penaye, J. and Poudjom Djomani, Y. H. 2004, Geodynamic evolution of the Pan-African belt in central Africa with special reference to Cameroon. *Canadian Journal of Earth Sciences,* **41**, pp. 73–85.

TREES, 1997, Tropical Ecosystem Environment Observation by Satellites (TREES). Vegetation Map of Central Africa. Ispra: World Resources Institute. http://www. globalforestwatch.org/english/datawarehouse/index.asp# (3.12.2008).

Ubangoh, R., Ateba, B., Ayonghe, S. N. and Ekodeck, G. E. 1997, Earthquake swarms of Mt Cameroon, West Africa. *Journal of African Earth Sciences,* **24**(4), pp. 413–424.

Villiers, J.-F. and Santoir, C. 1995, La végétation. In *Atlas régional Sud-Cameroun,* edited by Santoir, C. and Bopta, A. (Paris: Éditions ORSTOM), pp. 10–11.

Vita-Finzi, C. 1986, *Recent earth movements. An introduction to neotectonics,* (London, Academic Press), p. 226.

Wang, S., Chen, Z. and Smith, D.G. 2005, Anastomosing river system along the subsiding middle Yangtze River basin, southern China. *Catena,* **60**, pp. 147–163.

Watchman, A.L. and Twidale, C.R. 2002, Relative and ‚absolute' dating of land surfaces. *Earth-Science Reviews,* **58**, pp. 1–49.

Webster, R. 1965, A catena of soils on the Northern Rhodesian Plateau. *The Journal of soil science,* **16**(1), pp. 31–43.

Wiese, B. 1997, *Afrika. Ressourcen Wirtschaft Entwicklung,* (Stuttgart: Teubner, Studienbücher der Geographie—Regional), p. 269.

Wirthmann, A. 1994, *Geomorphologie der Tropen,* (Darmstadt: Wissenschaftliche Buchgesellschaft: Erträge), p. 222.

CHAPTER 4

Late Quaternary palaeoenvironments in Southern Cameroon as evidenced by alluvial sediments from the tropical rain forest and savanna domain

Mark Sangen

Institute of Physical Geography, Johann Wolfgang Goethe University, Frankfurt am Main, Germany

ABSTRACT: Extended fieldwork on Late Quaternary landscape evolution was undertaken across southern Cameroon to sample and elucidate the value and relevance of alluvial sedimentary archives for palaeoenvironmental research. Around 160 hand-corings reaching maximum depths of 550 cm were carried out on alluvial ridges and floodplains of major Cameroonian fluvial systems (Boumba, Ngoko, Ntem, Nyong and Sanaga). These multilayered, sandy to clayey alluvia contain sedimentary sequences and 'palaeosurfaces' (e.g., fossil organic horizons, relict swamps and mudds), which provide excellent additional proxy data for the reconstruction of palaeoenvironmental conditions in western equatorial Africa. Coring transects and sedimentary profiles document grain-size shifts and distinguishable sedimentary units in the stratigraphic record, which evidence fluvial-morphological adjustments of the fluvial systems to external forcing and (river-) intrinsic response and variability. Seventy-six ^{14}C (AMS)-dated samples from organic sediment and macro-remains (wood, leaves, etc.) embedded in these sedimentary units yielded Late Pleistocene to recent ages (uncalibrated ^{14}C-ages: 48 to 0.2 kyrs BP). The tentative interpretation of the alluvial record highlights hydrological, ecological and fluvial-morphological changes across Cameroonian fluvial ecosystems during the Last Glacial Maximum, the adjacent humidification in the course of the African Humid Period and the northern hemispheric Bølling-Allerød (Late Pleistocene-Holocene transition) as well as the Holocene. This multi-proxy approach of combining remote sensing methods with archaeobotanical, mineralogical, geomorphological and sedimentological analysis of numerous ^{14}C-dated alluvial stratigraphies offers excellent opportunities for palaeoenvironmental studies in those regions, where 'conventional' terrestrial proxy data archives (i.e., lakes, swamps, mires) are lacking or non-existent. The δ^{13}C-values (-35.5 to -18.0 ‰) of the dated samples primarily indicate the persistence of C_3-dominated gallery forests across the rivers ('fluvial rain forest refuges') despite several climatic fluctuations (aridifications e.g., Last Glacial Maximum around 20 kyrs BP, Younger Dryas 13–11 kyrs BP). Supported by earlier findings from lacustrine and hemi-pelagical sediment archives from Central Africa, the results from manifold river sections, in this essay primarily of the Ntem, Nyong and Sanaga Rivers, corroborate the synthetic reconstruction of the Late Quaternary climatic, hydrological and environmental evolution in southern Cameroon.

4.1 INTRODUCTION

4.1.1 Objectives and purpose of this study

The Rain Forest Savanna Contact ('ReSaKo') project focused on providing geoscientific evidence for past environmental change. The research groups' activities focused on examining the temporal and spatial coherence of altering land-use strategies, cultural innovations and climate change during the so-called 'First Millennium BC Crisis'. For this approach the ecologically sensitive ecosystem of the tropical southern Cameroonian evergreen to semi-deciduous rain forest and its adjacent savanna margin had been selected.

The ReSaKo project for the first time systematically identified (see contribution by J. Eisenberg in this volume) and sampled alluvial sedimentary archives in southern Cameroon. The main target of this contribution was to sample, analyse and interprete the hitherto unexplored alluvial sediments of southern Cameroonian rivers to contribute to a more sophisticated understanding of the nature and history of Late

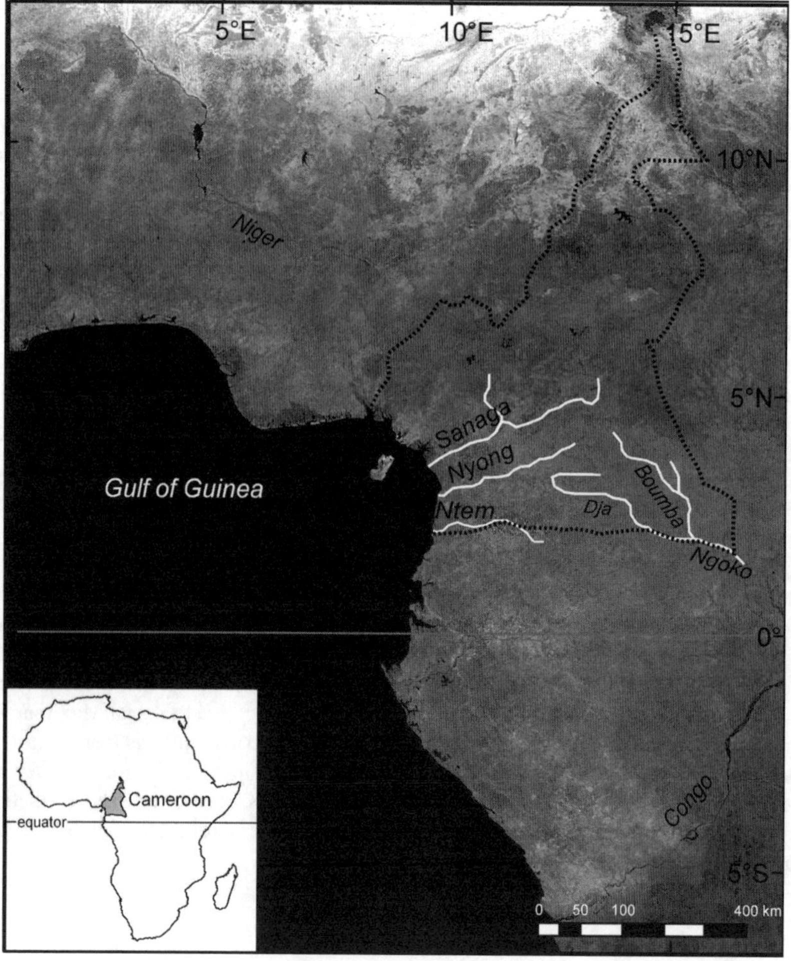

Figure 1. Overview map (SPOT 4-satellite image of GLC 2000) showing the location of Cameroon (broken line) and the surveyed fluvial systems Sanaga, Nyong, Ntem, Dja, Boumba and Ngoko.

Quaternary palaeoenvironments in this western Central African region. The following crucial questions were formulated:

- Is it possible to identify and sample in the context of a fluvial-morphological, palaeohydrological and palaeoenvironmental framework interpretable alluvial sediments across hitherto unexplored tropical fluvial systems of southern Cameroon?
- Can these alluvial archives offer any (new) evidence for the reconstruction of the Late Quaternary environmental conditions in the study area?
- How does the new evidence correlate with earlier findings from Central African terrestrial and marine (hemi-pelagic) sedimentary archives?
- How far can the results contribute to an overall interdisciplinary result (i.e., corroborate archaeological or archaeobotanical findings) of the DFG-Research-Unit 510?

Interpretation of the alluvial sedimentary archives, some of which show striking changes in grain-size distribution and embedded 'palaeosurfaces' (e.g., fossil organic horizons, relict swamps and mudds), can allow far-reaching conclusions on former climatic, hydrological and environmental conditions in the studied river catchment areas. In particular, such palaeohydrological interpretations can analyse how changing patterns of discharge and sediment fluxes can be induced by climate changes or other triggers, such as land-use, vegetation changes and the impact of human activity. Proxy data from alluvial sedimentary archives can contribute to enhanced understanding of major transformations of landscape settings, primarily of former fluvial environments.

Besides fluvial-morphological evidence, which supports reconstruction of past fluvial environments and fluvial system evolution, proxy data inherited in the alluvia (e.g., pollen, phytoliths, sponge spicules, clay minerals, diatoms, $\delta^{13}C$, etc.) corroborate a more holistic synopsis of palaeoenvironmental conditions in a region, where a significant data gap still exists. Supported by earlier findings from lacustrine and hemi-pelagical sediment archives from Central Africa (e.g., Gasse et al., 2008 and references therein), results from manifold river sections (here primarily of the Ntem, Nyong and Sanaga Rivers), supplement the synthetic reconstruction of the Late Quaternary climatic, hydrological and environmental evolution in southern Cameroon. Although initially the interdisciplinary DFG-Research Unit 510's focus was the Late Holocene and especially the 'First Millennium BC Crisis', the sampled alluvial sequences provided even records reaching back into the Late Pleistocene (up to 53,000 cal. yrs BP). In this way, abundant information was also gained on conditions during the Last Glacial Maximum, the Late Pleistocene-Holocene transition (northern hemispheric 'Bølling–Allerød') and the Early as well as Middle Holocene.

4.1.2 Methods

After extensive analysis and interpretation of satellite imagery (LANDSAT ETM+, ASTER; compare also contribution by J. Eisenberg in this volume) to identify suitable coring sites across alluvial ridges of meandering to anabranching/anastomosing rivers of southern Cameroon, extended fieldwork campaigns were organised and conducted to sample the selected alluvia. The studied rivers flow through recent extensions of tropical rain forest (Boumba, Ngoko, Ntem) and across rain forest-savanna border zones (Nyong und Sanaga); given the rivers' location, their alluvial sedimentary records should yield important information on past landscape and environment dynamics and evolution, as well as climate and vegetation, in the respective catchment areas.

Sampling was carried out with an Eijkelkamp Edelman hand-corer, reaching maximum depths of around 6 m. Eventually, a 50 cm long and 3 cm wide percussion probe was used to recover fine-grained (clayey) cores; this probe turned out to be very useful for sampling water-logged sediments. In total, ca. 160 corings in the alluvial basins of the Boumba, Ngoko, Ntem, Nyong and Sanaga Rivers were undertaken during dry seasons of 2005–2008. After texture and soil colour were determined in the field, samples were analysed at the laboratory of the Institute of Physical Geography in Frankfurt. Here pH (solved in 0.1 n KCl, following Meiwes et al., 1984), grain sizes (following Köhn), soil colour (wet and dry, Munsell Colour Charts), organic material (OM) and carbon contents (C_{tot}; with LECO EC–12), nitrogen (N_{tot}; following Kjeldahl), dithionite and oxalate solvable quantums of iron (Fe_d and Fe_o) and manganese (according to Mehra and Jackson, 1960; using AAS Perkin Elmer Analyst 300) were determined.

During coring and sediment treatment in the laboratory, macro-remains, charcoal and organic sediment from sedimentary units, fossil organic horizons and palaeosurfaces were collected for ^{14}C (AMS) radiocarbon dating and determination of stable carbon isotopic composition values (δ^{13}C). These data provided preliminary (maximum) ages of the alluvial sediments as well as information on former vegetation composition (C_3/C_4 species; cf. Schwartz et al., 1986; Schwartz, 1992; Giresse et al., 1994; Runge, 2002). In total, ca. 70 samples from different locations and depths were extracted and analysed by the '*CEntro di DAtazione e Diagnostica* (CEDAD)' of Salento University in Lecce (Italy), the AMS ^{14}C-laboratory of the Friedrich-Alexander-University in Erlangen, Nuremberg (Germany) and Beta Analytic Inc. in Miami, Florida (USA). For calibration of ^{14}C (AMS) ages, CalPal (www.calpal-online.de), which is based on calibration-curve CalPal2007_Hulu (Hughen et al., 2006), was used (Table 1). Besides the analysis of grain-size distributions, stratigraphic features and pedological as well as sedimentological characteristics of the alluvial sediments, a further range of prominent

Table 1. Dated sediments from southern Cameroonian fluvial systems.

Sanaga

core	site	depth (cm)	dated material	^{14}C (AMS) BP	cal. yrs BP	δ^{13}C (‰)	lab.-no.
C01	Dizangué	200–218	organic sediment	128.5 ± 0.9 pMC	recent	–19.6	LTL2104A
C06	Dizangué	360–365	organic sediment	9321 ± 90	10,515 ± 132	–27.2	LTL2106A
C13	Bélabo	320–340	organic sediment	11,840 ± 60	13,749 ± 137	–19.6	Beta-278642
		390–400	organic sediment	16,117 ± 200	19,294 ± 325	–28.2	LTL2107A
C14	Bélabo	335–340	organic sediment	1766 ± 50	1693 ± 72	–29.7	LTL2108A
C15	Bélabo	381	wood, etc.	1196 ± 45	1130 ± 58	–28.7	LTL2109A
C16	Ngoktélé	240–260	leaves, etc.	14 ± 100	108 ± 125	–24.2	LTL2110A
C19	Sakoudi	213	organic sediment	3894 ± 50	4327 ± 72	–29.2	LTL2111A

(*Continued*)

Table 1. (*Continued*)

Nyong

core	site	depth (cm)	dated material	^{14}C (AMS) BP	cal. yrs BP	δ^{13}C (‰)	lab.-no.
B01	Njock	200–225	organic sediment	2134 ± 41	2161 ± 100	−27.3	Erl-8941
B03	Makak	160–180	wood	−2105 ± 34	recent	−29.8	Erl-8944
		235–245	leaves, etc.	−4116 ± 31	recent	−29.4	Erl-8946
B05	Lipombe II	100–120	wood	707 ± 38	637 ± 48	−28.2	Erl-8943
B06	Donenda	180–200	wood, etc.	4093 ± 47	4656 ± 117	−29.7	Erl-8945
B07	Dehane	280–300	wood	400 ± 40	428 ± 72	−28.7	Erl-8942
C21A	Benana	255–265	leaves, etc.	1946 ± 50	1902 ± 56	−27.7	LTL2112A
C23	Akonolinga	150–160	organic sediment	28,358 ± 300	32,788 ± 411	−20.1	LTL2113A
C28	Akonolinga	320–340	organic sediment	14,160 ± 60	17,395 ± 244	−24.1	Beta-282934
		340–360	organic sediment	16,360 ± 70	19,605 ± 265	−24.9	Beta-282935
		420	organic sediment	42,940 ± 1500	46,680 ± 1817	−19.6	LTL2114A
exposure	Akonolinga	230	organic sediment	902 ± 45	832 ± 61	−24.2	LTL2117A
C30	Ayos	280	leaves, etc.	724 ± 45	679 ± 28	−25.8	LTL2115A
exposure	Ayos	180	organic sediment	1508 ± 50	1419 ± 65	−18.0	LTL2116A
C31	Mengba	220	organic sediment	11,107 ± 90	13,011 ± 138	−28.9	LTL3171A
		260–280	organic sediment	13,357 ± 60	16,295 ± 418	–	LTL2118A
NY05	Ebabodo	280	leaves, etc.	732 ± 30	684 ± 12	−22.0	LTL3178A
NY11	Tetar	280	organic sediment	31,904 ± 300	35,961 ± 482	−33.5	LTL3179A
NY13	Tetar	180	organic sediment	18,858 ± 100	22,702 ± 248	−18.1	LTL3180A
		180–200	organic sediment	11,840 ± 60	13,749 ± 137	−19.1	Beta-283401

Boumba

core	site	depth (cm)	dated material	^{14}C (AMS) BP	cal. yrs BP	δ^{13}C (‰)	lab.-no.
B02	Ouesso	360–376	organic sediment	2728 ± 60	2846 ± 59	−35.5	LTL3176A
B03	Mankako	300–320	organic sediment	> 45,000	–	−29.6	LTL3177A

(*Continued*)

Table 1. (*Continued*)

Ngoko

core	site	depth (cm)	dated material	^{14}C (AMS) BP	cal. yrs BP	δ^{13}C (‰)	lab.-no.
–	Mokounounou	480–485	organic sediment	898 ± 50	829 ± 64	−30.0	LTL3170A
N01	Mokounounou	550	organic sediment	938 ± 45	856 ± 52	−28.0	LTL3169A
N02	Mokounounou	393	wood, etc.	1041 ± 45	973 ± 41	−28.4	LTL3172A
N04	Mokounounou	470	wood (*Raphia*)	892 ± 40	827 ± 62	−26.0	LTL3173A
N05	Mokounounou	465	organic sediment	1228 ± 35	1167 ± 64	−28.1	LTL3174A
N06	Mokounounou	545–550	organic sediment	1126 ± 45	1046 ± 59	−29.4	LTL3175A

Ntem

core	site	depth (cm)	dated material	^{14}C (AMS) BP	cal. yrs BP	δ^{13}C (‰)	lab.-no.
L02	Meyo Ntem	100–120	wood	1066 ± 53	995 ± 51	−27.0	Erl-8249
		160–180	wood	1104 ± 52	1024 ± 53	−29.2	Erl-8250
C02	Nyabibak	380–400	wood etc.	5379 ± 51	6162 ± 95	−29.0	Erl-9573
L05	Meyo Ntem	220–240	leaves etc.	908 ± 50	835 ± 61	−29.4	Erl-8251
L08	Meyo Ntem	300–320	wood	45,596 ± 4899	50,851 ± 5663	−31.4	Erl-8252
–	Aloum II	78–88	organic sediment	8402 ± 67	9408 ± 75	−30.6	Erl-9574
C11	Aloum II	120–140	wood etc.	6979 ± 58	7822 ± 76	−30.1	Erl-9575
C13	Meyos	60–80	organic sediment	4081 ± 56	4646 ± 122	−29.9	Erl-10366
		120–140	wood	14,020 ± 106	17,271 ± 229	−28.1	Erl-9567
		140–160	wood	17,570 ± 141	20,981 ± 336	−26.5	Erl-9568
		160–180	wood	18,719 ± 161	22,389 ± 377	−26.5	Erl-9569
		180–200	organic sediment	17,197 ± 132	20,647 ± 322	−28.8	Erl-9570
		250–263	organic sediment	18,372 ± 164	22,011 ± 341	−27.6	Erl-9571
L14	Abong	340–360	wood	48,230 ± 6411	53,096 ± 6876	−29.6	Erl-8254
L17	Meyos	280–300	wood, leaves	14,263 ± 126	17,467 ± 265	−27.3	Erl-8253
		280–300	wood, leaves	13.690 ± 100	16.781 ± 251	−24.2	LTL2103A
		380–400	wood, leaves	14.516 ± 80	17.658 ± 248	−23.1	LTL3181A
L18	Nyabessan	140–160	leaves etc.	587 ± 64	596 ± 46	−29.1	Erl-8270
		200–220	leaves etc.	2337 ± 55	2389 ± 73	−31.9	Erl-8271

(*Continued*)

Table 1. (*Continued*)

L19	Nyabessan	320–340	wood etc.	2189 ± 52	2216 ± 78	−29.5	Erl-8272
C20	Nyabessan	430–440	organic sediment	2479 ± 43	2571 ± 106	−28.8	Erl-9576
L22	Nyabessan	140–160	wood etc.	3894 ± 57	4324 ± 79	−28.1	Erl-8273
L24	Nnémeyong	80–100	wood	441 ± 46	482 ± 39	−30.6	Erl-8266
		120–140	wood	671 ± 52	623 ± 46	−28.7	Erl-8267
L25	Nnémeyong	160–180	wood	21,908 ± 302	26,239 ± 583	−27.0	Erl-8268
		220–240	wood	22,398 ± 316	26,978 ± 631	−29.1	Erl-8269
L27	Akom	100–120	wood, leaves	443 ± 57	453 ± 72	−26.4	Erl-8261
		160–180	leaves	427 ± 52	440 ± 74	−27.6	Erl-8262
C27	Nyabessan	270–280	organic sediment	3829 ± 46	4257 ± 87	−28.5	Erl-9577
L30	Tom	140–160	wood etc.	1381 ± 49	1309 ± 30	−28.2	Erl-8260
L32	Nkongmeyos	60–80	charcoal	217 ± 46	196 ± 103	−28.0	Erl-8255
C32	Nkongmeyos	145–150	organic sediment	954 ± 39	865 ± 50	−28.9	Erl-9572
L34	Nkongmeyos	100–120	wood	435 ± 51	448 ± 70	−27.4	Erl-8256
L36	Aya Amang	140–160	leaves	4341 ± 60	4945 ± 75	−27.2	Erl-8263
		200–220	wood, leaves	4514 ± 45	5176 ± 94	−27.3	LTL3182A
L37	Aya Amang	320–340	wood	30,675 ± 770	34,962 ± 702	−30.8	Erl-8264
		360–380	wood, leaves	31,253 ± 240	35,220 ± 377	−27.1	LTL3183A
L38	Aya Amang	220–240	wood	5306 ± 64	6096 ± 85	−31.4	Erl-8265
		280–300	wood, leaves	5282 ± 45	6079 ± 79	−26.8	LTL3184A
L40	Anguiridjang	280–300	organic sediment	2180 ± 50	2209 ± 81	−26.9	LTL3185A
		360–380	wood etc.	2339 ± 52	2393 ± 65	−27.9	Erl-8259
L46	Nkongmeyos	80–100	wood etc.	8291 ± 89	9277 ± 127	−27.5	Erl-8599
		120–140	wood etc.	10,871 ± 99	12,846 ± 101	−28.3	Erl-8258
L49	Nkongmeyos	180–200	organic sediment	10,775 ± 144	12,744 ± 155	−30.1	Erl-8257

proxy data could be retrieved from the alluvial records. After initially finding and analysing pollen from one site in the Ntem's interior delta (Nyabessan; Ngomanda et al., 2009), research on clay minerals (Weldeab et al., 2011), phytholiths, sponge spicules (Sangen et al., 2011) and diatoms (minor results yet; François Nguetsop, personal communication) from the alluvia considerably contributed to the formulation of palaeoenvironmental reconstructions.

4.1.3. State of the art

Late Pleistocene to recent evolution of palaeoenvironmental and hydro-climatic conditions in western Central Africa

One of the earliest and most profound theories respecting palaeoenvironmental conditions in western Central Africa was developed by De Ploey (1964, 1965) and Van

Zinderen Bakker (1967, 1969, 1975). Among other things, their studies led to the definition of several arid (*Maluékien, Léopoldvillien, Kibangien B*) and humid phases (*Njilien, Kibangien A*) for the Central African domain since the Late Pleistocene. These correspond to hypothermal ('displuvial') and hyperthermal ('isopluvial') hydro-climatic conditions (Bernard, 1962; Giresse, 1978), which prevailed during cyclic changes in solar radiation (Milanković cycles, especially precession cycles, 19–21 kyrs) and consequent weakenings/strengthenings of the West African monsoon (e.g., Rossignol-Strick, 1983; Prell and Kutzbach, 1987; DeMenocal et al., 2000; Vidal and Arz, 2004; Weldeab et al., 2007, 2008, 2011). The imprints of the cyclic climatic changes have been conserved in the sedimentary records of Central Africa. The most far-reaching (up to 700,000 yrs. BP) information was gained from pelagic and hemi-pelagic deep-sea cores containing stratigraphic palaeoenvironmental proxy data (e.g., diatoms, phytoliths, pollen, clay minerals; Abrantes, 2003; Gasse et al., 2008). Perhaps the most significant data has come from sedimentary records derived from deep-sea fans of Central Africa's major rivers, the Congo (e.g., Schneider et al., 1997; Marret et al., 2001, 2006), Niger (Pastouret et al., 1978; Zabel et al., 2001; Adegbie et al., 2005) and, the Sanaga and Nyong in southern Cameroon (Weldeab et al., 2005, 2007).

Most detailed (high-resolution) information on local palaeoenvironmental conditions in monsoonal western equatorial Africa has been gained through palynological studies on lacustrine sediment archives, such as those of the Barombi Mbo (Maley and Brenac, 1998), Bambili (Stager and Anfang-Sutter, 1999) and Ossa (Reynaud-Farrera et al., 1996; Nguetsop et al., 2004) lakes in Cameroon, and several sites in southern Congo (Sinnda, Coraf, Songolo, Ngamakala, Bilanko and Kitina; e.g., Elenga et al., 1994, 2004; Vincens et al., 1998, 1999) and Gabon (Ngomanda, 2005; Giresse et al., 2008), although these records generally show Holocene ages and rarely date back to the Late Pleistocene (Barombi Mbo up to 30 kyrs BP). For southern Cameroon the most interesting results were found by the ECOFIT-programme (Servant and Servant-Vildary, 2000). In a wider context, findings from Lake Bosumtwi (Talbot et al., 1984; Maley, 1991) in Ghana, Lake Sélé in Benin (Salzmann and Hoelzmann, 2005) and Lake Chad (Maley, 1981) should be mentioned. Overall the results corroborate the occurrence of several major climatic (e.g., alterations of seasonality, aridifications) and ecological ('savannisation', cf. Kadomura, 1995, 2000; retreat of rain forest ecosystems, cf. Maley, 1991, 2001, 2002) changes, especially during the Last Glacial Maximum (LGM), Younger Dryas and several shorter Holocene episodes (around 8.2, 4 and 2.8 kyrs BP ['First Millennium BC Crisis']; e.g., Maley, 2001; Barker et al., 2004; Elenga et al., 2004; Gasse et al., 2008).

However, research on fluctuations in fluvial activity and fluvial-morphologic processes in Central Africa since the Late Pleistocene are still very limited (Thomas, 2000, 2008), although the work of Runge (2001, 2002) and Neumer (2007) in the Central African Republic, Thomas and Thorp (1980, 1995, 2003) and Thorp and Thomas (1992) in Sierra Leone and Ghana, as well as Preuss (1986, 1990) in the Congo-Zaire basin have shown that tropical African alluvia are exceptionally valuable for palaeoenvironmental reconstructions. Earlier studies from Australia and southern America showed the great potential of alluvia for palaeoenvironmental research (e.g., Stevaux, 1994, 2000; Nanson et al., 1993, 2008; Nanson and Price, 1998; Nott and Price, 1999; Latrubesse and Franzinelli, 2002, 2005; Latrubesse, 2003, 2008; Latrubesse et al., 2005). In contrast to the by very local phenomena depicted lacustrine and palustrine sediment archives, the value of alluvia for reconstructions of former hydro-climatic and environmental conditions is related to the fact that they bear witness to changing run-off, erosion and sedimentation behaviour of fluvial systems and, therefore, to ecosystems on wider, i.e., catchment/river basin, scales. And, although, interpretation of alluvial records can be hampered by complexity and uncertainty (e.g., interbedding of alluvial and colluvial material, iron crusts, stone-lines, hiatus; e.g., Schumm, 1977; Thomas, 2000, 2004; Lewin and Macklin, 2003), nonetheless, they provide an excellent proxy data source for wide-reaching modifications in landscape and ecosystem evolution. Nevertheless, it still remains a challenge to decipher the complex effects of climate, water balance, vegetation, tectonics and human activity on the deposition of alluvial stratigraphies (Holbrook and Schumm, 1999; Thomas, 2000, 2004; Vandenberghe, 2003).

4.2 STUDY AREAS, FIELD SITES AND OBSERVATIONS ON ALLUVIA

4.2.1 Sanaga catchment

4.2.1.1 Geographical characteristics

The Sanaga catchment covers around 25% of Cameroon's surface (133,000 km²) between 3°32' and 7°22'N as well as 9°45' and 14°57'E. About 200 km² of the source area, the major tributaries of the Djérem and Lom Rivers on the *Surface de Meiganga* (*Surface Post-Gondwanienne*), are located in the Central African Republic. The waters of the Sanaga start their 976 km long journey on the Adamaoua plateau (1000–1200 m asl.). This peneplain surface, also called *Surface Intérieure* (interior surface or inner plateau; Segalen, 1967), is the eastern extension of the Cameroon Volcanic Line (CVL; *Dorsale Camerounaise*), and is characterized by highly eroded surfaces with deeply weathered ferralitic soils (Figure 2). The Sanaga catchment is in the N and W limited by the *Surface Supérieure* (superior surface; Segalen, 1967), as part of the *Ligne du Cameroun* volcanic chain (i.e., CVL), including the active volcano *Mont Cameroun* (4095 m asl.) and *Mont Oku* (3008 m asl.). To the south, the catchment is joined by the Congo Craton transition zone, which is built up of metamorphites and intrusiva. The river follows a marked shear zone from E to W (*Faille de la Sanaga/ Sanaga fault*), which evolved during the Precambrian Pan-African orogenesis (Ngako, 2006; Toteu et al., 2004). This is a southward displacement of the Central Cameroon Shear Zone (CCSZ), a tectonic shear zone stretching in WSW-ENE direction through central Cameroon (Ngako et al., 2003). The source of the Djérem (at 1150 m asl.) is situated on the southern flank of the Adamaoua plateau, near the city of Ngaoundéré (7°19'N, 13°34'E, 1212 m asl.). The river then flows into Lake Mbakaou near Tibati (6°27'N, 12°37'E, 874 m asl.), and finally drains in southern to south-eastern direction untill it unifies with the second source river, the Lom, near Goyoum (5°12'N, 13°22'E, 640 m asl.); it here continues as Sanaga River. The Lom originates from the south-eastern flank of the Adamaoua plateau near Meiganga (6°31'N, 14°17'E, 1027 m asl.) at around 1200 m asl., and drains, strongly meandering, in southern to south-western direction. The vegetation of the Adamaoua plateau and the CVL (*Monts Mbouda* and *Oku*, as well as Bamiléké plateau) is characterized by subalpine prairie (*prairie afro-alpine*), bush-, tree- and thornbush savannas as well as grassfields (Cameroon grassfields). Achoundong (2006) describes the vegetation between 800 and 1200 m asl. (upper catchments of Djérem, Lom and Mbam Rivers) as Guinean-Sudanian altitudinal savanna (*'savane soudano-guinéenne d'altitude'*). The region around *Mont Oku* and the Bamiléké plateau is characterized by intense land-use (*'paysage agricole domestiqué'*). The middle and lower catchments of the Sanaga tributaries are situated in the Guinean-Sudanian rain forest-savanna transition zone (*'savanes périforestières'*) with gallery forest mosaics between 4 and 5°N (savannas with *Pennisetum purpureum*, *Imperata* and *Hyparrhenia*, rain forests with Guinean tree species *Terminalia glaucescens, Albizia adianthiifolia, Albizia glaberrima* and *Albizia zygia*; Achoundong, 2006). To the south, and following the course of the Sanaga, this zone is attached by the recent

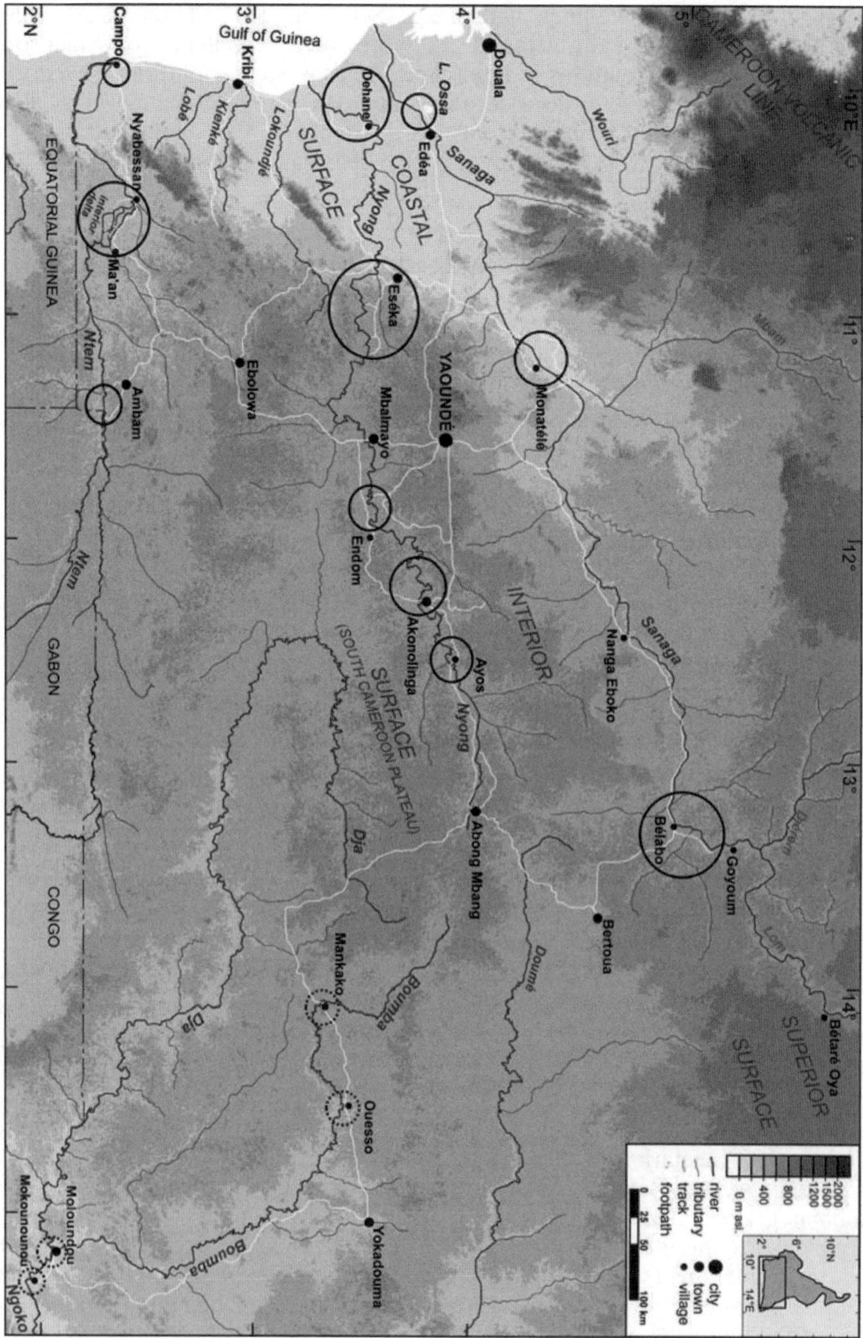

Figure 2. Map (SRTM 2000-data) showing peneplain surfaces according to Segalen (1967), relief, GPS-tracks (white lines), major rivers and investigated river reaches of the Ntem, Nyong and Sanaga Rivers (black circles) in the southern Cameroonian study area. In this abridged version, results from the investigated reaches of the Boumba and Ngoko Rivers (dotted black circles) are only occasionally discussed.

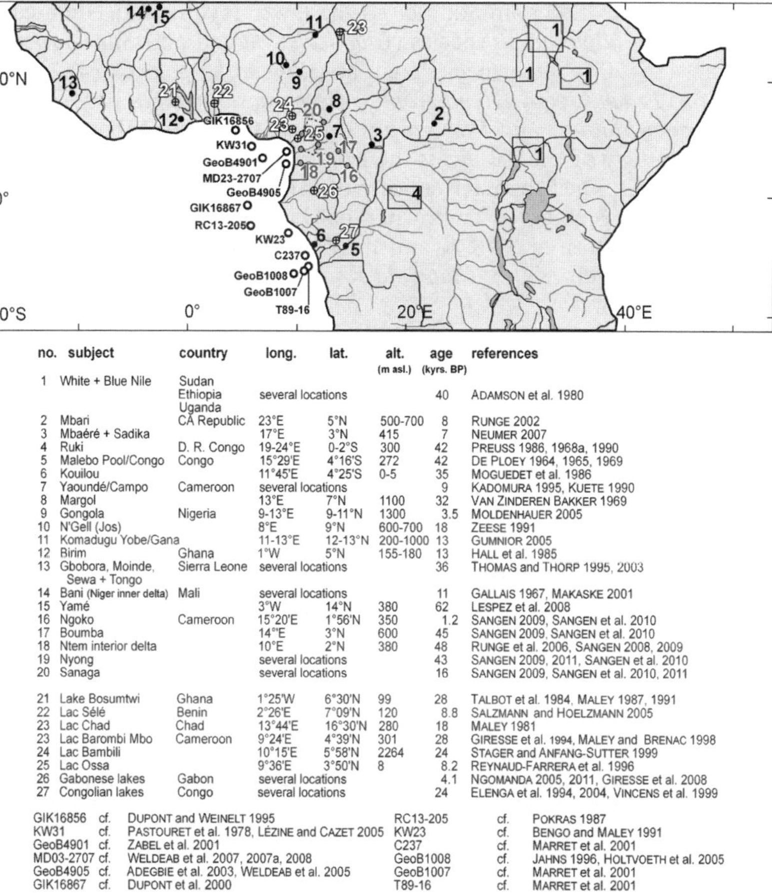

no.	subject	country	long.	lat.	alt. (m asl.)	age (kyrs. BP)	references
1	White + Blue Nile	Sudan Ethiopia Uganda	several locations			40	ADAMSON et al. 1980
2	Mbari	CA Republic	23°E	5°N	500-700	8	RUNGE 2002
3	Mbaéré + Sadika		17°E	3°N	415	7	NEUMER 2007
4	Ruki	D. R. Congo	19-24°E	0-2°S	300	42	PREUSS 1986, 1968a, 1990
5	Malebo Pool/Congo	Congo	15°29'E	4°16'S	272	42	DE PLOEY 1964, 1965, 1969
6	Kouilou		11°45'E	4°25'S	0-5	35	MOGUEDET et al. 1986
7	Yaoundé/Campo	Cameroon	several locations			9	KADOMURA 1995, KUETE 1990
8	Margol		13°E	7°N	1100	32	VAN ZINDEREN BAKKER 1969
9	Gongola	Nigeria	9-13°E	9-11°N	1300	3.5	MOLDENHAUER 2005
10	N'Gell (Jos)		8°E	9°N	600-700	18	ZEESE 1991
11	Komadugu Yobe/Gana		11-13°E	12-13°N	200-1000	13	GUMNIOR 2005
12	Birim	Ghana	1°W	5°N	155-180	13	HALL et al. 1985
13	Gbobora, Moinde, Sewa + Tongo	Sierra Leone	several locations			36	THOMAS and THORP 1995, 2003
14	Bani (Niger inner delta)	Mali	several locations			11	GALLAIS 1967, MAKASKE 2001
15	Yamé		3°W	14°N	380	62	LESPEZ et al. 2008
16	Ngoko	Cameroon	15°20'E	1°56'N	350	1.2	SANGEN 2009, SANGEN et al. 2010
17	Boumba		14°E	3°N	600	45	SANGEN 2009, SANGEN et al. 2010
18	Ntem interior delta		10°E	2°N	380	48	RUNGE et al. 2006, SANGEN 2008, 2009
19	Nyong		several locations			43	SANGEN 2009, 2011, SANGEN et al. 2010
20	Sanaga		several locations			16	SANGEN 2009, SANGEN et al. 2010, 2011
21	Lake Bosumtwi	Ghana	1°25'W	6°30'N	99	28	TALBOT et al. 1984, MALEY 1987, 1991
22	Lac Sélé	Benin	2°26'E	7°09'N	120	8.8	SALZMANN and HOELZMANN 2005
23	Lac Chad	Chad	13°44'E	16°30'N	280	18	MALEY 1981
23	Lac Barombi Mbo	Cameroon	9°24'E	4°39'N	301	28	GIRESSE et al. 1994, MALEY and BRENAC 1998
24	Lac Bambili		10°15'E	5°58'N	2264	24	STAGER and ANFANG-SUTTER 1999
25	Lac Ossa		9°36'E	3°50'N	8	8.2	REYNAUD-FARRERA et al. 1996
26	Gabonese lakes	Gabon	several locations			4.1	NGOMANDA 2005, 2011, GIRESSE et al. 2008
27	Congolian lakes	Congo	several locations			24	ELENGA et al. 1994, 2004, VINCENS et al. 1999

GIK16856	cf.	DUPONT and WEINELT 1995	RC13-205	cf.	POKRAS 1987
KW31	cf.	PASTOURET et al. 1978, LÉZINE and CAZET 2005	KW23	cf.	BENGO and MALEY 1991
GeoB4901	cf.	ZABEL et al. 2001	C237	cf.	MARRET et al. 2001
MD03-2707	cf.	WELDEAB et al. 2007, 2007a, 2008	GeoB1008	cf.	JAHNS 1996, HOLTVOETH et al. 2005
GeoB4905	cf.	ADEGBIE et al. 2003, WELDEAB et al. 2005	GeoB1007	cf.	MARRET et al. 2001
GIK16867	cf.	DUPONT et al. 2000	T89-16	cf.	MARRET et al. 2001

Figure 3. Overview of the most important western tropical African marine (white dots), lacustrine (dots with crosses) and fluvial/alluvial (black and grey dots) sedimentary basins mentioned in this study. Grey dots indicate 'ReSaKo'-project study areas.

rain forest-savanna border, which merges into the Congolian-Guinean semi-deciduous tropical rain forest. At Bélabo (4°55'N, 13°17'E; 623 m asl.), the Sanaga River enters the interior surface, which stretches into Gabon to the south and the Central African Republic in the east. It is limited by the *Surface Gondwanienne* to the north and the *Surface Côtière* (coastal surface; Tertiary/Quaternary origin, Segalen, 1967)) in the west. The transition to the coastal surface is located near the *Chutes de Nachtigal* (4°21'N, 11°38'E, 465 m asl.), 50 km north of Yaoundé. From here, the Sanaga changes its meandering morphology and exposes a braiding to anastomosing and mostly rectilinear pattern. Near Batchenga, the red ferralitic soils are replaced by distinct yellow coloured soils of the lower elevated landscape. Towards the coastal surface the soils are increasingly hydromorph, especially in the Sanaga and Wouri deltas, which are characterised by mangrove vegetation (Segalen, 1967; Olivry, 1986). Mangrove biotopes also occur at the estuaries of the Nyong and Ntem Rivers and merge into (Atlantic) coastal rain forest ('*forêt atlantique littorale*') with typical species *Lophira alata, Saccoglottis gabonensis* and *Coula edolis* (Achoundong, 2006). This is further inland associated with the tropical Biafra rain forest ('*forêt atlantique biafréenne*') between Edéa and

Monatélé. This type is also common across the lower Nyong and Ntem catchments and marked by *Caesalpiniaceae* species (*Calpocalyx heitzii* and *Sacoglottis gabonensis*) and higher (35–50 m) tree species such as *Anthonotha glaucescens, Aphanocalyx margininervatus, Brachystegia cynometroides, Desbordesia glaucescens, Erythrophleum ivorensis, Lovoa trichilioides* and *Pterocarpus soyauxii* (Tchouto Mbatchou, 2004).

Largest tributary of the Sanaga is the Mbam River having its source near Mayo Mbamti (1850 m asl.). The Mbam River north of Monatélé (4°23'N, 11°16'E; 418 m asl.) confluences with the Sanaga, after flowing 494 km to the south. The Sanaga basin has an average inclination of 1.18 ‰ and the river's longitudinal profile (Figure 4A) shows several steps, such as the *Chutes de Nachtigal* and the *Chutes d'Edéa* (largest

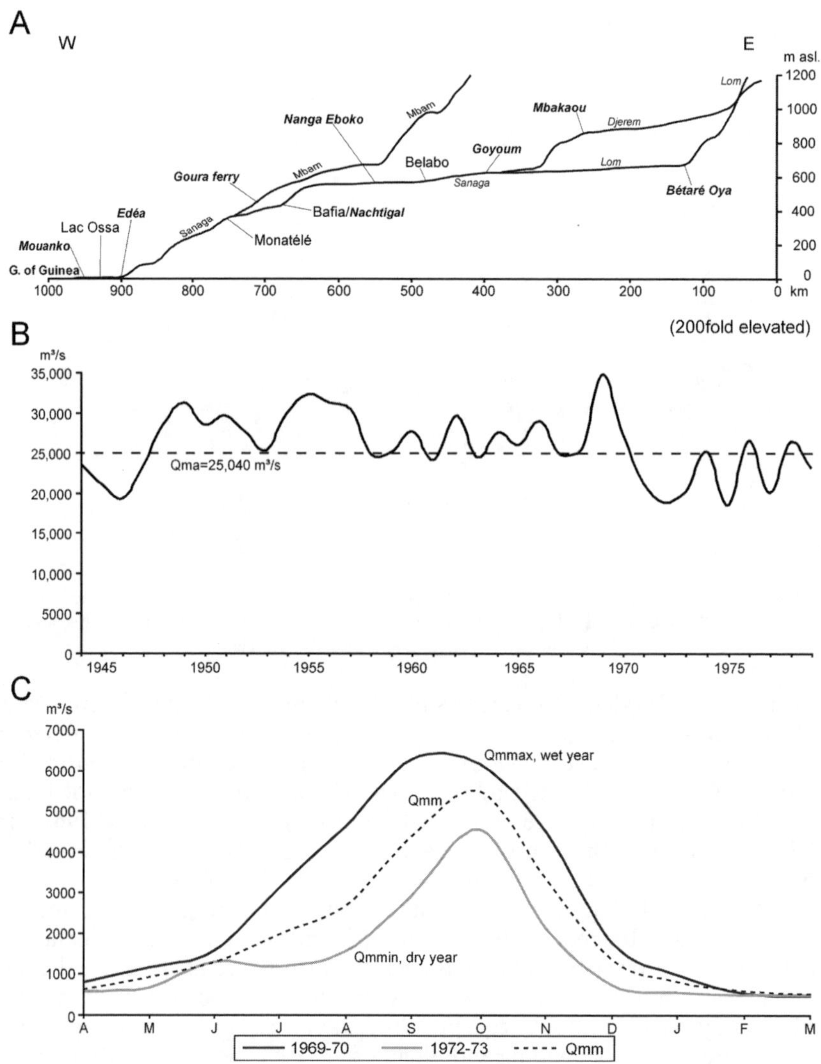

Figure 4. A Longitudinal profile of the Sanaga River, with its sources on the Adamaoua plateau and mouth near Mouanko, **B** cummulated yearly discharges at Edéa (3°48'N, 10°8'E; 31 m asl.) 1944–1979, **C** mean (Qmm, 1944–1979), maximum (Qmmax, 1969–70) and minimum (Qmmin, 1972–73) monthly Sanaga discharges at Edéa, 1944–1979.

hydropower station), where cataracts have formed and the various peneplain surfaces successively are transcended. Downstream of Edéa, the Sanaga is navigable for 67 km and free of further rapids, before draining into the Gulf of Guinea (at 3°33'N, 9°38'E; Olivry, 1986; Segalen, 1967).

4.2.1.2 Climatic and hydrological characteristics

The Sanaga crosses the savanna-rain forest border in E-W direction, which is nearly in accordance with the transition from semi-humid to humid tropical, or according to Rodier (1964) and Olivry (1986) tropical and equatorial climate. While in the source area (Ngaoundéré, Adamaoua plateau) and upper catchments of the Djérem (Tibati) and Lom (Meiganga) only one dry and rainy season occurs, in the middle (Nanga Eboko, Nachtigal, Bafia and Monatélé) and lower (Edéa) catchment areas two distinct dry and rainy seasons prevail, with rainfall maxima in May/June and September/October. Total rainfall depends on occurrence, length and intensity of the West African monsoon (WAM), while the long dry season (November-March) is dominated by the harmattan, hot aeolian sediment laden Sahara air masses, which are blown into the catchment area by the NE trade winds (Figure 5). Olivry (1986) estimated average yearly rainfall amounts at Ngaoundéré being 1658 mm, 1720 mm at Tibati, 1763 mm at Meiganga, 1584 mm at Bétaré Oya, 1674 mm at Goyoum (Bertoua), 1640 mm at Nanga Eboko, 1363 mm at Nachtigal and 1493 mm at Bafia. In the catchment area of the Mbam River, they reach 1897 mm at Mantoum and gradually increase towards the coast (2628 mm at Edéa). All stations display a very high inter-annual variability for the period of observation.

Accordingly, discharge amounts are subject to identical fluctuations having delayed peaks occurring during highest precipitations from August to October. Figure 4B shows the high variability of cumulated yearly discharges at the station Edéa, where the Sanaga has converged with all its major tributaries. Here, mean

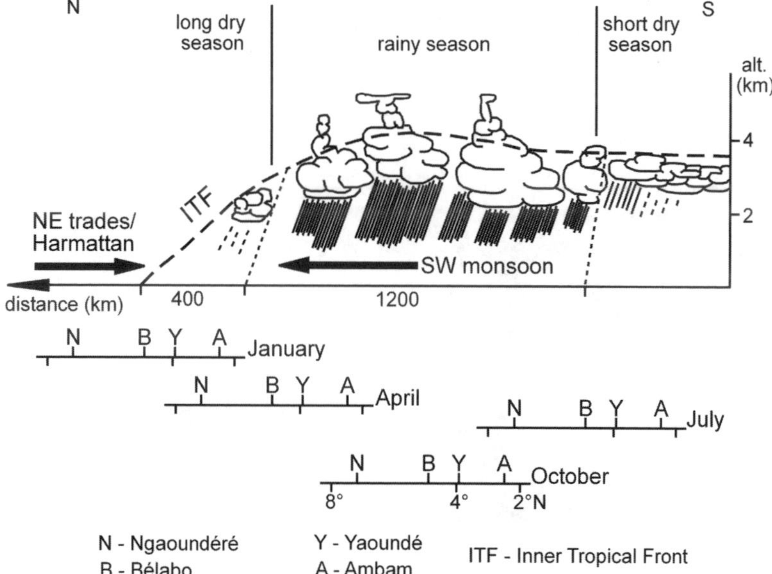

Figure 5. Schema of the weather zones in southern Cameroon (south of Ngaoundéré) during the course of the seasons (modified according to Olivry, 1986, p. 24).

annual discharges reach 30,000–35,000 m³/s in humid years. At the station Mbakaou (Djérem catchment area: 20,200 km²), south of Lake Mbakaou, a mean annual discharge (Qma) of 4654 m³/s was recorded between 1959 to 1980 and at Bétaré Oya, in the catchment area of the Lom (11,100 km²), 2081 m³/s between 1951 to 1981. The station Goyoum registered 10,141 m³/s between 1961 to 1970 and in the middle catchment area of the Sanaga the following discharges were measured: Nanga Eboko (catchment area: ~65,100 km²): 12,300 m³/s (1949–1970), Nachtigal (catchment area: ~65,100 km²): 12,907 m³/s, Goura ferry (lower Mbam catchment, 42,300 km²): 8468 m³/s (1951–1980) and Edéa (lower catchment area, 131,500 km²): 23,676 m³/s (1943–1980). During extreme humid years (i.e., 1969–70), discharge may exceed average amounts by 73% (32,014 m³/s), and in extreme arid years (i.e., 1972–73) they can fall below 17,251 m³/s (37% under average; Figure 4C). During individual months, the precipitation and discharge amounts may considerably rise above (up to 250%) or fall below (up to 70%) the average amounts, especially during the dry season.

4.2.1.3 Alluvial sediment archives

Alluvia of the upper Sanaga catchment

The upper Sanaga catchment is in this case defined as the region around the city of Bélabo (4°55'N, 13°17'E). Here mainly sandy deposits occur across strongly anabranching to braiding channel systems. Very often basement outcrops characterise the smaller branches, which regularly divert from the main channel. Coring was carried out near the settlements Mbaki II, Sakoudi, Bélabo, Ngok Etélé and Mbargué. All sampled river benches were covered by tropical rain forest, although the rain forest-savanna border stretches only several 100 m north of the Sanaga main channel. While near Bélabo a coring transect (cores C12–15) was conducted in order to find an assumed palaeochannel (identified by LANDSAT ETM+ satellite imagery), further corings to the west (Mbargué, C17–18 and Ngok Etélé, C16) and northeast (Sakoudi, C19 and Mbaki II, C20) were accomplished for prospection purposes.

Near Sakoudi (05°02'N, 13°20'E; 633 m asl.), in core C19, very fine-grained sandy to clayey, carbon-rich (C_{tot}: 1.39%, OM: 2.40% and N_{tot}: 0.084%) sediments were found in 200–215 cm depth, overlaying black mica basement (Figure 6). This

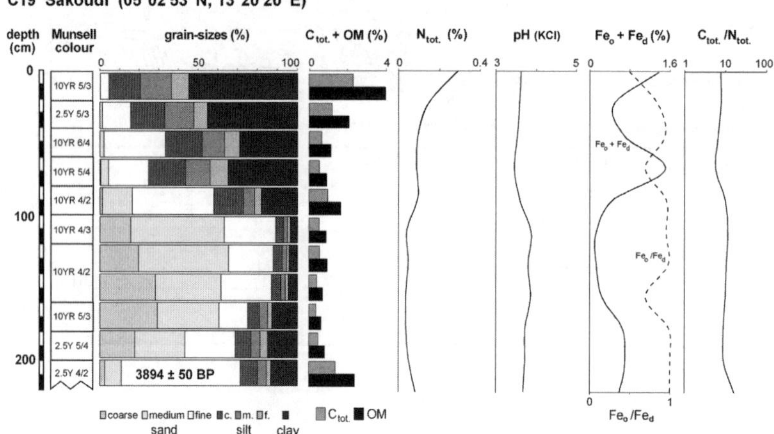

Figure 6. Sediment profile of core C19 near Sakoudi with ¹⁴C-age.

dark grey sediment layer (2.5Y 4/2) contained very dark macro-remains, which were dated to 3894 ± 50 ^{14}C yrs BP (δ^{13}C: −29.2 ± 0.2‰).

Additionally, pH (3.70) is very low and C/N reasonably high (16). This dark layer is covered by a much lighter fining-upward sequence with increasing sandy fractions and maxima of C_{tot}, OM and N_{tot} in the top layer.

Most significant sediments were found across the Bélabo transect (Figure 7B), which crosses a Sanaga anabranch (Figure 7A) and contains 4 corings, stretching over a distance of around 400 m. Whereas near the Sanaga anabranch and associated river

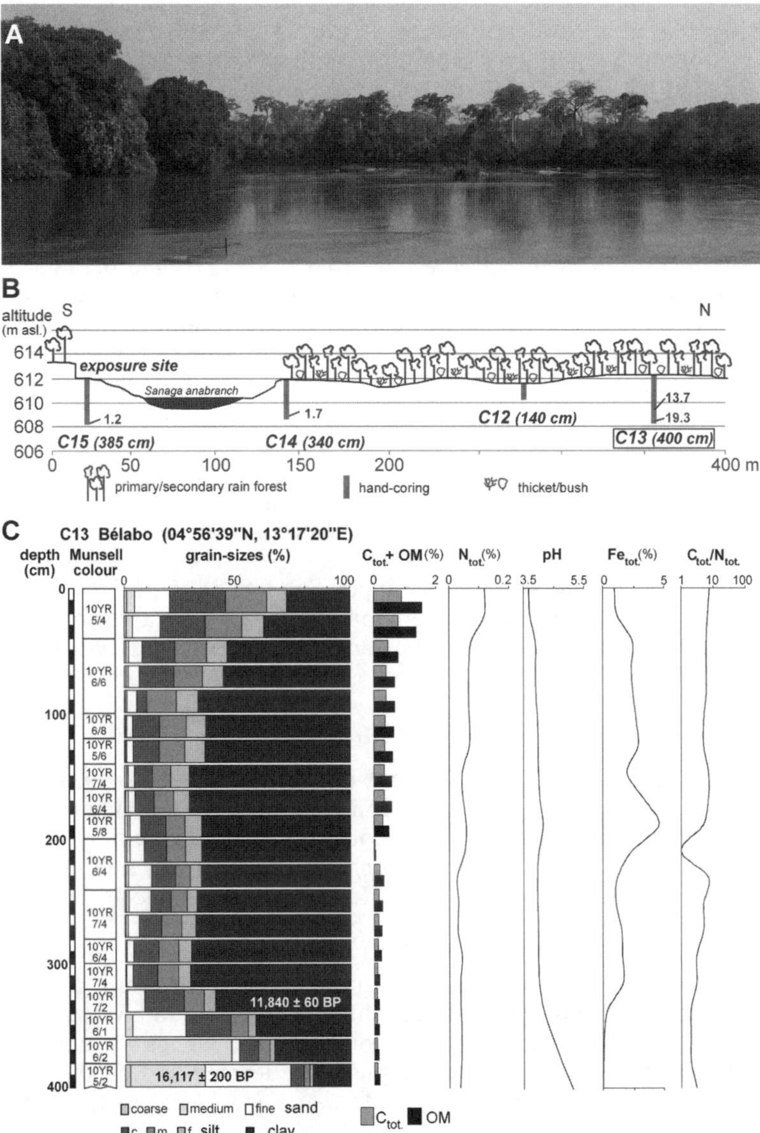

Figure 7. A Anabranch of the Sanaga (Photo: M. Sangen, January 2007) near the transect Bélabo (**B**, with hand-corings and cal. ^{14}C-data in kyrs BP) and sediment profile of core C13 Bélabo with ^{14}C-ages (**C**).

benches mainly sandy to silty fractions dominate, in the sampled sediments towards the assumed abandoned channel more silty to clayey deposits occur. At coring sites C14 and 15, dark organic sediments (10YR 3/2 and 4/2; C_{tot} up to 1.8%, OM 3% and N_{tot} 0.2% in 385–320 cm) having elevated C/N (13–19) and a pH around 4 were sampled, covered by increasing loamy to sandy deposits. Iron content of the sediments averages between 2–5% and reaches maxima in around 300 cm depth. While organic macro-remains from the base of core C15 (381 cm) were dated to a ^{14}C-age of 1196 ± 45 yrs BP (δ^{13}C: −24.3 ± 0.2 ‰), clayey-sandy organic sediment from the base of core C14 yielded a ^{14}C-age of 1766 ± 50 yrs BP (δ^{13}C: −23.6 ± 0.5 ‰).

Core C12 reached only 140 cm depth over basement, and those sediments were characterised as silty loam (140–80 cm; 10YR 5/5 and 5/6) covered by clayey loam (80–40 cm; 10YR 5/6 and 2.5Y 5.5/6), sandy loam (40–20 cm; 2.5Y 5/6) and silty-loamy sand (20–0 cm; 10YR 6/3). Core C13 reached 400 cm; Precambrian granite was here covered by clayey sand (10YR 5/1, 6/1 and 7/1 as well as organic particles having 10YR 2/1), clayey loam (360–340 cm) and silty clay (340–320 cm). These were overlaid by very clayey sediments in 320 to 140 cm (10R 4/8, 10YR 4/6, 6/6 and 5Y 8/1) and silty clay between 140 and 40 cm depth (10YR 7/1 and 7.5YR 5/8). C_{tot} and OM are very low near the base (0.10 and 0.17% in 400–380 cm) and gradually increase to 0.9 and 1.5%. Like in the other cores, N_{tot} (0.05 to 0.12%) and pH (3.5–5) are low; increased iron contents (around 2.5%) reached a maximum (5%) in ca. 200 cm depth. C/N is low (2–3) in the under- and overlying (4–8) stratum. Organic material from 400–390 cm was dated to a ^{14}C-age of 16,117 ± 200 BP (no δ^{13}C available).

Near Ngok Etélé (04°55'N, 13°08'E; 614 m asl.), where an anabranch of the Sanaga was identified, core C16 (S. 289) reached 293 cm over basement. Sandy sediments (2.5Y 4/4) are from 180–120 cm overlaid by loamy-sandy (10YR 4/3, 4/4 and 4/6) and sandy-loamy (10YR 5/6 and 6/6) deposits in a fining-upward sequence. Sediment characteristics are comparable to those found further upstream the Sanaga in the Bélabo region. Organic sediment from 260–240 cm yielded a recent age of 14 ± 100 BP (δ^{13}C: −24.2 ± 0.2 ‰).

At the final coring site of the upper Sanaga, Mbargué (04°55'N, 13°03'E; 600 m asl.), less suitable sediments were found. Nonetheless, a savanna island inside the gallery forest could be sampled, but only revealed very clayey-loamy sediments without datable fossil remains.

Alluvia of the middle Sanaga catchment (Monatélé)

Especially near the town of Monatélé (4°15'N, 11°12'E; 418 m asl.), some 150–200 km NE of Edéa, huge amounts of sand are conveyed from the braiding to anabranching Sanaga river bed (Figures 8A and B). Accordingly, over basement (muscovite-granite and migmatite of the Neo-Proterozoic *Nappe de Yaoundé*) very sand-dominated sediment profiles were found during corings. However, these sediments (cores C07–09) as well as those from corings at the sites Ossebe (C10, 220 cm; 04°12'N, 11°06'E; 351 m asl.) and Ntol (C11, 190 cm; 04°10'N, 11°04'E; 350 m asl.) were considered of minor interest for palaeoenvironmental studies. The Ossebe and Ntol corings revealed strong quartzous and micaceous sediments, and in all corings around Monatélé no datable material was found. Some km SW, near Nkon Ngok, the Sanaga was traversed to reach the rain forest-savanna border. Here, rapids have formed and on the river benches huge amounts of sand have been deposited (*Plages de la Sanaga*, Figure 8C).

Alluvia of the lower Sanaga catchment (Edéa)

Corings taken from the region Edéa (3°48'N, 10°08'E; 31 m asl.) were principally believed to reveal a more precise insight into the genesis of Lake Ossa.

Figure 8. **A** Rapids and basement outcrops at Nkon Ngok. On the right, thick sandy sediments can be recognized. **B** Sand exploitation at Monatélé, where the Sanaga has a very shallow and sandy river bed (Photos: M. Sangen, January 2007). **C** LANDSAT ETM+ scene 186–57 (19.10.2001) showing the study and reconnaissance region of Monatélé in the middle Sanaga catchment with tracking route (lines) and locations of cores C07–11.

The LANDSAT-TM-satellite scene from 26.04.2001 shows a small river bench separating the Sanaga from Lake Ossa (Figure 9A). The suspicious morphology gives rise to the assumption that Lake Ossa originates from a former Sanaga tributary. The scene additionally shows a very linear river section of the Sanaga having sand banks on the convex sides of the meanders; this hints to increased vertical accretion of sandy sediments. The river bed is bordered by up to 10 m high levees, which are flooded during the rainy season. These seasonal floods provide adjacent floodplains with fine-grained sediments.

Near the village of Dizangué (03°46'N, 09°59'E; 11 m asl.), a 6 core-transect (C01-C06) was sampled, stretching from the northern Sanaga river bench to Lake Ossa (Figure 9B). The river bench was covered by grasses and reeds and is seasonally inundated by floods. Core C01 (210 cm depth) revealed coarse clayey sand, dated to recent age (128.52 ± 0.90 pMC; $\delta^{13}C$: −19.6 ± 0.4 ‰), and covered by sandy, loamy clays and silty clays, showing oxidation and reduction, at the top. Core C02 reached the basement at 375 cm and also comprised of clays covered by loamy clay, clayey loam and sandy-clayey loam. In core C06 (540 cm), sandy-clayey sediments (10YR 5/6, 6/6, 6/4) prevailed until around 350 cm depth; they were overlaid by clayey and grey coloured (10BG 5/1, 6/1, 7/1; reduction processes) deposits. The units are separated by a distinct oxidation layer, and organic clay from 360–354 cm (Figure 9C) depth yielded a ^{14}C-age of 9321 ± 90 BP (δ^{13}C: –27.2 ± 0.1 ‰).

Core C05 reached 240 cm and contained increasing coarse and silty sandy sediments (10YR 7/6 and 7/8) covered by first loamy (7.5YR 6/8) and then more silty sands (10YR 4/2, 4/4; 7.5YR 5/8). In 170–160 cm, organic particles (10YR and 7.5YR 2/1) were mixed in a sandy matrix, but radiocarbon dating was not possible. This material was underlain by a ferritic, 2–3 cm thick hardpan (iron concentration layer).

In core C04 (200 cm), coarse sandy (2.5Y 8/1) sediments were overlaid by increasing clayey-loamy (10YR 5/8 and 6/8) and, finally, clayey (10YR 6/8 and 5Y 7/1) sediments. Again, the sandy sediments were seperated from the clayey by an iron concentration layer in 190 cm depth. In C03, taken directly on the shore of Lake Ossa, very coarse sand (> 43%) was sampled until around 70 cm depth (2.5Y 7/1 and 8/1).

4.2.2 Nyong catchment

4.2.2.1 Geographical characteristics

The Nyong catchment stretches between 2°48' to 4°32'N and 9°54' to 13°30'E, having a surface area of 27,800 km² (Olivry, 1986). The perennial Nyong River, having a length of 400 km, is the second largest river in Cameroon, and drains the Precambrian South Cameroon plateau (also interior surface or inner plateau, *Plateau Sud-Camerounais*)

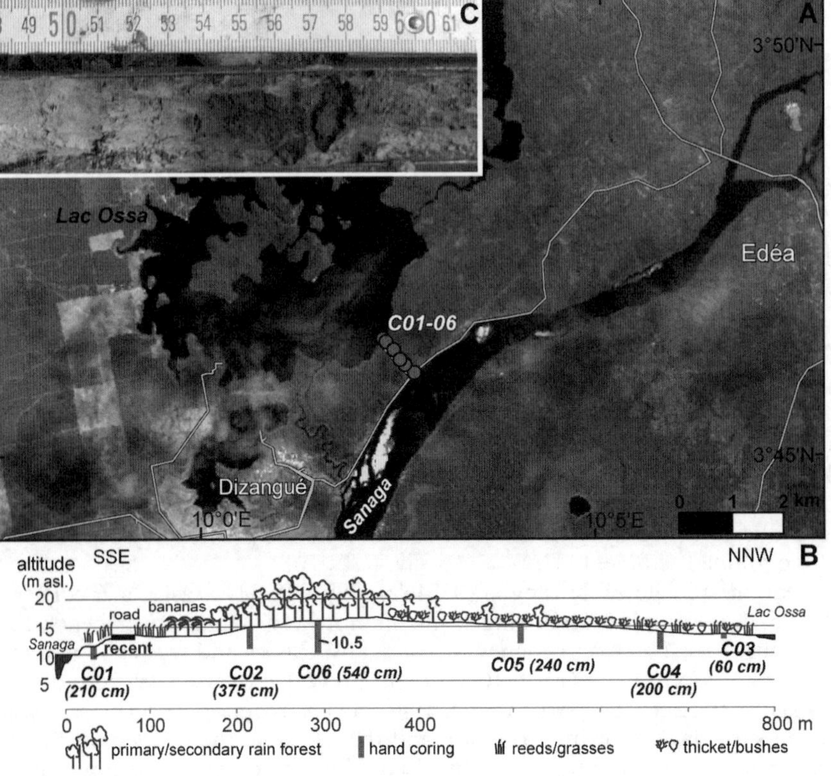

Figure 9. A LANDSAT ETM+ scene 186–58 (26.04.2001) of the Dizangué study area, **B** transect Diganzué (with hand-corings C01–06, physio-geographical settings and cal. ^{14}C-data in kyrs) and **C** fossil organic clayey sediments sampled in C06 (354–360 cm, Photo: M. Sangen, January 2007).

on heights between 600–760 m asl. Its source is located in the most eastern part of the catchment area (13°30'E, near Djouyaya at 690 m asl.) on the interior surface. This plateau marks the watershed divide between the Nyong catchment area and the basins of the Dja and Sangha, which drain into the Congo basin. The Nyong catchment area comprises the major cities of Abong-Mbang (3°59'N, 13°10'E; 657 m asl.), Ayos (3°53'N, 12°31'E; 654 m asl.), Akonolinga (3°47'N, 12°15'E; 643 m asl.), Yaoundé (3°52'N, 11°31'E; 730 m asl.), Mbalmayo (3°31'N, 11°30'E; 634 m asl.) and Eséka (3°41'N, 10°42'E; 146 m asl.) from E to W. The Nyong drains from ENE to WSW, and flows into the Gulf of Guinea west of Dehane (3°34'N, 10°07'E; 15 m asl.; see also 200 fold elevated longitudinal profile Figure 10A).

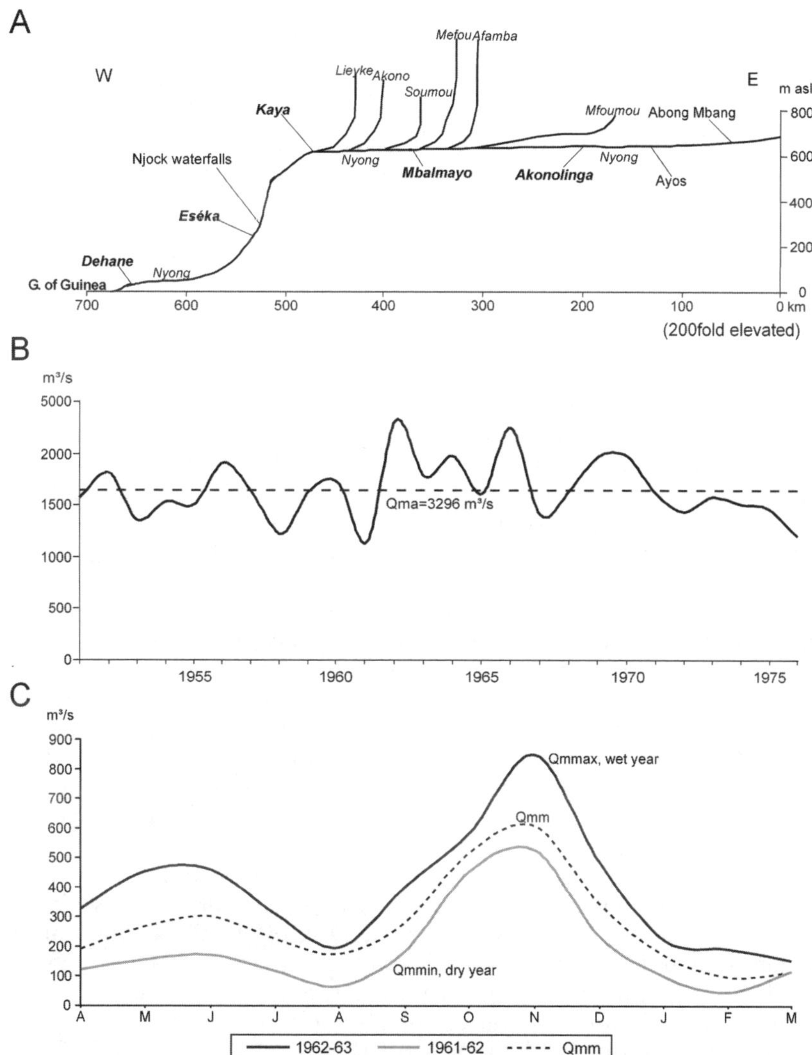

Figure 10. A Longitudinal profile of the Nyong River, having its source near Abong Mbang and mouth near Dehane, **B** cummulated yearly discharges at Eséka (3°38'N, 10°47'E; 423 m asl.) 1951–1976 and **C** mean (Qmm), maximum (Qmmax) and minimum (Qmmin) monthly Nyong discharges at Eséka, 1951–1976.

The geological formations of the South Cameroon plateau reach from the Lower (Congo Craton) to the Upper Precambrian, where ferralitisation is among the dominating pedogenetic processes. High precipitation and temperatures favour the development of red ferralitic soils, while in the valleys Quaternary alluvia and hydromorphic soils dominate. The Nyong catchment area is for the most part (middle and lower reach) covered by tropical rain forest and can be classified into three morphological units: a meandering upper reach having slightly undulating plains and low base slope (0.05–0.15%, between Abong Mbang and Akonolinga), wide valleys (up to 3 km) with seasonal swamp and flooded plains (between Abong Mbang and Mbalmayo), a rectilinear lower reach with steep slopes, rapids and waterfalls (Edéa region), and finally a deltaic estuary south-west of Dehane.

According to Villiers (1995), the seasonal swamp and flooded area is characterised by swamp forest (*Raphia regalis, Macaranga* sp. and *Sterculia subviolacea*) and aquatic prairie (*Echinochloa pyramidalis, Bacopa egensis, Ceratopteris cornuta, Cyclosurus interruptus, Cyperus* spp., *Carex* sp., *Ethulia conyzoides, Impatiens irvingii, Ipomoea cf. arisifolia, Ludwigia leptocarpa, Oxycaryum cubense, Panicum* sp., *Pentondon pentandrus* and *Polygonum* spp.); he refers to this area as '*dépressiones marécageuses périodiquement inondées*' with aquatic graminaceous plants and 25–30 m high swamp forest. During the 2007 field campaign, the valleys between Abong Mbang and Akonolinga were still flooded in the dry season (February), which seriously complicated the sampling of alluvia. Downstream these swamps (~1000 km²), near the town of Kaya (3°32'N, 11°05'E; 617 m asl.), the base slope rapidly rises and rapids as well as cataracts (*Chutes de Mpoume, Chutes de Manyanga, Chutes de Makai, Chutes de Milly, Chutes de Mouilla Mage* and *Chutes de Njock* from east to west and covering a distance of around 20 km) become most dominant.

Some 70 km southwest of Yaoundé, the Nyong transcends most of these cataracts between the towns of Kaya and Eséka while crossing the peneplain step between the interior surface and the coastal lowland, therby falling around 300 m. Across the peneplain step and the north-western boundary area of the craton, amphibolite and quartzite of the Neo-Archaic basement are strongly present. In the upper and middle catchment area, the Nyong crosses mica schist and gneiss of the *Nappe de Yaoundé*; the river follows these structures from Mbalmayo until around 50 km downstream. Here is the distinctive border between Neo-Proterozoic rocks of the *Nappe de Yaoundé* and the Congo Craton (Maurizot, 2000; Toteu et al., 2004).

It also marks the transition between two different landscape types. From here, the Nyong traverses more or less disturbed Cameroonian-Congolian tropical evergreen rain forest ('*forêt Camerouno-Congolaise*') with typical species *Uapaca paludosa, Afrostyrax lepidophyllus, Baphia pubescens, Beilschmiedia louisii, Cryptosepalum congolanum, Drypetes paxii, Fernandoa fernandii, Heisteria trillesiana, Irvingia excelsa, Manilkara letouzeyi, Millettia laurentii, Pentaclethra eetveldeana, Pericopsis alata, Tessmannia africana* and *Vincentella* sp. (Villiers, 1995).

From Eséka to Dehane, base slope is strongly reduced, and the river bed is marked by Precambrian basement, reducing the occurrence of alluvial sediments to a minimum. Not until the coastal lowland around Dehane, abundant alluvia of remarkable thickness can be found. The Nyong demonstrates a structural morphology, which is affected by rectilinear stretches and several abrupt flow-direction alterations. This can be linked to the Southwest Cameroonian Shear Zone, which was like the Sanaga Shear Zone, formed during the Panafrican orogenesis (Ngako et al., 2003). That structural influence is considerably reduced when the Nyong enters the Quaternary deposits of the coastal surface near Dehane. In the framework of this research, especially the reaches between Abong Mbang, Ayos and Akonolinga on the interior surface and

Eséka as well as Dehane on the coastal surface were in detail studied. Further fieldwork was done across several abandoned channels near Endom (Edjom).

4.2.2.2 Climatic and hydrological characteristics

Taking into account recent IRGM-CRH data and studies undertaken by Olivry (1986) and Suchel (1987), mean annual rainfall amounts are estimated to have reached about 1556 mm between the years 1930 to 2002. Precipitation across the Nyong, like over the whole of southern Cameroon, decreases from the coast (3000 mm) towards the hinterland and the South Cameroon plateau (1000 mm). In Kribi (some 35 km south of the mouth of the Nyong at Dehane) precipitation averages 2836 mm, in Eséka 2150 mm and in Abong Mbang 1674 mm. Between Makak and Dehane, rainfall varies between 1800 to 3000 mm. In strong correlation with rainfall, highest discharges occur (with some delay) around June and November. The largest tributary of the Nyong is the 233 km long Soo River, which has a catchment area of 3120 km² and drains into the Nyong near Mbalmayo. Further important tributaries of the Nyong are: the Kom River (catchment area 695 km², length 106 km), Long Mafok (1520 km², 166 km), Mfomou (1260 km², 181 km), Afamba (485 km², 90 km), Ato (740 km², 117 km), Mefou (840 km², 121 km), Mefou at Nsimalen (425 km², 90 km), Mefou at Etoa (235 km², 66 km), Soumou (820 km², 116 km), Kama (445 km², 104 km), Akono (610 km², 118 km), Liyeke (505 km², 104 km) and Kelle (2770 km², 274 km).

Accordingly, discharge amounts rapidly increase towards the coast and river mouth. Olivry (1986) has estimated a mean annual discharge of around 7.8 m³/s (1940–1952) for the source area near Abong Mbang (965 km²). At Akonolinga (catchment area: 8350 km²) it increases to ~93 m³/s (1953–1977) and at Eséka, where most tributaries have confluenced with the Nyong, the catchment area of around 21,600 km² represents a mean annual discharge of ~275 m³/s (1937–1976). Lastly, at Dehane this increases (catchment area 26,400 km²) to ~445 m³/s (1951–1977). Here again it must be stressed that in the Nyong catchment area the precipitation and discharge amounts are prone to high variability and can, in individual months, distinctly exceed (up to 290%) or fall below (50%) average amounts (Figures 10B and C).

4.2.2.3 Alluvial sediment archives

Alluvia of the upper Nyong catchment (Abong Mbang)

In the upper catchment and source area of the Nyong River near Abong Mbang, Ayos and Akonolinga extended palaeoenvironmental archives were expected. From its source near Abong Mbang (03°59'N, 13°10'E; 694 m asl.) until Mbalmayo (some 150 km downstream) the Nyong possesses a very wide (2–3 km), shallow and swampy floodplain, which is flooded during the rainy season (Figure 11). Inclination is very

Figure 11. River Nyong at Abong Mbang (M. Sangen, January 2007, see also Figure 8 contribution J. Eisenberg).

low (0.15 ‰) and altitude difference between Abong Mbang and Ayos (~80 km distance) amounts to only 3 m. The floodplain is home to a typical aquatic prairie flora (e.g., *Raphia regalis, Macaranga* sp. and *Sterculia subviolacea*).

Coring was not possible close to the river and two cores from Abong Mbang II (C21; 04°01'N, 13°06'E; 693 m asl.) and Zende (C22; 04°01'N, 13°06'E; 692 m asl.) only reached 140 and 100 cm, respectively. The cores revealed sandy-loamy sediments lacking fossil organic material and remains, and hence, are of minor interest to our studies.

Nyong floodplain at Ayos

Around Ayos (03°54'N, 12°31'E; 693 m asl.), the floodplain is limited to the N and S by up to 20 m high terraces of the *Ayos-Series* (Eno Belinga, 1984) comprising metamorphous rocks (crystalline schist, mica schist, phyllite and quartzite) of the *Nappe de Yaoundé*.

Similar to the Abong Mbang region, floodplain vegetation is characterised by aquatic prairie. The great interest of the Ayos research area relates to the existence of several abandoned and/or seasonal channels, which, during the dry seasons are not filled with waters of the Nyong. At several locations corings were made in the beds of

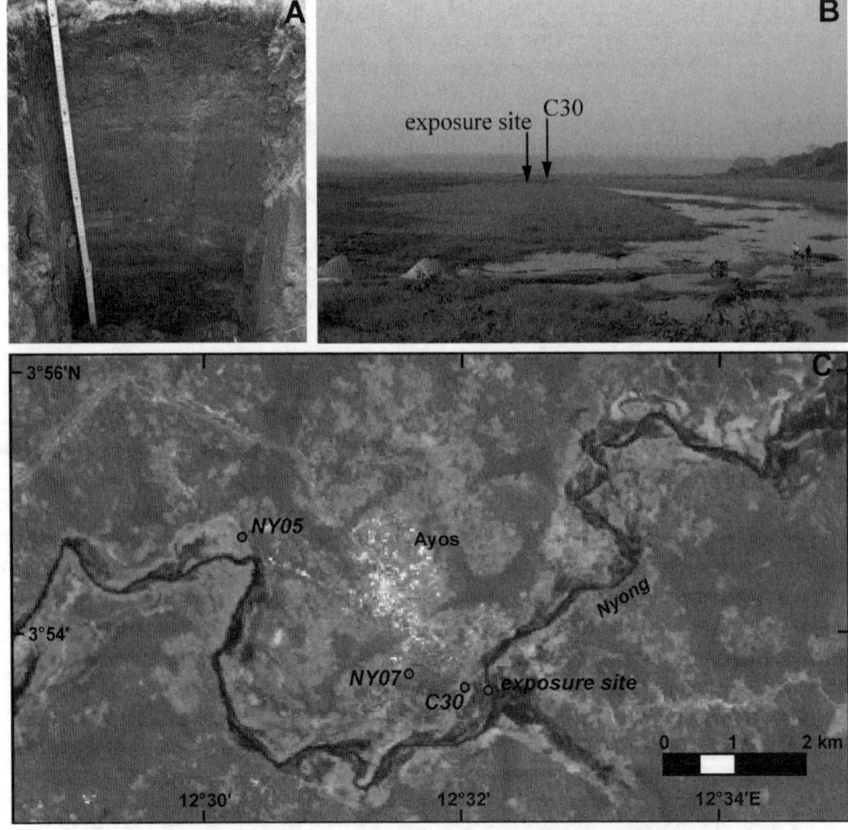

Figure 12. A Exposure site at Ayos, **B** the Nyong River at Ayos with position of core C30 and exposure site. In the front: exploited sand from the river bed (Photos: M. Sangen, January 2007). **C** LANDSAT ETM+ scene 185–57 (16.01.2002) showing the Ayos study area, position of the exposure site and locations, where abandoned (N07, C30) and nearly abandoned channels (NY05) were sampled.

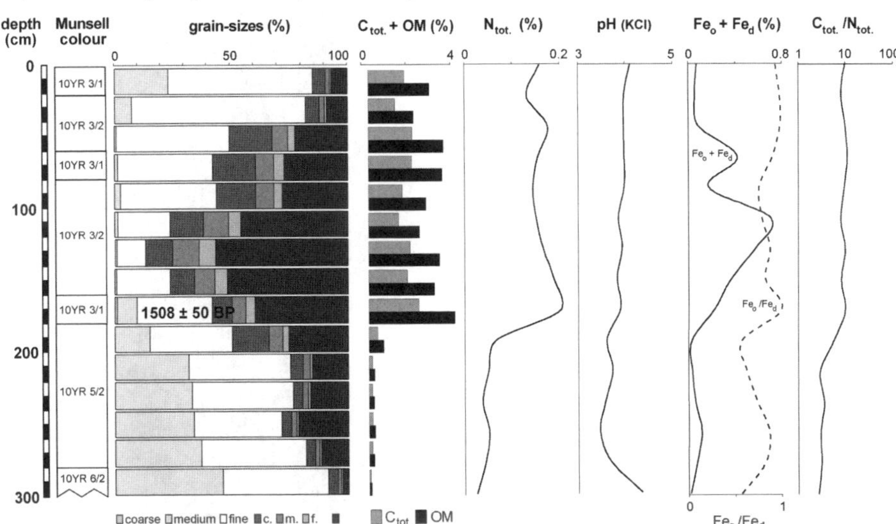

Figure 13. Sediment profile 'Ayos exposure site' with ^{14}C-age.

certain of these channels. Of most interest was core C30 (Figure 12B and C). It reached 280 cm depth and contained very organic sediments at 280–260 cm (C_{tot}, OM and N_{tot}: 18, 30 and 1.07%) of a very dark colour (10YR 2/1) and elevated C/N (20). This organic sediment layer was dated to 724 ± 45 ^{14}C yrs BP (δ^{13}C: -25.8 ± 0.3 ‰). It was covered by a very sandy layer, which, once again, was followed by very dark organic sediments until around 140 cm depth. Here, very high C_{tot} (between 24 to 6.5%), OM (41–11%) and N_{tot} (1.289–0.415%) occurred. These values rapidly decrease towards the surface, where a coarsening upward sequence was found (up to 57% fine sandy texture and 25% coarse silt). The pH fluctuates around 4 and the sediment only slightly lightens towards the surface (10YR 3/2).

In a distance of around 15 m from core C30 an exposure site in the 2.5–3 high river bench of the Nyong gave an excellent insight into the stratification of the Nyong deposits (Figure 12A). The profile showed sandy sediments between 300 and 200 cm depth, which decreased in a fining upward sequence towards 120 cm. A clayey organic layer in 180–160 cm yielded a ^{14}C-age of 1508 ± 50 yrs BP (δ^{13}C: -18.0 ± 0.3 ‰). Above 140 cm a coarsening upward sequence was found with again increasing sandy deposits.

By analyzing ASTER-satellite data (image from 12.03.2007) further abandoned and seasonal channels of the Nyong near Ayos were identified. Here sediments of lesser thickness were cored with coarser sandy deposits near their base. At the locations NY05 (Ebabodo) and NY07 (see Figure 12C) sediments comparable to C30 were found. Also at these locations coarsening upward sequences were identified with increasing sandy textures. In NY05 the sediments yielded a similar ^{14}C-age (732 ± 30 yrs BP; δ^{13}C: -20.0 ± 0.4 ‰) as in core C30 in comparable depth (280 cm) and sediment matrix.

Nyong floodplain at Akonolinga

Some 25–30 km downstream, the Nyong passes the town of Akonolinga (03°46'N, 12°15'E; 671 m asl.). Like at Ayos, there is a bridge to cross the Nyong and a laterite gravel road that traverses the floodplain (Figure 14). The Nyong meanders through the

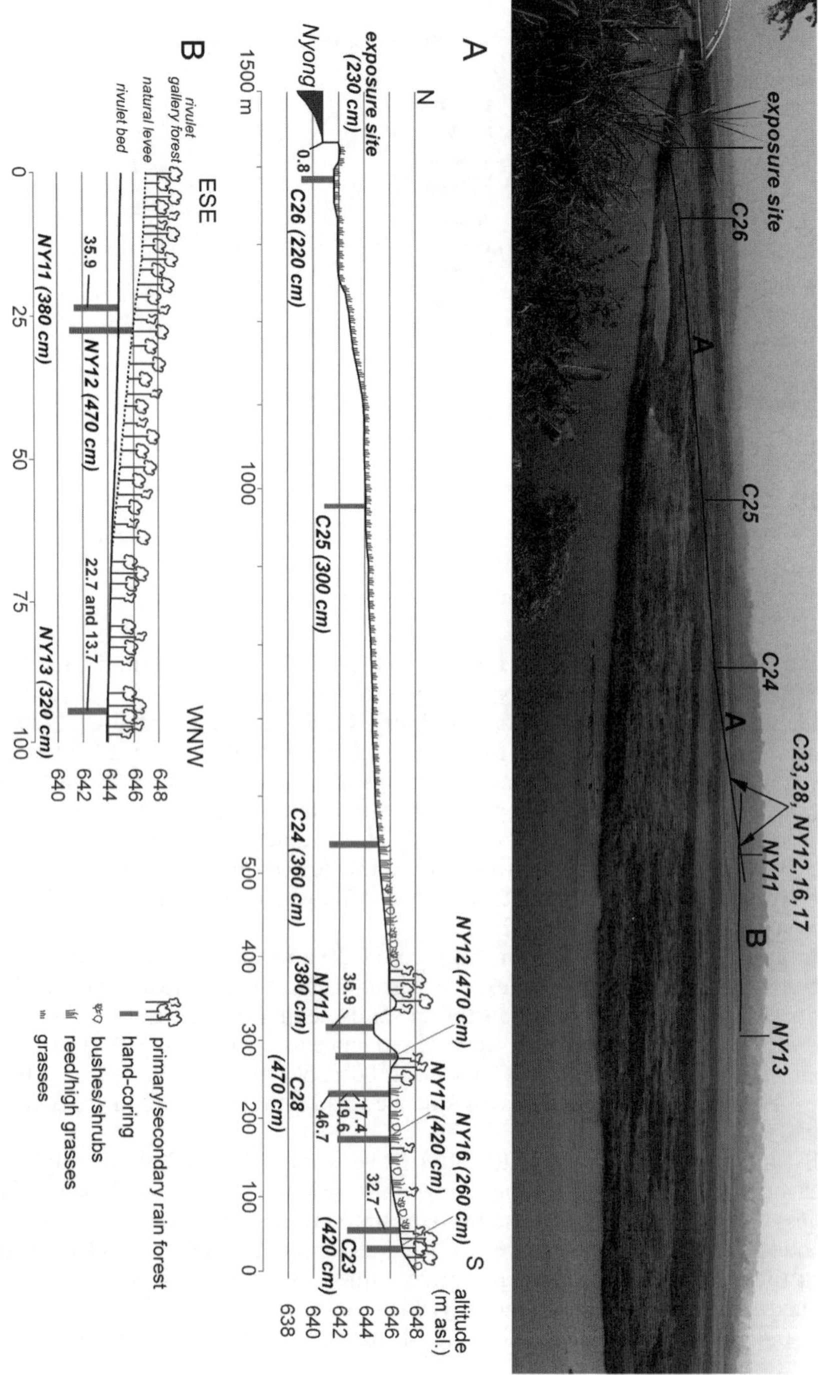

Figure 14. Overview of the Akonolinga research area (Photo: M. Sangen, 2008), showing transects through the Nyong floodplain (A) and across the Tetar rivulet (B) as well as physiogeographic settings, hand-coring sites and calibrated ^{14}C-data in kyrs BP (for depths in C28 see Table 1).

floodplain and is joined by a seasonal tributary called Tetar. To the west of the road, a transect was cored from S to N (Figure 14).

From the southern Nyong river bench to the Tetar tributary, two corings (C23 and C28) were made over a distance of 300 m. While the river bench is covered by tropical rain forest and cacao trees, towards core C23 swampy brushwood and shrubs prevail. On the 2 m broad and 1.5 m high benches of the 5 m wide Tetar River bed, gallery forest has formed. Towards the northern Nyong River terrace and the town of Akonolinga, aquatic prairie and savanna (with up to 1 m high grasses) grow; these vegetations, like in the Ayos region, are regularly burned in the dry season. On this passage (around 1200 m) cores (C24–26) were taken every 400–500 m and finally an exposure site was investigated across the Nyong River bank. During the rainy season (September-November) the floodplains at Ayos and Akonolinga are inundated with 1–2 m of water.

Core C23, at the southernmost point of the transect, reached a depth of 420 cm. Over sandy sediments at the base (420–380 cm) increasingly silty and loamy-clayey deposits occurred (380–240 cm). These are followed by, once again, sandy and silty deposits until around 200 cm (for a sedinment profile, see Sangen et al., 2010, p. 175). Then clayey sediments were cored between 180 to 120 cm, which were very concentrated between 160 to 140 cm. This layer was dated to a ^{14}C-age of 28,358 ± 300 yrs BP (δ^{13}C: −20.1 ± 0.3 ‰).

About 120 m further north, core C28 (420 cm) was taken close to the Tetar tributary. Here, dark (10YR 4/2) sandy (~2% coarse, ~16% middle and 30–35% fine sand) to clayey basal Late Pleistocene deposits were dated to 42,940 ± 1500 ^{14}C yrs BP (δ^{13}C: −19.6 ± 0.5 ‰). They are overlain by increasingly sandier LGM deposits (10YR 5/2), dated to 16,360 ± 70 ^{14}C yrs BP (δ^{13}C: −24.9 ‰) in the upper boundary (360–340 cm) and abruptly increasing clayey (≥ 80%) alluvia (10YR 6/2) from 340–320 cm, dated to 14,160 ± 60 ^{14}C yrs BP (δ^{13}C: −24.1 ‰). Like in the other cores from the Akonolinga transect, C_{tot}, OM and N_{tot} are very low in deeper sections of the cores and pH ranges from 3–4. C/N fluctuates around 5–15 (Figure 15C).

A further 100–150 m north of the Tetar tributary, core C24 was taken in a central position of the Nyong floodplain, the savanna vegetation of which had been recently burned and so the floodplain was denuded. At this site, clayey sands were found until 360 cm depth covered by silty clay from 260 cm upwards. Again, mostly grey colours (10YR 6/2 and 7/2; 2.5Y 7/1 and 8/1) were identified, intermixed with sections showing oxidation (10R 5/8, 10YR 7/8 and 2.5Y 7/8). No fossil, organic sediments were found. Another 500 m more towards the recent Nyong channel (dry season) C25 was cored. At the base (300–220 cm) clayey sands of a dark grey colour (2.5Y 7.5/4) were uncovered followed by sandy clays from 220–140 cm, which were grey (10YR 6/1, 6.5/1 and 7/1) and again contained elements of oxidation (10YR 6/8 and 10R 5/8). Towards the surface, first loamy clay (140–80 cm) and then silty clays occurred. Again, no datable fossil, organic deposits were found.

The last core of the transect, C26, was taken very close to the main Nyong channel and reached 220 cm depth. It consisted of mainly sandy sediments (50% middle sand and 75% fine sand) and revealed a fining upward sequence towards the surface. Like to former cores, no interesting phenomena for palaeoenvironmental research could be identified.

Finally, an exposure site across the Nyong river bench was scratched and inspected. Within the mainly (90%) sandy deposits, fine-grained, organic layers were discovered between 220–200, 140–120 and 80–60 cm depth. They showed dark colours (10YR 2/1 and 3/1), high C_{tot} (1.50%, 2.79% and 1.80%), OM (2.59, 4.81 and 3.10%) and N_{tot} (0.121, 0.226 and 0.160%). The C/N fluctuates from around 10–20 and pH

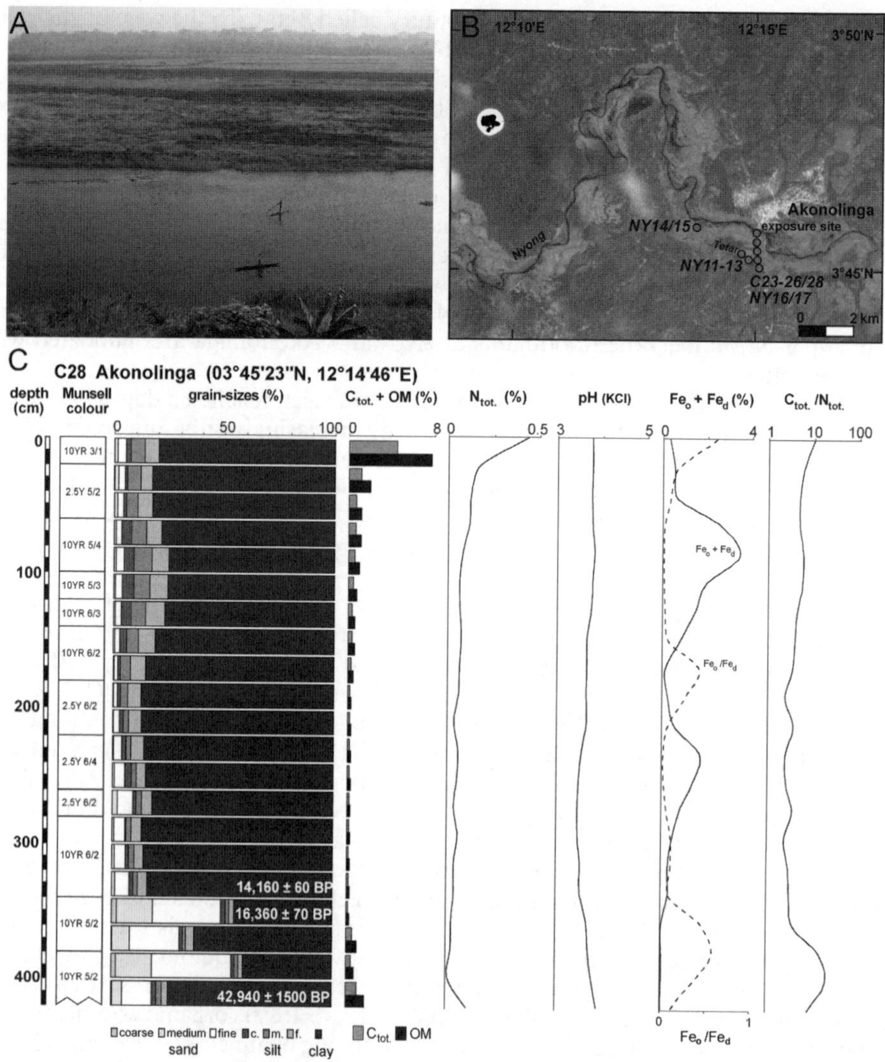

Figure 15. **A** The Nyong at Akonolinga and sandbank formation (Photo: M. Sangen, 2008). **B** LANDSAT ETM+ scene 185–57 (16.01.2002) showing the Akonolinga study area and location of hand-corings and exposure site. **C** Sediment profile and ^{14}C-data of core C28 near Akonolinga.

was about 4. Iron content was very low (ca. 0.1%). Sediment from the deepest organic layer (220 cm) was dated to a ^{14}C-age of 902 ± 45 yrs BP (δ^{13}C: −24.2 ± 0.5 ‰).

Between C23 and C28, another coring (NY16) was made during fieldwork the following year. Core NY16 reached 400 cm depth but did not provide any fossil organic material. Additionally, two corings were undertaken in the Tetar rivulet bed (NY11 and NY13, Figures 16A and B).

Core NY11 reached 340 cm and brought coarse sands to the surface (340–280 cm), which were covered by silty to clayey, organic sediments between 180–120 cm (Figure 16C). From 80 cm onwards, the deposits again show increasing proportions of sand. Sandy to clayey organic deposits from 280 cm depth were dated to a ^{14}C-age of 31,904 ± 300 yrs BP (δ^{13}C: −33.5 ± 0.4 ‰).

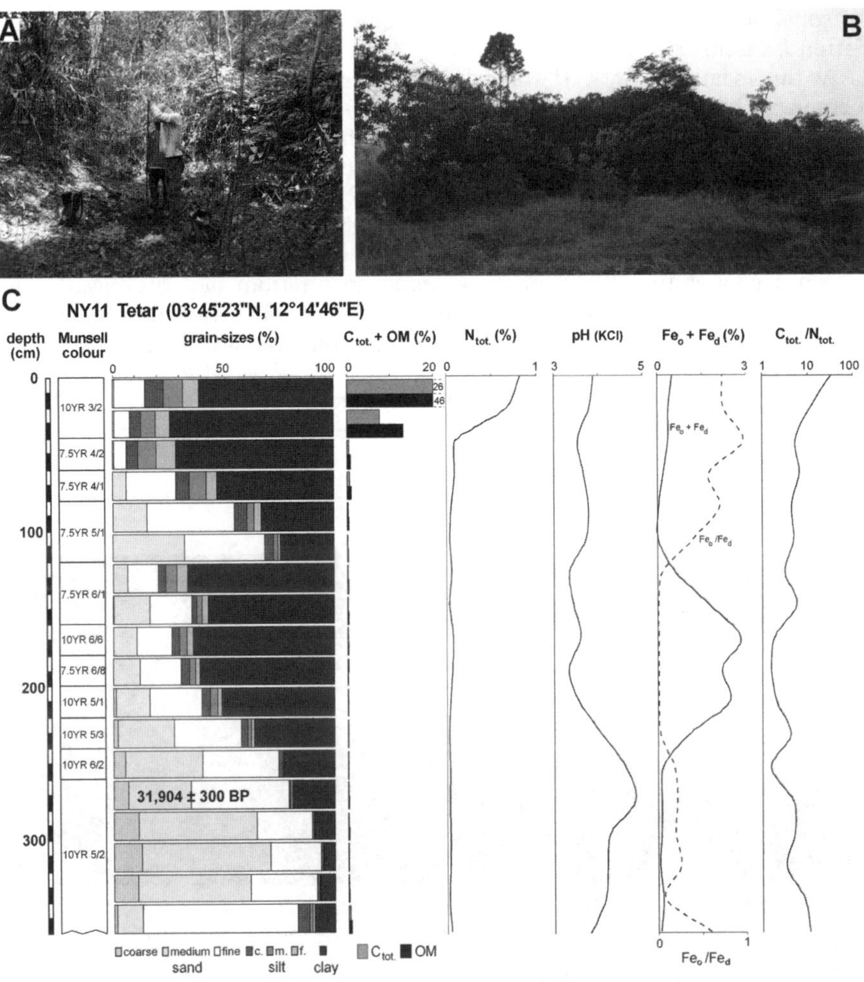

Figure 16. A Location of coring NY11 in the Tetar rivulet bed, **B** location of NY13 in the rivulet gallery forest (Photos: M. Sangen, February 2008) and **C** sediment profile of NY11 Tetar with ^{14}C-age.

Some 100 m west of NY11, NY13 was taken; between 320 to 180 cm depth dark (10YR 3/1, 3/2), sandy-clayey (~20% middle sand, ~30% fine sand and ~30% clay), and organic sediments were unearthed. They last were, from 180 cm, covered by increasing clayey (70–93%), lighter (10YR 4/1) organic (1–37% OM) deposits. These revealed a very low pH of around 3.5. At the transition from sandy to clayey deposits (ca. 200–180 cm depth), organic sediments yielded ^{14}C-ages of 18,858 ± 100 (δ^{13}C: −18.1 ± 0.2 ‰) and 11,840 ± 60 yrs BP (δ^{13}C: −19.1; for a sediment profile see Sangen et al., 2010, p. 176).

Corings from abandoned Nyong channels between Endom and Mbalmayo

The last station across the Nyong upper catchment was located at the Mengba floodplain, near the village Edjom (03°32'N, 11°58'E). Here, the Nyong has developed a much smaller floodplain than at Akonolinga or Ayos.

The satellite image (Figure 17) shows bowlike structures near Mengba, where the Nyong changes its flow direction from W to N. These structures were identified as seasonal or abandoned channels of the Nyong (see also Figure 10 Chapter 3, contribution J. Eisenberg).

At this location, core C31 was taken 500 m west of the recent Nyong channel; this core was 280 cm (Figure 17). At the base (280–220 cm) coarse-grained sediments (6% coarse, 13% middle, 60% fine sand, 10% silt, 11% clay) occurred, which were covered by increasingly fine-grained and clayey (70–80%) deposits. The former are marked by dark colours (10YR 2/1 and 10YR 3/2), high carbon content and low pH (3.5). The upper sediment layers were of much lighter colour (10YR 4/2, 5/2 and 6/3) and revealed much lower contents of C_{tot}, OM and N_{tot}. C/N is high at the base (17–20) and then decreases to 5–10. Sandy organic sediment from the core base yielded a ^{14}C-age of 13,357 ± 60 yrs BP (δ^{13}C not available). From the transition layer between sandy and clayey deposits (220 cm) clayey organic material was dated to a ^{14}C-age of 11,107 ± 90 yrs BP (δ^{13}C: –28.9 ± 0.1 ‰).

At Nkolmvondo, another abandoned channel (meander loop or oxbow) was cored at three different locations (M01–03; Figure 17). Unfortunately, the cores did not deliver any suitable results.

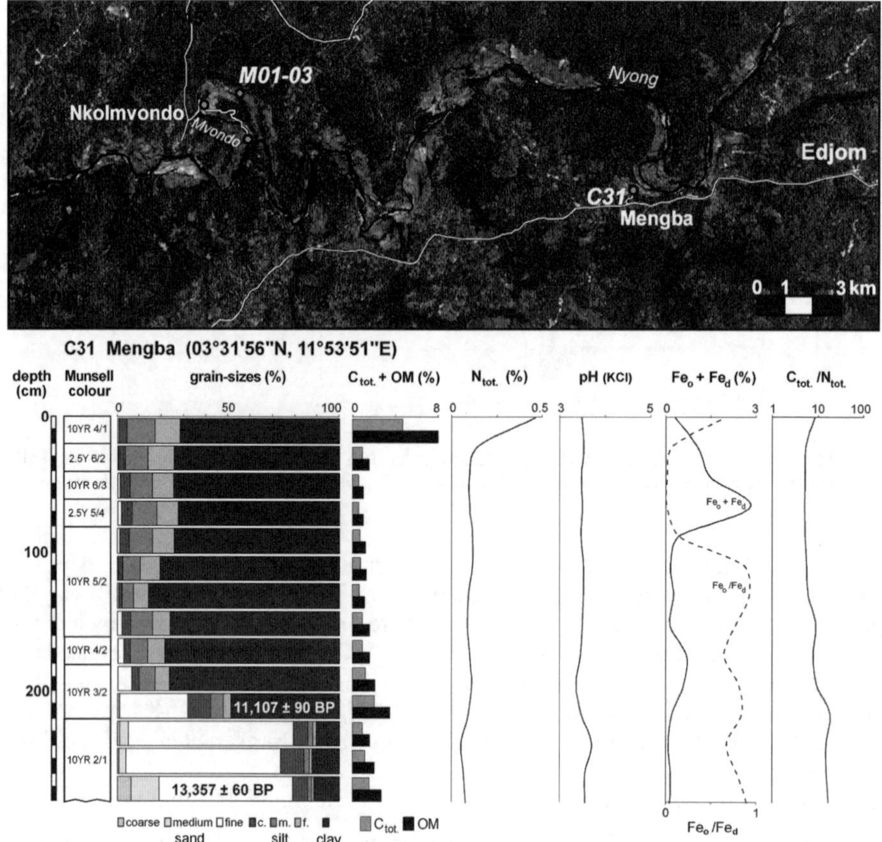

Figure 17. LANDSAT ETM+ scene 185–58 (18.03.2001) showing the Mengba and Nkolmvondo study area with reconnaissance route and corings from the abandoned channels of the Nyong (C31, M01–03) and sediment profile of C31 near Mengba with ^{14}C-ages.

Cores from the middle Nyong catchment area near Eséka

The middle catchment area of the Nyong was investigated between the towns of Eséka and Makak. The satellite image of the Eséka region (Figure 18B) shows a very linear river physiognomy of the Nyong, which is strongly associated to underlying geological complexes. The river course shows several erratic and acute changes. Only at a few locations does the river branch out. In this region, the Nyong traverses the peneplain step of the South Cameroon plateau by numerous cataracts and waterfalls. One of the latter, the *Chutes de Njock* (Figure 18A) was investigated ca. 5 km south of Eséka, near the settlement of Njock (03°34'N, 10°47'E). At this place, the Nyong drains the peneplain step by the three-stepped waterfall of Njock. No suitable cores for study, i.e., containing alluvial sediments, could be found.

Above the waterfalls, core B01 was taken directly on the river bench. At several places the basement (amphibolic gneiss of the Neo-Archaic to Palaeo-Proterozoic socket, Maurizot, 2000; *Unité du Nyong et de l'Ayna* after Ngako, 2006) cropped out of the river. Below the waterfalls, sesquioxid-hardened peddles were identified, like those found at the *Chutes de Dehane* and *Chutes de Memvé'élé* (Ntem interior delta).

Samples from B01 were cored on a 2.5 m high river bench, some 2 m from the river. At the base (250–225 cm) dark (2.5Y 4/2) sandy alluvium was found, which merges into silt and clay with decreasing depth (Figure 18C). Fossil macro-remains from the basal layer (225 cm) were dated to a ^{14}C-age of 2134 ± 41 yrs BP (δ^{13}C: –27.3 ‰). While C/N fluctuates between 10 to 22, pH reaches 4–5.4.

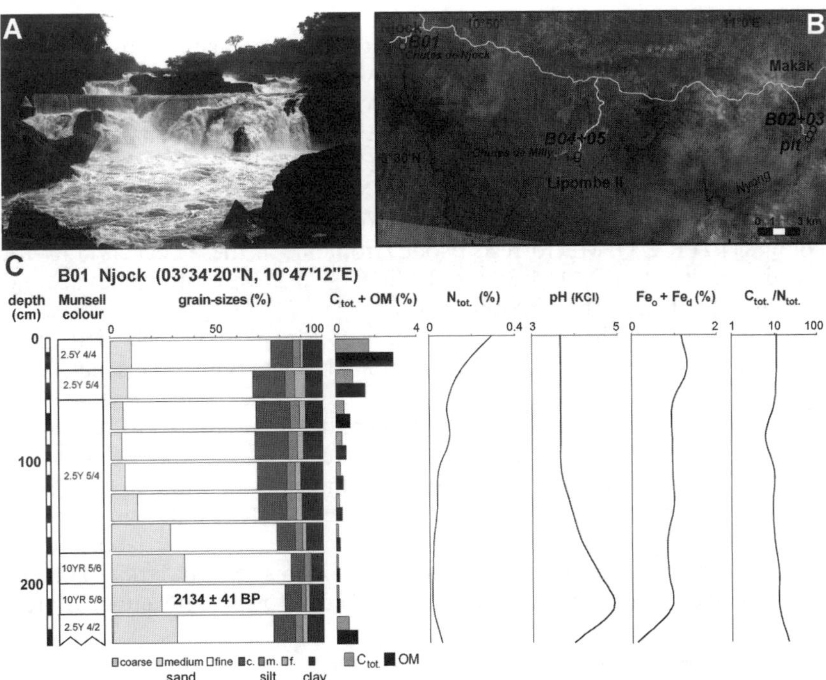

Figure 18. A The Njock waterfalls, where the Nyong River crosses three steps of gneiss basement. Core B01 was sampled above the waterfalls (Photo: M. Sangen, February 2005). **B** LANDSAT ETM+ scene 186–58 (26.04.2001) showing the Eséka study area and corings. **C** Sediment profile B01 from near Njock with ^{14}C-age.

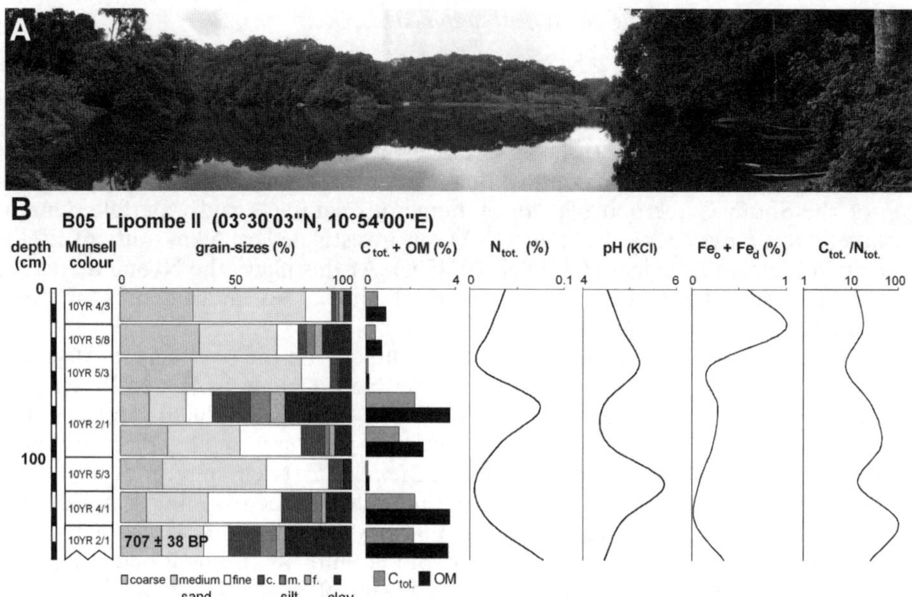

Figure 19. Location (**A**) and sediment profile (**B**) of core B05 Lipombe II. Above the rapids a flooded area has emerged (Photo: M. Sangen, February 2005).

Some 10 km SE of Njock, further corings were taken near the village of Lipombe II (03°30'N, 10°54'E; 575 m asl.). Across a series of cascades (*Chutes de Milly*), the Nyong has created a very shallow floodplain (Figure 19A).

Cores B04 and B05 (around 140 cm depth) revealed very coarse-grained sandy sediments and coarsening upwards sequences (Figure 19B). C_{tot} (2.18 to 2.19%), OM (3.76 to 3.78%), N_{tot} (0.078 to 0.074%) and C/N (109 to 47) reach relatively high values. The profiles contain pocket-like concentrations of fossil organic sediments having dark colours (10YR 2/1). Macro-rests (wood) from one of these pockets (140–120 cm) were dated to a ^{14}C-age of 707 ± 38 yrs BP (δ^{13}C: –28.2 ‰). With values around 4–5, pH was somewhat higher than in other cores.

Ten kilometres further to the east, south of Makak (03°30'N, 11°03'E; 564 m asl.), sands from the Nyong river bed are exploited. Some pits in the sandy sediments were investigated and one coring (B02, 300 cm depth) was done. Organic sediments were sampled from 245–235 cm and 180–160 cm depth and dated to recent ^{14}C-ages (1979–1980 AD and 1966–1967 AD) with δ^{13}C-values of –29.8 and –29.4 ‰.

Core B02 exhibited mainly sandy alluvial sediments in a fining-upward sequence. Middle sand fractures decrease from 70% at the base to around 20% in the upper layers. Fine sand, silt and clay fractures distinctly increase. All other laboratory results on these sediments were rather inconspicuously; for this reason dark (10YR 3/2 and 10YR 4/3) fossil organic sediments found between 300 to 240 cm were not dated.

Cores from the lower Nyong catchment near Dehane

Between the villages Donenda and Dehane north of Kribi (Figure 20A), the lower catchment area of the Nyong was investigated for suitable alluvial sediments. At Donenda (03°20'N, 10°02'E; 15 m asl.), core B06 was taken from the south bench of the Nyong at a site covered with rain forest (Figure 20C). Core B06 reached 220 cm

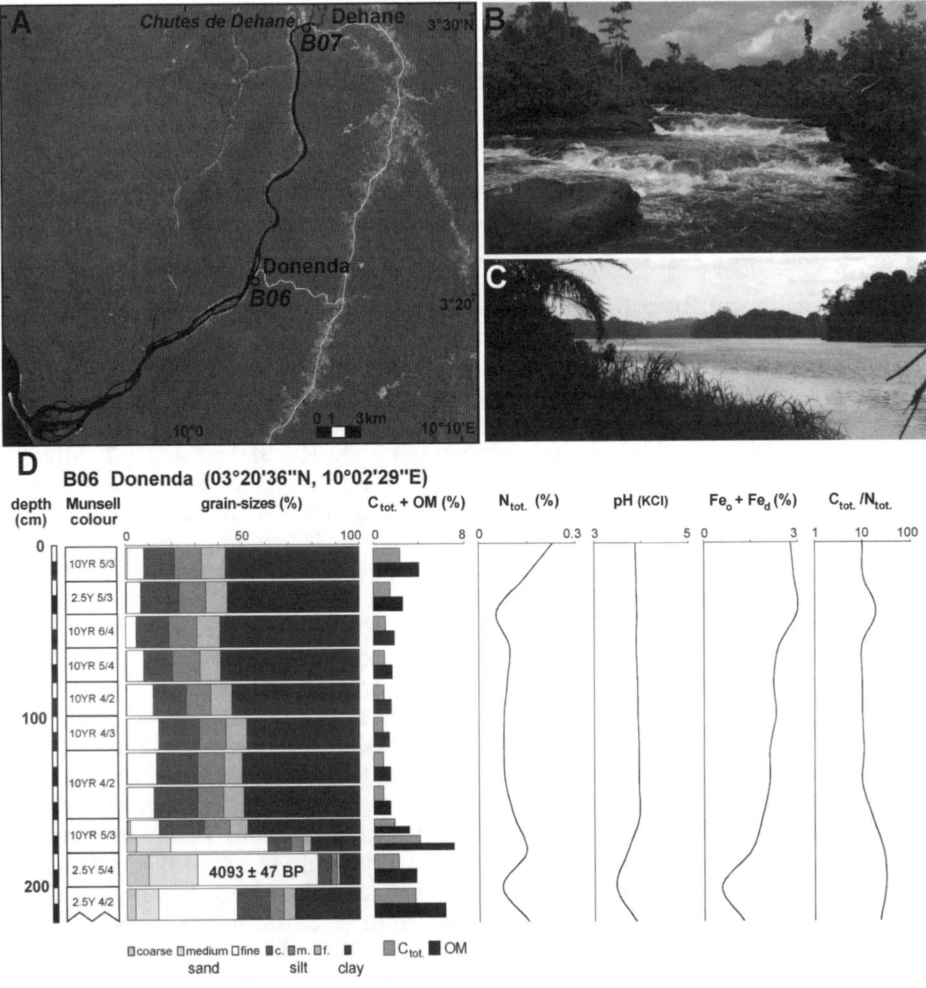

Figure 20. A LANDSAT ETM+ scene 186–58 (21.02.2001) showing the Dehane study area and location of corings (B06 and B07). B The waterfalls at Dehane (*Chutes de Dehane*), where sesquioxide-cemented pebbles were found. C The Nyong River at Donenda (Photos: M. Sangen, February 2005). D Sediment profile B06 near Donenda and 14C-age.

and revealed dark (10YR 2/1, 4/1) and mainly coarse-grained (5–10% coarse, 10–20% middle, 40–50% fine sand and 5–15% coarse silt) sediments having a high proportion of organic material between 220 to 170 cm depth (Figure 20D). These sediments are covered by brownish (10YR 4/6), fine-grained alluvia (~30% silt and 55% clay). At the base, relative high values C_{tot}, OM and N_{tot} (4.09, 7.05 and 0.147%) were recorded and pH fluctuated around 4 with a minimum of 3.47 (200–180 cm). Here, also, high C/N was found (24–30) and organic macro-remains yielded a ^{14}C-age of 4093 ± 47 yrs BP (δ^{13}C: –29.7 ‰).

About 10–12 km downstream, the Nyong cascades over a 10 m high step near Dehane (03°34'N, 10°07'E; 15 m asl.), this cataract is called *Chutes de Dehane* (Figure 20B). Around 100 m downstream the cataract, a coring was made on a bench covered with bamboo. Core B07 reached 340 cm depth and also showed dark

(2.5Y 3/2 and 4/3), sandy organic sediments near the base (340–260 cm) with, again, relative high values C_{tot}, OM and N_{tot} (1.18%, 2.03% and 0.064%). Above the base, more brownish colours (2.5Y 5/6 and 6/4) prevail and again pH fluctuates around 4. Macro-remains (wood) from 300–280 cm depth was dated to a ^{14}C-age of 400 ± 40 yrs BP ($\delta^{13}C$: –28.7 ‰).

4.2.3 Ntem catchment

4.2.3.1 Geographical characteristics

The perennial Ntem River has a catchment area of 31,000 km², which spreads over the countries of Equatorial Guinea, Gabon and much (ca. 75%) of Cameroon. The Ntem finds its source in Oyem, Gabon, at ca. 700 m asl. The catchment area in the N and W is bordered by the catchment areas of the Nyong, Lokoundjé, Kienké and Lobé rivers, and in the NE by the Ayina, a tributary of the Ivindo (which drains into the Ogoue), and in the SW by the catchments of the Rio Benito and the Mbia in Equatorial Guinea. The Ntem's main branch is 460 km long and drains in a south-western direction; it traverses the Archaic rocks of the Ntem-Complex basement (*Complex du Ntem/Unité du Ntem*, 2.9–2.5 Ga; Ngako, 2006), comprised of charnock-ite, gneiss, granite and quartz, which became exposed during peneplain genesis. On the surrounding peneplain, plateaus to the north and south of the Ntem basin have evolved heights of over 1000 m asl. (ex. *Mont Tembo* in Gabon is 1200 m asl.).

The upper Ntem reach is situated on the South Cameroon plateau and, like the Nyong between Abong Mbang and Akonolinga, characterized by slow flowing swampy tributaries at altitudes about 560–600 m asl. From its source in Gabon (Oyem, 700 m asl.) to the Cameroonian border (~580 m asl.), base slope averages ~2 ‰ and decreases (0.27 ‰) towards the interior delta near Ma'an (518 m asl.), a major study area of this research. In this section, the Ntem absorbs its major tributaries the Kom (Mfam), Mboua, Kie, Kyé, Mgoro, Nlobo, Nyé, Rio Bolo and Rio Guoro. Before uniting with the Nlobo and Kom (which has already further upstream absorbed the Mboua and Mfam), the Ntem receives the discharges of several Gabonese tributaries: the Nye (and Boleu), Kie, Sossolo and Aya. Upstream from Ngoazik, the Kyé flows into the Ntem, and downstream from Ngoazik (at ca. 535 m asl.), the Ntem confluences with the major tributaries from the northern catchment area (in Cameroonian territory): at 215 km from the rivers mouth in the west, the Ntem takes up the Mgoro river, and at 150 km the Mvila. This last is an important tributary, originating in the Ebolowa region, where it united with the Seng (catchment area around Assoseng: 440 km²). The Ntem River's longitudinal profile analogues with that of the Nyong's and is reflected in the general morphology and hydrological regime of both catchment areas. However, the hydrological regime for the Ntem catchment exhibits an equatorial pattern, having distinct dual rainfall and run-off maxima, whereas the Nyong and Sanaga catchment areas belong to the tropical climate zone. Given these longitudinal profiles, both river basins have considerable hydro-electrical potential, and there is growing interest in exploiting cataract sites across the basins for hydro-power such as at the *Chutes de Njock*, along the Nyong, and the *Chutes de Memvé'élé*, along the Ntem.

Before its confluence with the Mvila, the Ntem fans out into an anastomosing river system having multiple branches. This so-called interior delta is situated in the sub-prefecture of Ma'an in the district of the *Vallée du Ntem*; it extends from 2°14'N and 10°39'E in a north-western direction towards the waterfalls of Memvé'élé (2°24'N, 10°23'E), located close to Nyabessan. The delta has a length (SE-NW) of around 35 km and a maximum width (NE-SW) of some 10 km (total surface area of 210 km²).

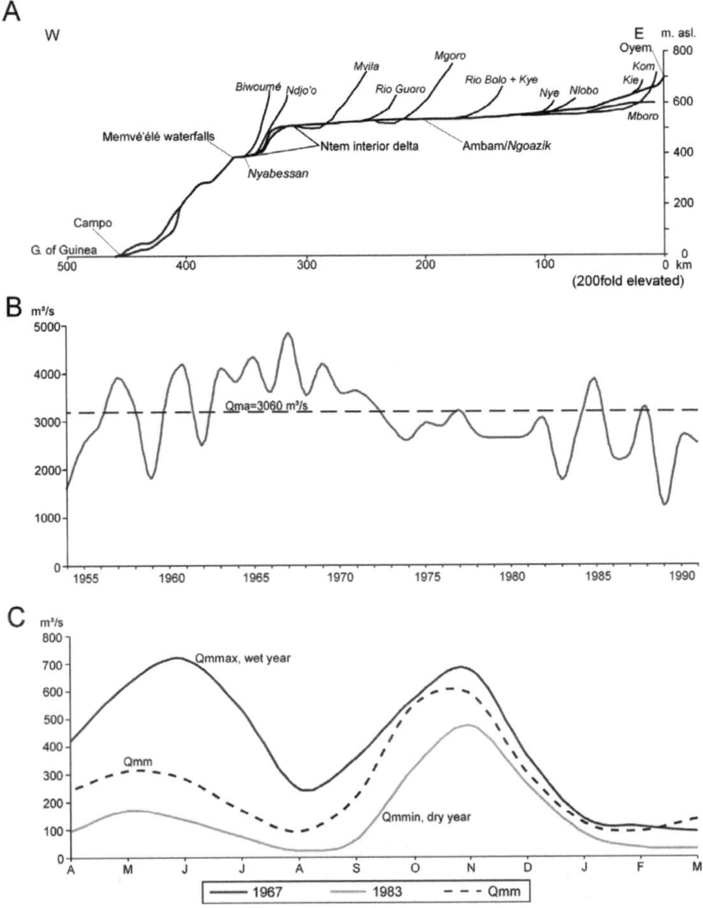

Figure 21. A Longitudinal profile of the Ntem River, with its source at Oyem and mouth at Campo Beach; **B** cummulated yearly discharges at Ngoazik (2°18'N, 11°18'E; 535 m asl.) 1954–1991; **C** mean (Qmm), maximum (Qmmax) and minimum (Qmmin) monthly Ntem discharges at Ngoazik, 1954–1991.

It divides the slightly inclined upper reach of the Ntem from the stronger one in the lower catchment area. After crossing the interior delta, the Ntem falls from 518 m asl. to 405 m asl. (inclination: 4.5 ‰). After passing the town of Nyabessan (385 m asl.), where the anastomosing system reunites into a single branch, the Ntem receives the discharges from the Biwomé (catchment area: 402 km²) and Ndjo (376 km²), which originate from the *Monts Kenle* (1020 m asl.) and Bingalawoa (1059 m asl.) mountains. After traversing the waterfalls of Memvé'élé (~384 m asl.), inclination increases to 5 ‰. The Ntem then crosses up to 80 m deep gorges and falls around 200 m over a distance of 40 km. The hydrography of the Ntem changes after passage through the Nkoltom- and Nkolebengue mountain ranges (500–1059 m asl.), because of the considerable rain-shadow effect. The Ntem becomes forked into two branches (Bongola and Ntem) and a series of cataracts before reaching the coast and Gulf of Guinea. These two branches circumfluent the island of Dipikar (ca. 16 × 40 km), and reunite around 8 km before the Ntem's mouth where it is known as the Rio Campo. Here, the

Ntem traverses younger Cretaceous and Quaternary deposits (see also Ntem River longitudinal profile, Figure 21A).

Letouzey (1985) and Tchouto Mbatchou (2004) describe vegetation in the Ntem catchment area as evergreen lowland rain forest (mostly *Caesalpiniaceae*) and mixed semi-deciduous forest and on the seasonally flooded plains, swamp forests with *Raphia* spp. *Uapaca guineensis* and *Gilbertiodendron dewevrei* are dominant. In reality, vegetation in large areas of the Ntem catchment and especially inside the interior delta is secondary (rain) forest because of continued and increasing anthropogenic disturbance in the region (e.g., shifting cultivation, mono-cultures [bananas, cacao]). Widespread species are *Ceiba pentandra, Elaeis guineensis, Musanga cecropioides* and *Terminalia superba* (Tchouto Mbatchou, 2004).

According to Segalen (1967), in south-western Cameroon and the Ntem catchment area there are three common soil types: yellow to red tropical clay soils (*'sols ferrallitiques jaunes sur les roches acides'*), red to brown tropical clay soils over gneiss and granite (*'sols ferrallitiques rouges sur les roches acides'*) and alluvial soils (*'sols alluviaux'*).

4.2.3.2 Climatic and hydrological characteristics

The Ntem catchment is located in the equatorial climate zone and rivers show a pronounced equatorial regime having two discharge maxima during the short (April to June) and long (September to November) rainy season (Figure 21C). In this region, the WAM is very influential and provides continuous heavy rainfall during the short and long rainy seasons, which diminishes towards the hinterland. Mean annual rainfall in the Ntem catchment area totals around 1695 mm; 1675 mm was recorded at the meteorological gauging station at Nyabessan (2°24'N, 10°24'E; 385 m asl.) having a catchment surface of 26,350 km^2 and 1640 mm at the Ambam/Ngoazik station (1951–1992). At Campo average annual rainfall can exceed 3000 mm and mean annual temperature is about 2° C. Mean annual discharges, like rainfall, show high annual variability, cumulating between 1000 to 1600 m^3/s (Olivry, 1986). Like in the Nyong and Sanaga catchments, this variability becomes most significant when considering monthly rainfall and discharge data, and may exceed the average amounts by 290% and fall below 60%. These phenomena can be amplified by periodically recurring ENSO-events, causing extreme dry (e.g., 1958, 1961, 1983) as well as wet (1960, 1962, 1985) years every 2–3 years (Figure 21B). Although the hydrologic regime is characterized by two discharge maxima in May and November, with total maxima in November, this typical Gulf of Guinea rainfall and run-off pattern may be disrupted such that rainfall and discharges during the short rainy season exceed those of the long. Maley (1997) contends that the ENSO-events and the varying SST's in the Gulf of Guinea are responsible for these perturbations, as they influence humidity transfer between the ocean and the African continent and, thereby, rainfall and discharge behaviour. A mean annual discharge of around 441 m^3/s could be estimated for the period 1954–1991, though Olivry (1986) indicates ~382 m^3/s for 1978–1991. At the Ngoazik gauging station (2°18'N, 11°18'E; 535 m asl.; catchment area: 18,100 km^2), some 150 km upstream from Nyabessan, mean discharge amounted to ~260 m^3/s between 1954–1976.

4.2.3.3 Alluvial sediment archives

Ntem interior delta

The Ntem interior delta research area is located near the village of Ma'an and stretches from 2°14'N and 10°39'E to the NW, where it is limited by the waterfalls of Memvé'élé (2°24'N, 10°23'E) close to Nyabessan (Figure 23A). It has a length (SE-NW) of around

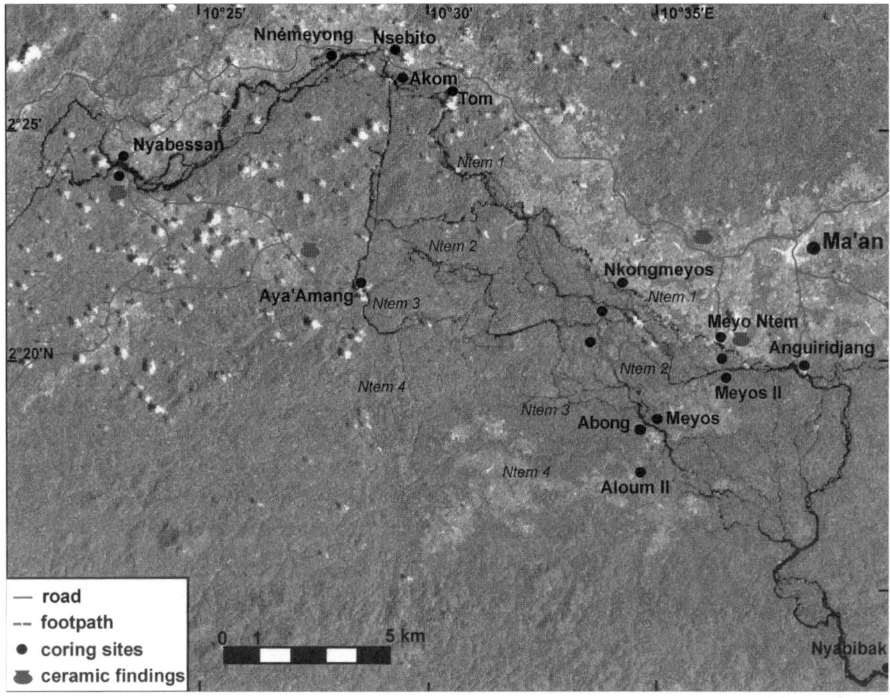

Figure 22. LANDSAT ETM+ scene 186–58 (21.02.2001) showing the Ntem interior delta study area and the location of sampled sites. The four major anastomosing channels of the Ntem were denominated as Ntem 1–4.

35 km and a maximum width (NE-SW) of 10 km. Over a surface of nearly 210 km^2 the Ntem fans out into an anastomosing river system with 4 main branches (Ntem 1–4). In order to explore the manifold spectra of this sediment archive, transects were set up through the interior delta. The main branches were crossed by piroges, and cores taken on the northern and southern river benches. During the dry seasons of the years 2005 and 2006 extended fieldwork campaigns were carried out in the interior delta. Eighty-eight cores and 4 pits were made at 16 different, central as well as marginal locations of the interior delta. Figure 37 shows the physiognomy of the anastomosing Ntem and the investigated localities.

Anguiridjang

At Anguiridjang, a location situated to the E of the first and most northern Ntem branch (Ntem 1), 4 corings were made, two on the northern and two on the southern river bench.

The alluvial sediments found at Anguiridjang were mainly composed of sand (80–95%). The four cores from this location (L39–42) are of minor interest, except core L40, which shows a fining-upward sequence and reached 400 cm depth. The other cores only reached minor depths over basement (Archaic noritic and amphibolic gneiss, which have been reworked during the Panafrican genesis; Maurizot, 2000). The sediments in general show grey colours (10YR 5/1 and 4/1). In core L40, between 400–300 cm depth dark (10YR 3/1 and 4/1) organic sediments were found. In 380–360 cm maximum C_{tot} (1.61%), OM (2.81%) and N_{tot} (0.061%) occurred. The lowest pH was measured here, with 3.73 in 380–360 cm,

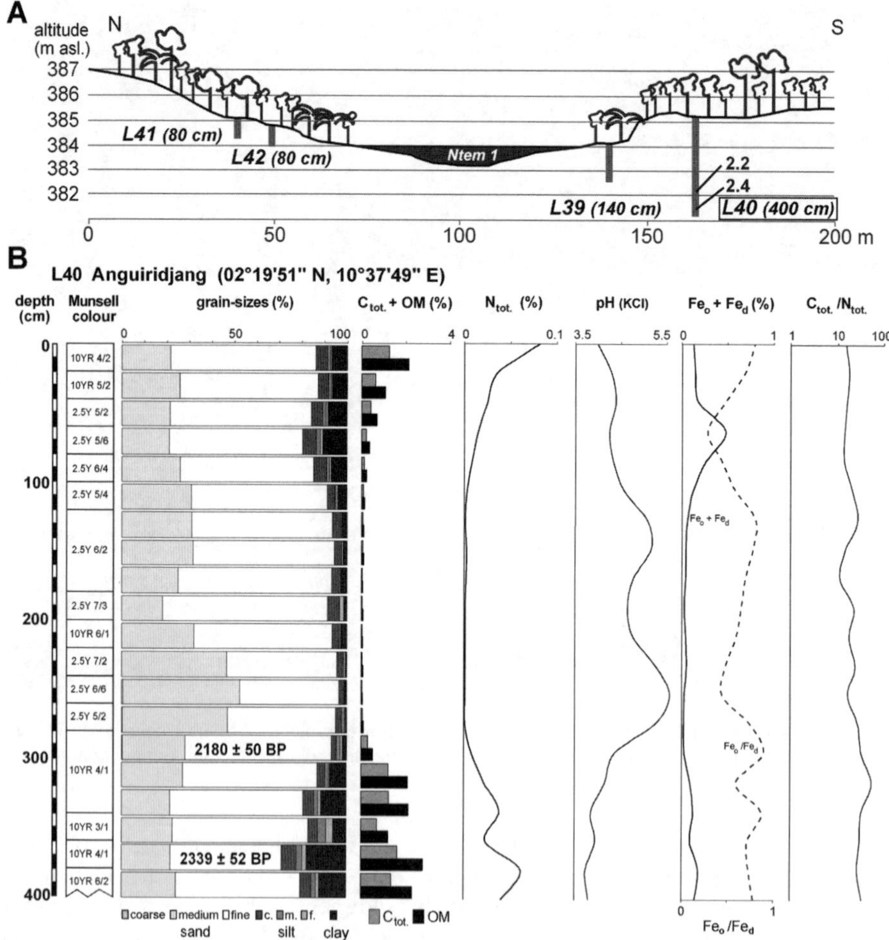

Figure 23. A Physio-geographical settings and corings as well as cal. ^{14}C-data (ages in kyrs BP) at Anguiridjang. **B** Sediment profile L40 from near Anguiridjang with ^{14}C-data.

and in higher layers this increased to ca. 5.7. Also relatively high C/N was found with values of 26–54. Sandy, organic sediment having macro-remains from 380–360 and 320–300 cm yielded ^{14}C-ages of 2339 ± 52 (δ^{13}C: −27.9 ‰) and 2180 ± 50 yrs BP (δ^{13}C: −26.9 ± 0.3 ‰), respectively (Figure 23B).

Transect Meyo Ntem-Abong

Between the settlements Meyo Ntem and Aloum II, a transect of 37 cores over a distance of 17 km was laid (see Figure 24). Thereby, four channels of the anastomosing Ntem river system were crossed. In order to study the sediment stratigraphies in detail, 3 pits (Meyo Ntem, Meyos and Meyos II) were dug. Preferably, sediments were sampled from periodic (seasonal) channels, natural levees, crevasse splays, abandoned channels, terraces (river benches in the dry and rainy season), backswamps and floodplains. For a better understanding and a more detailed description of all locations, the transect was divided in 4 parts: Meyo Ntem across the first (most northern) Ntem

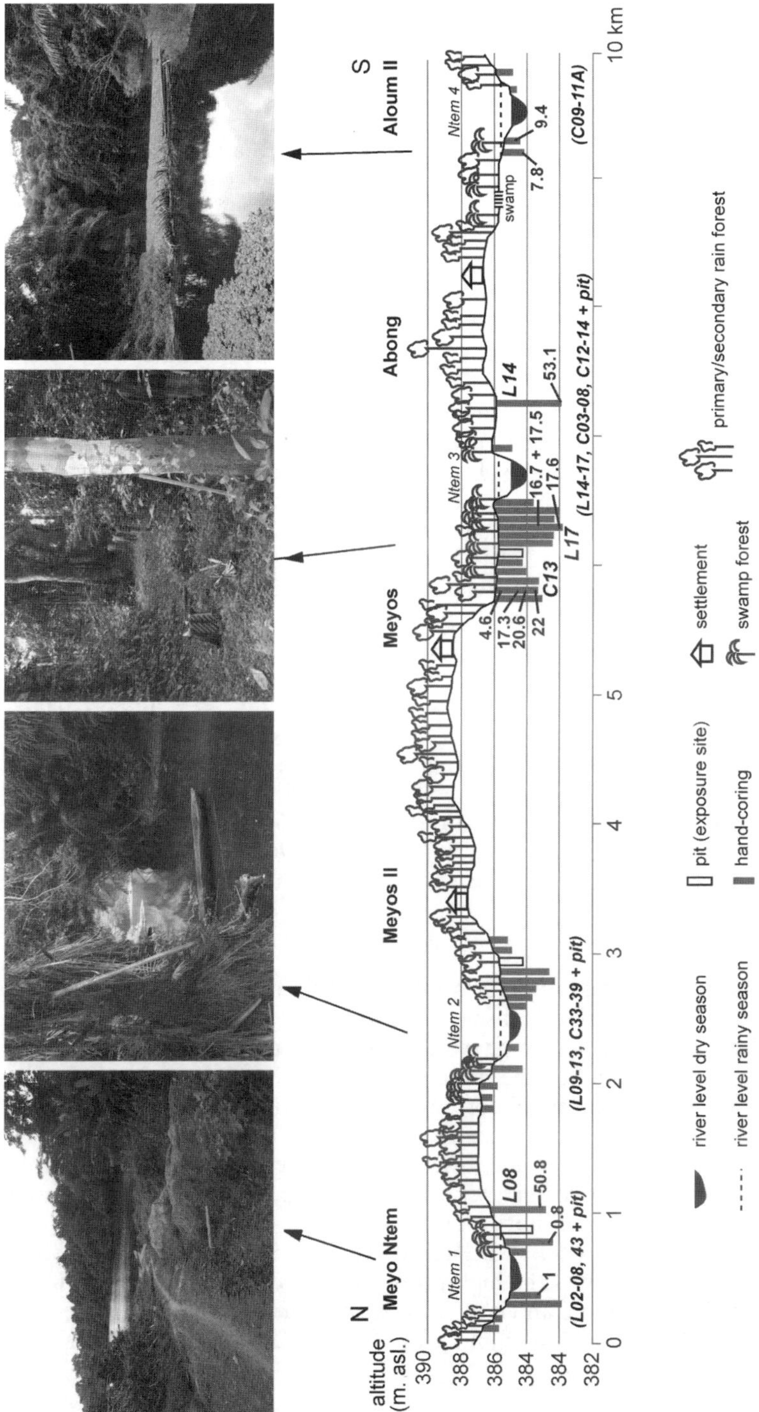

Figure 24. Transect Meyo Ntem-Aloum II (corings with cal. ^{14}C-ages in kyrs BP) and physio-geographical settings; Photos: M. Sangen, 2005).

channel (Ntem 1), Meyos II at the second (Ntem 2), Meyos at the third (Ntem 3) and, finally, Aloum II at the fourth (Ntem 4).

Transect Meyo Ntem

At Meyo Ntem, 9 cores (L01–08 and L43) were taken and one pit was dug. On both river benches two terrace levels were identified. Both river benches are covered by primary to secondary rain forest, which merges into swamp forest near the river. Two further cores, L03 and L06, were taken close to the nearby rapids. Here, the basement crops out of the river in the dry season. These rapids were subject of extended research on river bed genesis (lineaments etc.; compare also contribution J. Eisenberg). The pit (250 cm depth) was dug in the transition zone from the upper to the lower terrace and very clearly demonstrates the stratigraphy of the sediments at the Meyo Ntem site having a typical upper horizon of oxidized sediments followed by a reduction horizon and finally a fossil organic horizon having a high content of macro-remains (see Sangen, 2008, p. 88). In general, the sediments had a higher fraction of sand and a lesser fraction of silt, with the exception of L07 and 08, which manifested the presence of silty to clayey overbank deposits. The cores all show fining upward sequences and dated macro-remains yielded quite young ^{14}C-ages in some cores (1066 ± 53 yrs BP [δ^{13}C: −29.2 ‰] and 1104 ± 52 yrs BP [δ^{13}C: −27.0 ‰] in L02 and 908 ± 50 yrs BP [δ^{13}C: −29.4 ‰] in L05) and an amazing old Late Pleistocene age in L08 (45,596 ± 4899 yrs BP, δ^{13}C: −31.4 ‰). This last core (L08) was the most promising core from this location. It was cored on the upper terrace, which represents the bankfull stage during the rainy season.

Core L08 showed dark (2.5Y 3/1 and 4/1) fossil organic sediments at the base (340–300 cm) with increasing C_{tot} 0.74–1.32%, OM 1.28–2.28% and N_{tot} 0.009–0.015%, high C/N (80–87) and an acid pH (3.51–3.90). These were covered by grey (2.5Y 5/2 and 7/2) sandy deposits having decreasing C_{tot}, OM and N_{tot}, lower C/N (around 10) and higher pH (4).

Towards the surface, the fractions of clay (35–40%) and silt (25–30%) slightly increase as well as does iron content (10YR 6/5, 6/6; 2.5Y 6/6, 6/4; see also Sangen, 2008, pp. 90 and 91).

Transect Meyos II

Across the northern and southern river bench of the second Ntem channel (Ntem 2) no fossil organic sediments/horizons were found during coring (Figure 25). Five (L09–13) of the 12 cores from this location were sampled, while the other 7 (C33–39) were only documented. At the end of the transect, another pit was dug. Basement or saprolite was reached at the base, in all cores (except L10 and C34), and in the pit, at about 80–150 cm depth.

The sediments could be identified as natural levee and crevasse splay sediments. Besides the fact that two terraces can be distinguished, the main stratigraphy at Meyos II was an oxidation-horizon over a saprolite-layer and basement (metamorpic amphibolic and noritic gneiss) or laterite. Another indicator of the highly weathered state of the sediments was the high proportion of dithionitic iron (> 90%) and their colour (10YR 5/6, 5/8, 6/6 and 6/8). Towards the surface, the colours become increasing brown (10YR 5/4, 6/2, 6/4) and grey (10YR 5/2, 7/2). C/N varies around 10 and pH lies between 3.5 and 5.

Transect Meyos

At the Meyos site, 11 cores and one pit were made (Figure 26A). Coming from the village Meyos, one descends into a floodplain having two significant depressions, which

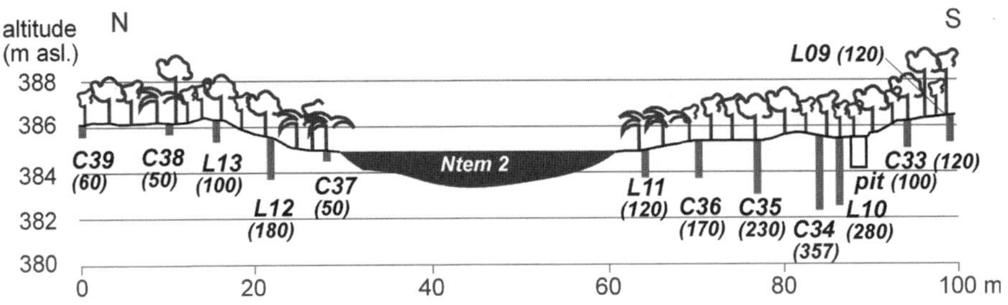

Figure 25. Corings and physio-geographic settings at Meyos II.

are separated by an E-W stretching ridge (gneiss basement near the surface). Towards the S, the depressions are limited by a natural levee and the third Ntem channel (Ntem 3). In both depressions, continuous (up to 1 m thick) palaeosurfaces were found between depths of 2 to 4 m. In core L17 a fossil gyttja (7.5YR 3/1, 5YR 2/1) was found in 380–300 cm depth, covering basement (gneiss) and a saprolitic layer (Figure 26B), indicative of a formerly swampy environment (probably backswamp). From 380–340 cm a sandy stratum was incorporated in the palaeosurface, while further upwards the sandy fractions are increasingly displaced by silt. Macro-remains from the core base (400–380 cm) yielded a ^{14}C-age of 14,516 ± 80 yrs BP (δ^{13}C: –23.1 ± 0.2 ‰). Over the palustrine sediments, undecomposed macro-remains (7.5YR 3/1) occur between 300–280 cm, which were dated to a ^{14}C-age of 14,263 ± 126 yrs BP (δ^{13}C: –27.3 ‰).

Here, the highest values for C_{tot} (9.90%), OM (17%), N_{tot} (0.134%) and C/N (74) were found. Moreover, there was high iron content here (1.64%, 0.72% Fe_o and 0.92% Fe_d) and very low pH (2.28). Above 280 cm, fine-grained, clayey-silty sediments (10YR 5/1, 5/2 and 6/2) were deposited (erratic increase of clay [50–60%] and silt [20–30%]). In 120–100 cm, iron reached a maximum of iron (2.73%), probably indicating a transition layer (10YR 5/8; Figure 26C).

Core L16 was taken on a natural levee bordering Ntem 3. It reached 300 cm and also contained dark (7.5YR 3/1) organic sediments, although no macro-remains were found. To the north and south of L17, corings were made every 10 m. This led to similar findings achieved in cores C03 (352 cm depth, palaeosurface between 352 and 255 cm), C04 (330 cm, 330–230 cm), C08 (320 cm, 320–250 cm) and C12 (395, 395–240 cm). Cores C05 and C06 reached saprolite and basement (gneiss) in 160 and 175 cm depth, respectively. In a dug pit, oxidation and reduction layers as well as adjacent fossil organic horizon were exposed and documented.

In the northern depression (cores C07, C13 and C14) another dark (10YR 2/1–4/1) palaeosurface was detected. In C07 (250 cm depth) it occurred between 250 to 175 cm, in C13 (265 cm) between 265 to 140 cm and in C14 (270 cm) finally between 270 to 170 cm. Similar to the southern depression, the palaeosurface contained thin sandy layers in the lower sections of the cores and layers of macro-remains (undecomposed vegetal litter). Again, very low pH (around 3), high C_{tot} (2–5%), OM (4–10), N_{tot} (0.05–0.1%) and C/N (30–80) were determined. In C13, macro-remains from six different organic layers were dated and δ^{13}C determined (Figure 27).

Abong

The cores L14 and 15 were taken on the southern river bench of Ntem 3, some 500 m from the village of Abong. In L14, mainly sandy textures (80%) were found in the lower core (400–260 cm) with brownish colours (10YR 6/5, 6/7 and 6/8). The values for C_{tot}

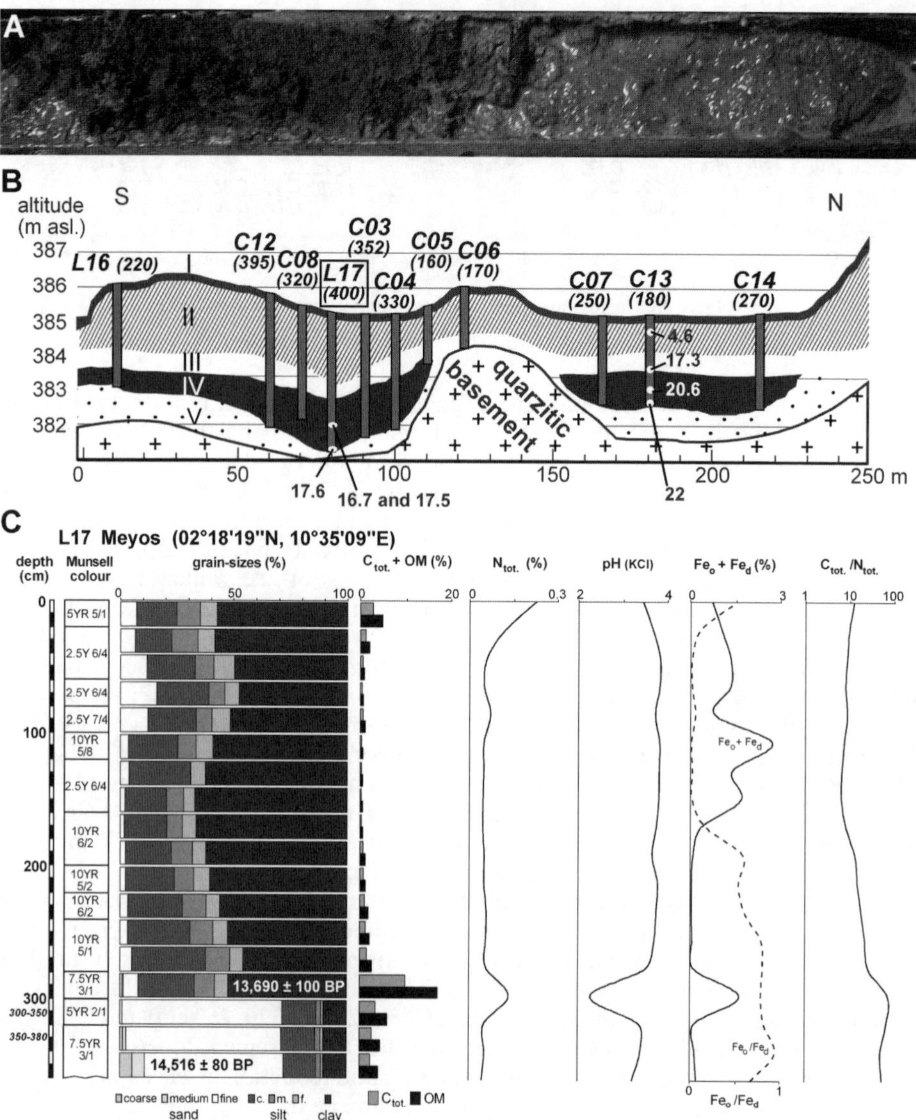

Figure 26. A Organic macro-remains and saphrolite found at Meyos (here base of core C04, 310–330 cm), **B** overview of corings (depth in cm, ages in cal. kyrs BP) and stratigraphy of alluvial sediments at Meyos: Over quarzitic basement (gneiss) with saprolite cover (V) a palaeosurface (IV), with ^{14}C-ages between 18 and 14 kyrs BP (22–17 cal. kyrs BP), sediments with reduction symptoms (III) and oxidation symptoms (II), as well as a humuc top layer were discovered (I). **C** Sediment profile L17 near Meyos.

and OM only reached 0.5%. Also, N_{tot} (0.002–0.012%) and iron contents (0.17–0.74%) were very low, while pH fluctuated around 4. In 360–340 cm (10YR 5/5), where C/N reached 87, macro-remains were found in a coarse-grained sandy matrix. They were dated to a ^{14}C-age of 48,230 ± 6411 yrs BP (δ^{13}C: –29.6 ‰). From 260 cm upwards, silt and clay textures (30–40%) strongly increased. From 260–180 cm, an iron concentration (1–2%) horizon was distinguished. While pH continuously decreased (until 3.64), C_{tot} and OM slightly increased towards the surface (Figure 28).

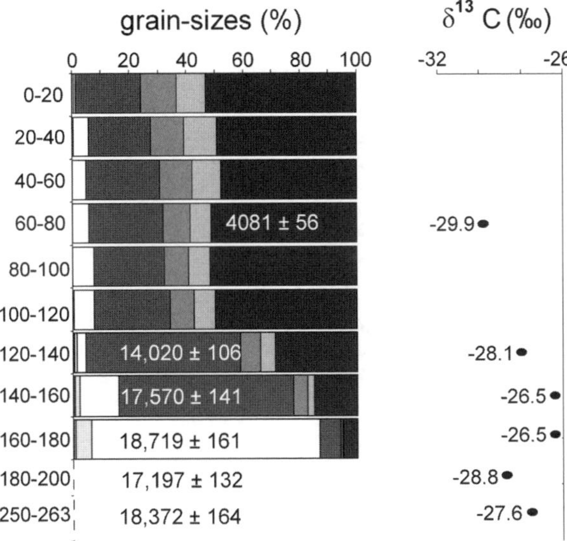

Figure 27. Grain-sizes, ^{14}C-data (in yrs BP) and corresponding δ^{13}C from core C13.

Core L15 only reached 120 cm over basement (gneiss) and showed a coarsening-upward sequence. The reddish (2.5Y 6/4, 6/6) to brown (2.5Y 5/4, 4/3) sediments revealed barely interesting laboratory results, nor was fossil organic material found. Behind the village of Abong there is a paved footpath to Equatorial Guinea, which can be used to cross a very swampy area. Some 1.6 km away from Ntem 3, there was another, very small channel, which was designated Ntem 4.

Aloum II

On the southern river bench of Ntem 4, cores C09 and 10 were cored. On the northern side, cores C11 and C11 A as well as a further coring to sample material for dating were carried out (Figure 29B). While C09 and C10 reached basement (110 cm) and laterite (10YR 3/6; 17 cm), respectively, core C11 turned up fossil organic sediments (Figure 29A).

Core C11 shows a fining-upward sequence between 140 to 80–60 cm (2.5Y 6/2 and 6/4) with 30% clay, 30% silt and 30% fine sand (Figure 29C). While pH varies between 3.5 and 4, C_{tot} (1.5–5.4%), OM (2.50–9.3%) and N_{tot} (0.1–0.3%) reach relatively high values. C/N varies between 9 and 22 and iron content is very low. Macroremains (wood) from the base of C11 yielded a ^{14}C-age of 6979 ± 58 yrs BP (δ^{13}C: −30.1 ‰). Organic sediment from another core next to C11 (88–78 cm depth) gave a ^{14}C-age of 8402 ± 67 yrs BP (δ^{13}C: −30.6 ‰).

Transect Nkongmeyos

Some 2 km NW of Meyo Ntem another 5 km long transect of 15 cores was carried out near the village of Nkongmeyos. During work on this transect, all three main channels of the anastomosing Ntem (Ntem 1–3) were crossed and integrated (Figure 30). The northern bench of Ntem 3, a seasonal floodplain, was cored using a very small-scale transect because a palaeosurface was found in ca. 80–140 cm depth, hence a very detailed study of the stratigraphy of the alluvial sediments was required.

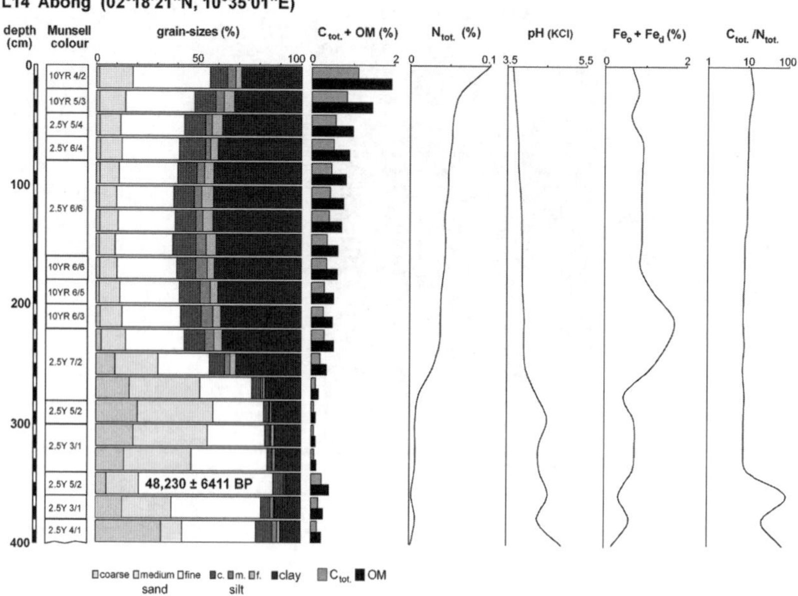

Figure 28. Sediment profile of L14 near Abong with ^{14}C-age (modified from Sangen 2011, p. 217).

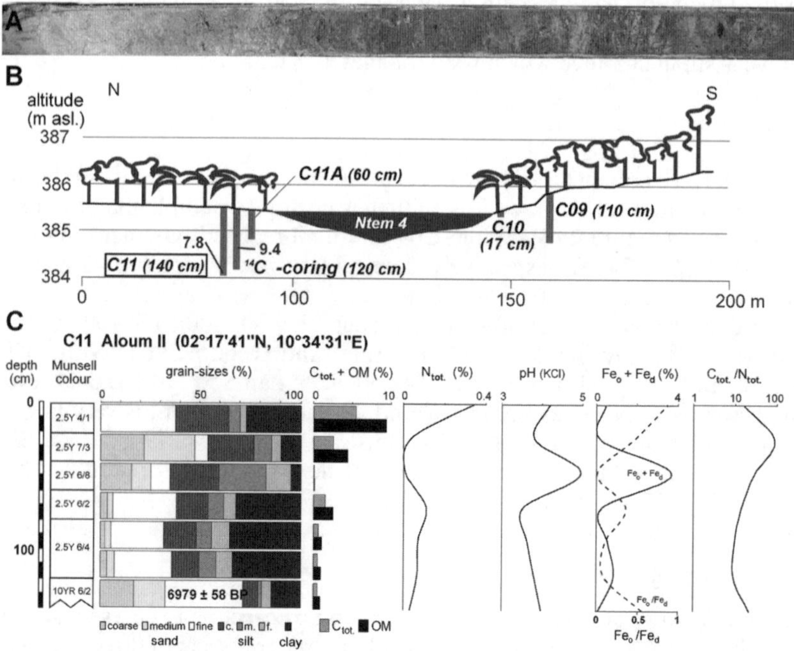

Figure 29. A Sediment core (90–140 cm) from C11 at Aloum II showing increasing fossil organic sediments towards the base (Photo: M. Sangen, February 2006), **B** corings (with depths and cal. ^{14}C-ages in kyrs BP) and physio-geographic settings at Aloum II. **C** Sediment profile C11 near Aloum II.

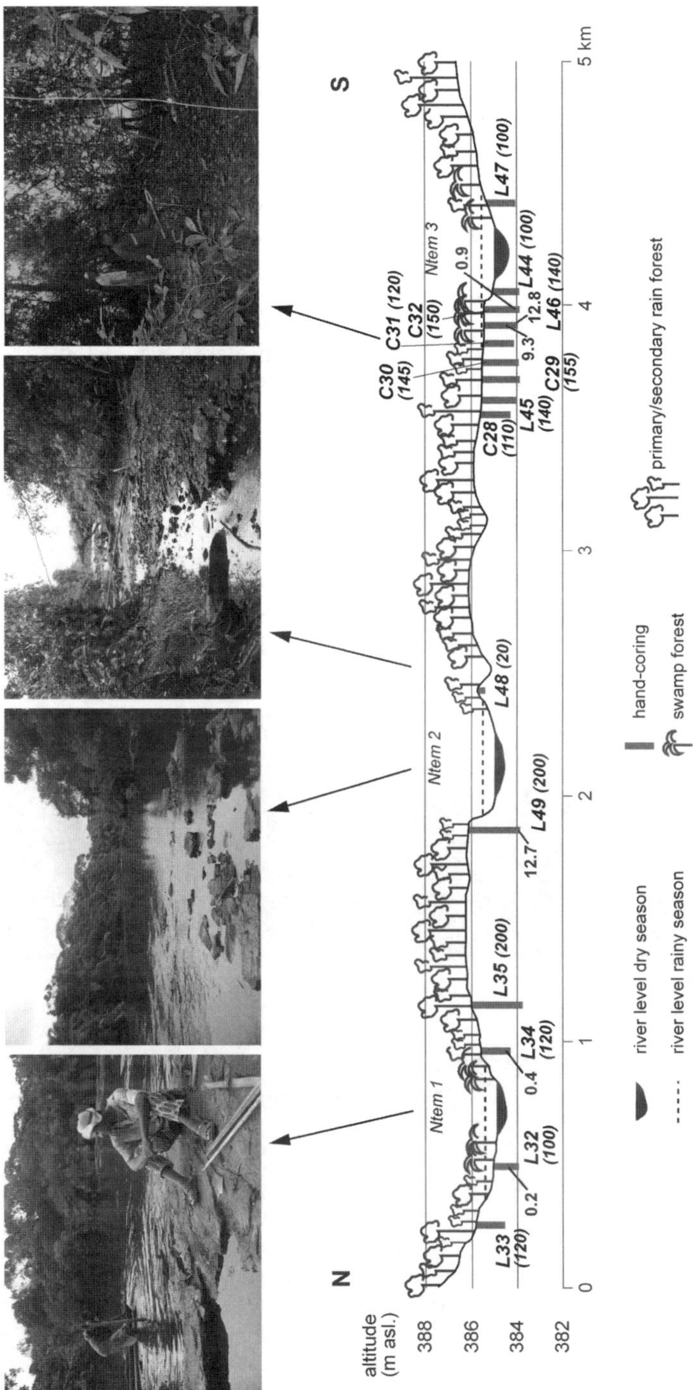

Figure 30. Transect Nkongmeyos showing corings (depths in cm) and cal. ^{14}C-ages in kyrs BP. The transect contains 15 corings stretching over 5 km and three anastomosing main channels of the Ntem River (Photos: M. Sangen, February 2005 and 2006).

Nkongmeyos Ntem 1

From the village of Nkongmeyos, Ntem 1 was reached after a ca. 2 km walk through secondary tropical rain forest and cacao plantations. Before L32 was cored on the northern river bench of Ntem 1, rapids and outcropping basement (gneiss) were identified some 50 m upstream (Figure 31A).

Further to the west, 2 cores were taken from the northern and southern river bench (Figure 31B). On the northern river bench two cores (L32, 33) were taken from an identified river terrace. Core L32 showed a coarsening-upward sequence with mainly sandy fractions.

Values for C_{tot}, OM and N_{tot} were very low (< 1%), similar to all cores taken from this location, and pH fluctuated between 4–5. However, charcoal from dark (2.5Y 4/2, 4.5/2 and 4.5/3) sandy sediments in 80–60 cm depth was dated to a ^{14}C-age of 217 ± 46 yrs BP (δ^{13}C: −28.0 ‰).

In L33 (120 cm depth) similar sediments were found. On the southern river bench L34 was (like L32) taken under *Raphia*-palm vegetation.

Core L34 (Figure 31C) was taken on a lower terrace and L35 on an upper one (river level during rainy season). At the base (120 cm) of L34, dark (2.5Y 3/1) sandy sediments with maxima for C_{tot} and OM (1.17 and 2.02%), low pH (3.8) and high C/N (41) contained wood, which was dated to a ^{14}C-age of 435 ± 51 yrs BP (δ^{13}C: −27.4 ‰). Core L35 reached the highest depth (200 cm) at this location and showed the same characteristica as L34. However, macro-remains (wood fragments) from the base of this core were not dated.

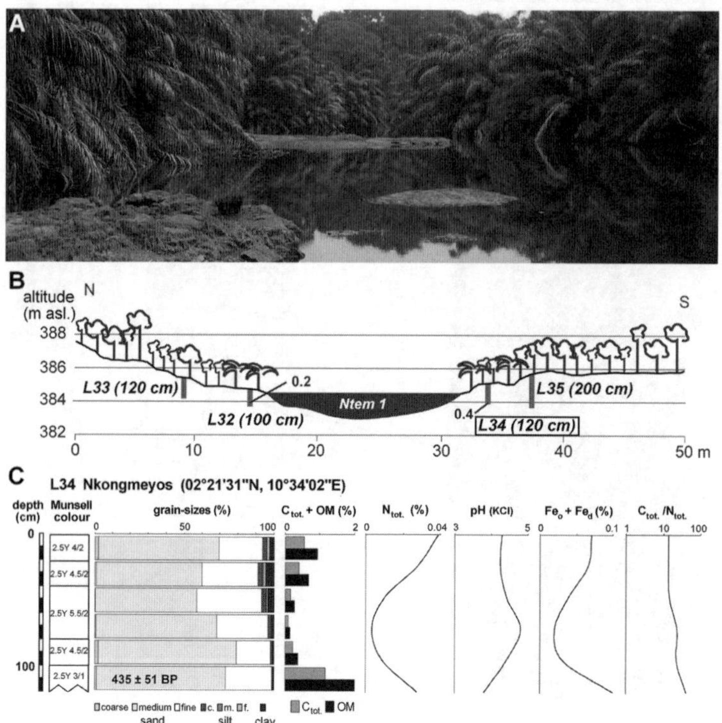

Figure 31. A Rapids and outcropping basement at Nkongmeyos Ntem 1 (Photo: M. Sangen, February 2006). **B** Corings (with depths and cal. ^{14}C-data in kyrs BP) and physio-geographical settings at Nkongmeyos Ntem 1. **C** Sediment profile L34 at Nkongmeyos Ntem 1.

Nkongmeyos Ntem 2

Near the coring locations of Ntem 2, rapids, small cataracts and basement (gneiss) occurred in the river channel. A NE-SW stretching basement step (gneiss) crosses the river bed, which continues towards Ntem 1 and 3 (lineament; see Figure 31, contribution J. Eisenberg). On the northern river, the outer (cut) bank was cored. Core L49 reached 200 cm (Figure 30) and the brown (2.5Y 5/2) to grey (2.5Y 6/2, 10YR 6/2) sediments from the base first show a fining- and then a coarsening-upward sequence. At the base, C_{tot} (0.30%), OM (0.52%) and N_{tot} (0.012%) were comparable low, as in cores across Ntem 1. However, in the dark sediment, having increased C/N (25) and low pH (3.56), a relative old ^{14}C-age of 10,775 ± 144 yrs BP (δ^{13}C: −30.1 ‰) was found. On the southern river bench, a coarse-sandy point bar was identified, which had an adjacent seasonal rivulet and floodplain, separated again by a basement step (gneiss). Core L48 had to be abandoned at 20 cm depth because the coarse sand could not be sampled by the Edelman-corer. Both river benches were covered by tropical rain forest.

Transect Nkongmeyos Ntem 3

Eight cores (C28–32, L44–46) were taken from the small-scale transect of the northern river bench floodplain and one reconnaissance core (L47) from the southern, which ended on basement (gneiss) at 100 cm depth (Figure 32B). All cores were carried out in tropical rain forest, which merges with swamp forest (mainly *Raphia* sp.) towards the river (Ntem 3). The transect was established after the discovery of fossil organic horizons in cores L44 and L46 suggested a continuous palaeosurface at this location. These dark (10YR 3/1 and 5/1), sandy (20–30% middle and 60% fine sand) fossil organic sediments (Figure 32A) were found in 80–90 cm depth. In L46 (Figure 32C) they occurred between 140 (base) to 80 cm and in combination with high C_{tot} and OM (0.75–1.47 and 1.29–2.53%), low pH (3.06–3.16), high C/N (60–102) and low iron content (0.14–0.17%). In 80–60 cm, this is followed by an iron concentration layer (2.5Y 6/6) and, subsequently a fining-upward sequence (20–30% silt and 15–30% clay). Macro-remains (wood) yielded ^{14}C-ages of 8291 ± 89 yrs BP (80–100 cm; δ^{13}C: −27.5 ‰) and 10,871 ± 99 yrs BP (120–140 cm; δ^{13}C: −28.3 ‰).

Similar sediments (10YR 4/2 and 3/1; 30% middle sand, 60% fine sand) were obtained from core L44, located directly on a 40 cm high step towards the dry season river level. However, this core only reached 100 cm because the waterlogged sediments could not be sampled.

In L45 no fossil organic sediments were found. Over basement (noritic and amphibolic gneiss) and a thin saprolitic layer, a sandy fining-upward sequence was found (10YR 6/2, 6/4). In this core, C_{tot} (0.15–0.36%), OM (0.26–0.62%) and N_{tot} (0.021–0.060%) were low, pH fluctuated around 4 and C/N varied between 6 to 11.

The sediments of further corings (C28–32) were only sampled to verify the existence of the palaeosurface and not analysed in the laboratory. In C28, basement was reached after only 110 cm, and overlaid by pisolites and an iron concentration layer (10YR 5/8). Similar results were found in C29, which reached 155 cm. Nevertheless, a thin fossil organic layer was discovered between 90 to 80 cm; in core C30 this was thicker and found in 145–100 cm, over saprolite and gneiss.

Over basement, the sandy, fossil organic layer containing macro-remains was again found in core C31 between 120 to 60 cm (10YR 5/1 and 5/2). Here, it was more coarse-grained than in the previous samples, pH was around 3.5 and C/N varied between 16 to 20. Finally, the palaeosurface was once again found in C32 (150 cm) between 145 and 95 cm. In C32, the palaeosurface was overlaid by more fine-grained, silty to

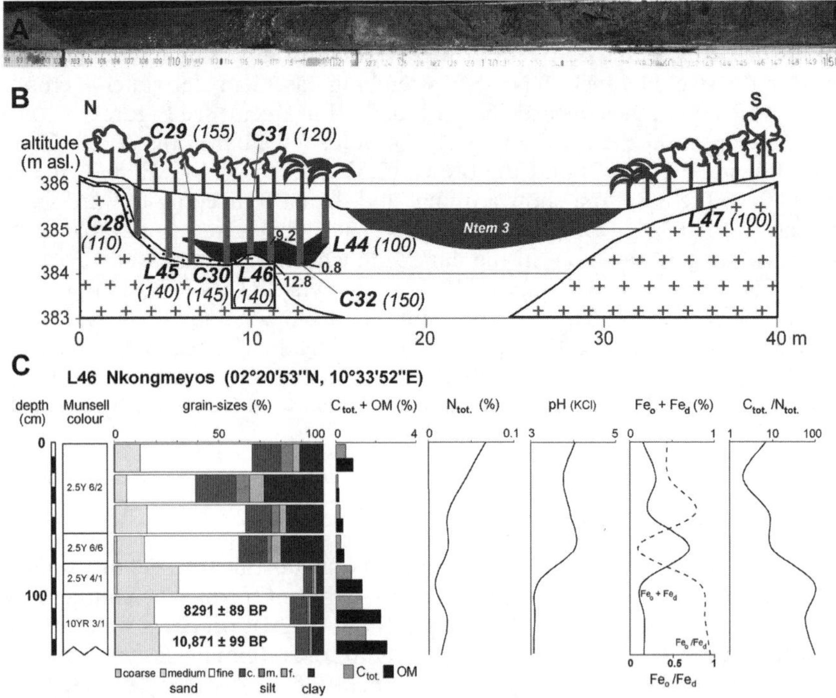

Figure 32. A Sandy, fossil, organic alluvial sediments from Nkongmeyos Ntem 3. Here from core C32, 100–150 cm depth (Photo: M. Sangen, February 2006). **B** Corings with depths in cm and cal. ^{14}C-ages in kyrs BP as well as physio-geographic settings at Nkongmeyos Ntem 3. Between C29 and L44 a continuous palaeosurface of around 30–50 cm thickness was found over saprolite and basement (gneiss). **C** Sediment profile of core L46 at Nkongmeyos Ntem 3 with ^{14}C-ages.

clayey sediments having oxidation characteristics between 95 to 15 cm and a humic top layer (Ah) in 5–0 cm depth. A quite young ^{14}C-age of 954 ± 39 yrs BP (δ^{13}C: -28.9 ‰) was found at the base of the fossil organic macro-remains (150–145 cm).

Aya'Amang

Three cores were sampled at the Aya'Amang site (L36–38; Figure 33B), located in the southwesternmost part of the research area, where Ntem 3 changes its flow direction from E to N. All corings took place in tropical rain forest.

From the village of Aya'Amang, there was a considerable distance by foot over hilly terrain and primary forest to reach the coring sites. The site is characterized by basement outcrops along the river channel, which were partially covered by sand banks. On the western river outer (cut) bank, a river terrace could be identified, on the eastern inner bank, two. In core L38 (300 cm depth; Figure 33C), taken on the terrace of the western river bench, very fine-sandy (80–90% fine sand), dark (10YR 4/1, 5/1, 5/2 and 5/3) deposits were found. C_{tot}, OM and N_{tot} were low (0.53, 0.91 and 0.011%), C/N high (28–72) and pH was acid (3.26–3.82). Above 160 cm, a fining-upward sequence was found having a rapidly increasing clay fraction (40–50%) and brown colour (2.5Y 6/4 and 6/5). From the two samples with maximum OM (300–280 and 240–220 cm) macro-remains yielded ^{14}C-ages of 5282 ± 45 yrs BP (δ^{13}C: -26.8 ± 0.5 ‰) and 5306 ± 64 yrs BP (δ^{13}C: -31.4 ‰).

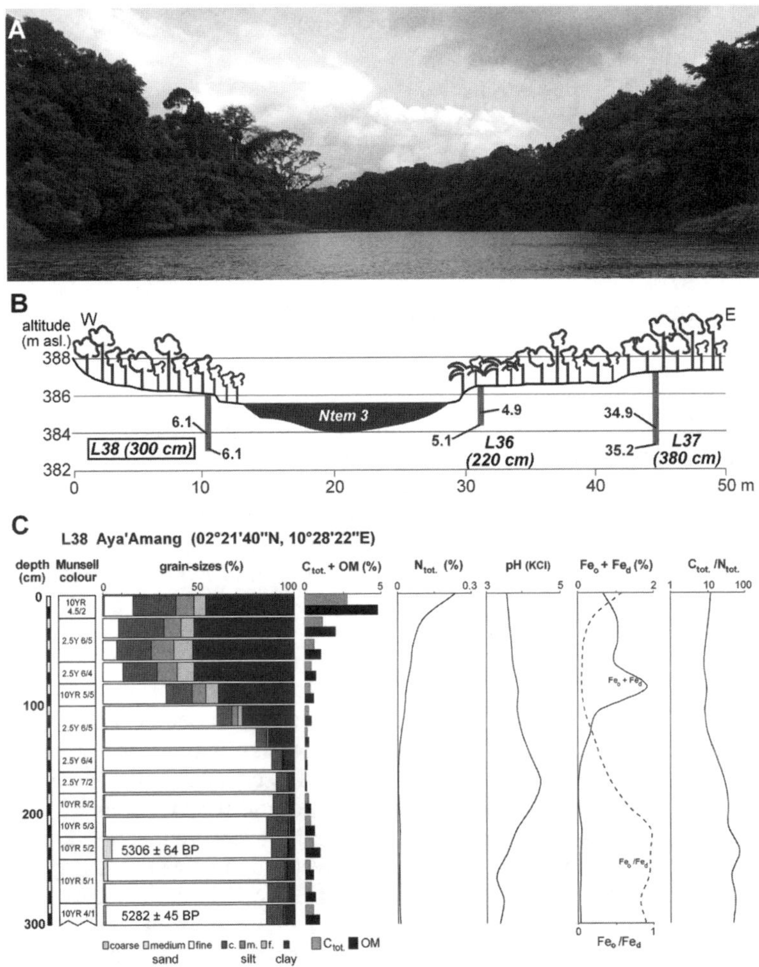

Figure 33. A Northward view on Ntem 3 at Aya'Amang and densely vegetated river banks (Photo: M. Sangen, January 2005). **B** Corings with depths in cm and cal. ^{14}C-ages in kyrs BP as well as physio-geographical settings at Aya'Amang. **C** Sediment profile L38 Aya'Amang with ^{14}C-data.

On the eastern inner river bank, core L36 was taken 100–120 cm above the dry season river level. Again, dark (10YR 3/1) sandy (12–15% middle sand, ~80% fine sand) fossil organic deposits were sampled from the base (220 cm) up to 140 cm depth. Values for C_{tot}, OM and N_{tot} increased: 0.85–1.30%, 1.47–2.24% and 0.020–0.024%, respectively. Additionally, low pH (3.39–3.91) and high C/N (43–57) were found. Organic sediment from the upper and lower layer of the fossil organic deposits was dated to ^{14}C-ages of 4514 ± 45 yrs BP (δ^{13}C: −27.3 ± 0.1 ‰) and 4341 ± 60 yrs BP (δ^{13}C: −27.2 ‰).

Some 15 m further to the east, another river terrace was seen and interpreted as a rainy season river bench. Here, core L37 reached 380 cm depth (see Sangen, 2008, p. 90). Dark (7.5YR 2.5/1, 3/1 and 4/1), sandy (70–80% fine sand, 15–20% coarse silt) fossil organic deposits were found up to 300 cm depth. These contained increased C_{tot} (0.48–2.07%), OM (0.83–3.57%) and C/N (84–106), but lower N_{tot} (0.006–0.023%) as

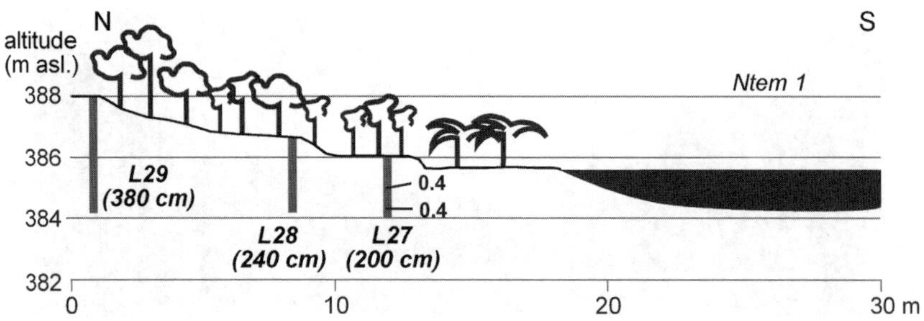

Figure 34. Corings with depths in cm and cal. ^{14}C-data in kyrs BP as well as physio-geographical settings at Akom.

compared to the previous core L36. Again, pH was very low (2.99–4.40). The fossil organic deposits are followed by a fining-upward sequence (~10–45% silt, ~20–35% clay) with greyish (7.5YR 6/1, 10YR 7/1) and yellow-brown colours (10YR 6/3, 6/4 und 2.5Y 6/3, 6/4). Macro-remains (wood) from the lower (380–360 cm) and upper (340–320 cm) end of this palaeosurface were dated to Late Pleistocene ^{14}C-ages of $31,253 \pm 240$ yrs BP ($\delta^{13}C$: -27.1 ± 0.2 ‰) and $30,675 \pm 770$ yrs BP ($\delta^{13}C$: -30.8 ‰).

Tom

Near the village of Tom, on the northern river bench of Ntem 1, only two cores were taken because at this location the river could not be crossed. Core L30 was taken directly behind a series of rapids (outcropping gneiss), which are very pronounced between Asseng and Tom. Core C30 reached only 160 cm and contained mainly sandy sediments (35–65% fine sand, 22–57% middle sand and 0.5–5% coarse sand). At the base (10YR 4/1), C_{tot} (0.41%), OM (0.71%) and C/N (82) were increased, iron (0.02%) and N_{tot} (0.005%) were low and pH comparatively high (4.48). An organic macro-remain (wood) yielded a ^{14}C-age of 1381 ± 49 yrs BP ($\delta^{13}C$: -28.2 ‰).

Core L31 was taken from an avulsion trunk channel and reached only 120 cm depth because the deeper water-logged sediment could not be sampled. The sandy (30–62% fine sand, 14–37% middle sand and 6–31% coarse sand), dark (2.5Y 3.5/1, 4/1.5 and 5/2) sediments showed very low carbon and OM content and were not considered interesting for the purpose of this study.

Akom

Around 1 km NW of Tom, a transect of three cores was carried out near the village of Akom (Figure 34).

At this location two terrace levels were identified and sampled. Cores L27 and 28 contained coarse- to fine-grained sandy sediment deposited in fining-upward sequences. In comparison to other locations, pH is relatively high (4.21–6.35). Although the sediments were also characterized by dark colours (10YR 4/1, 4/2 and 5/2), values for C_{tot}, OM and N_{tot} were very low. C/N is rather high with 30–77. However, the only dated samples (leaf fragments and wood) yielded relatively young ^{14}C-ages of 427 ± 52 yrs BP (180–160 cm; $\delta^{13}C$: -27.6 ‰) and 443 ± 57 yrs BP (120–100 cm; $\delta^{13}C$: -26.4 ‰).

Between Akom and Nnémeyong, a series of rapids, flowing over significant steps in the gneiss basement, were sighted and investigated near Nsebito (Figure 35A). Like at Meyo Ntem these structures belong to a complex geological system and fluvial system genesis (see Chapter 3.3.3, contribution J. Eisenberg).

Nnémeyong

Near the village of Nnémeyong, Ntem 1 was twice sampled on the northern and southern river bench. While on the northern bank only one terrace was identified and sampled, the southern bank revealed two terraces, which were both sampled (one core each). Similar to most of the sites inside the interior delta, the cores were taken in tropical rain forest, which merges with swamp forest (*Raphia* sp.) near the river bank (Figure 35B and C).

The basal layers (260–180 cm; 10YR 6/4) in L26 mainly contained coarse-grained sandy deposits (10–24% coarse and 43–59% middle sand). From 180 cm onwards

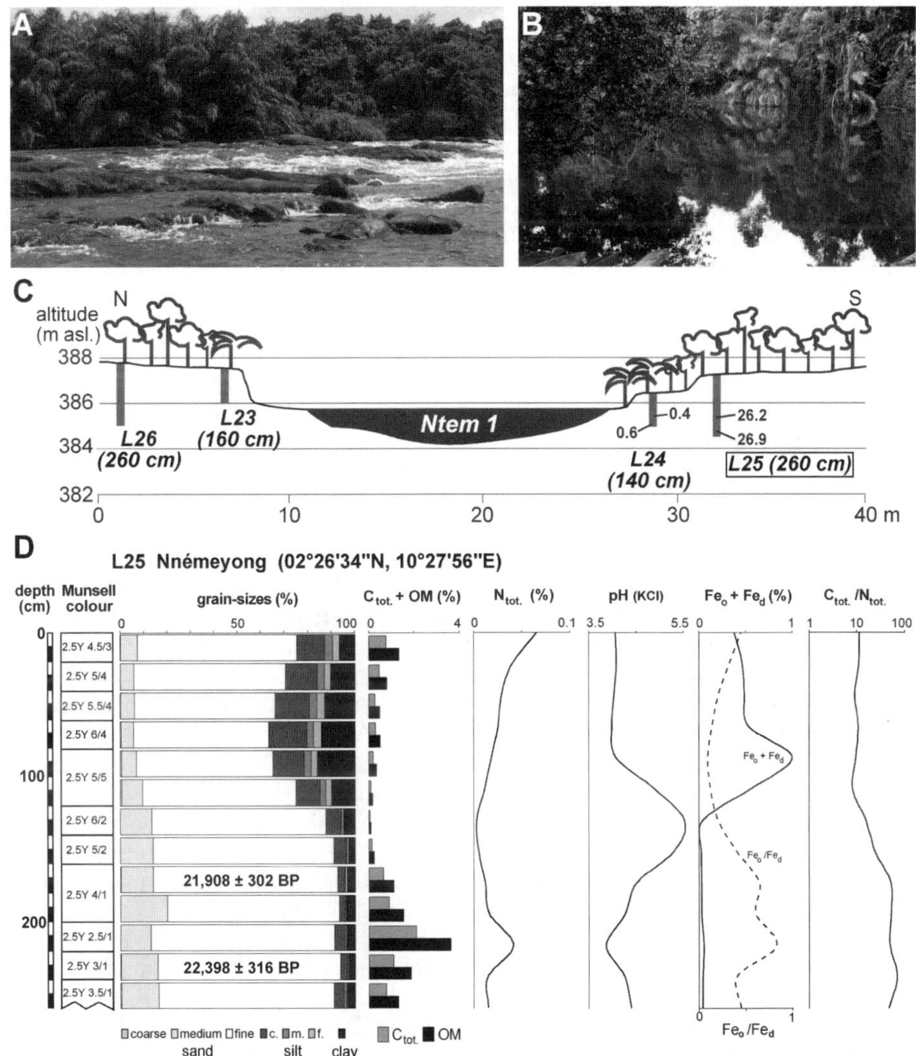

Figure 35. **A** Rapids at Nsebito. **B** Coring site Nnémeyong (Photos: M. Sangen, February 2005). **C** Corings with depths in cm and cal. ^{14}C-data in kyrs BP as well as physio-geographical settings at Nnémeyong. **D** Sediment profile of L25 near Nnémeyong with ^{14}C-data.

coarse-grained textures diminished and fine-grained textures strongly increased in a fining-upward sequence (18–28% fine sand, 8–15% silt and 23–38% clay); C_{tot}, OM and N_{tot} were quite low around the base and increased at the surface (0.13–1.21%, 0.22–2.09% and 0.013–0.098%). C/N varied between 8 to 13 and pH between 3.8 to 4.

In L23, taken directly from the river bank, 2 m above river level (dry season), similar results were found. C_{tot}, OM and N_{tot} are slightly higher (0.24–3.23%, 0.41–5.57% and 0.021–0.208%). C/N fluctuates between 10 to 15 and pH between 3.31 to 4.01. After grey colours (2.5Y 7/2, 6/2 and 6/3) near the base, brown colours became more dominant in the upper core.

On the lower terrace of the southern river bank (80 cm above river level) L24 reached 140 cm depth. At the base (10YR 3/1 over 3/2) again sandy textures dominated (18–31% middle and 54–72% fine sand), which decreased in a fining-upward sequence towards the surface (2–5% middle and 26–59% fine sand, 18–28% silt and 17–44% clay). C_{tot} and OM were elevated (1.55–2.73% and 2.67–4.71%), while N_{tot} (0.057–0.092%) was quite low. C/N reached values between 10 to 33 and pH fluctuates between 3.59 to 3.89. Macro-remains (wood) from 140–120 and 100–80 cm yielded ^{14}C-ages of 441 ± 46 yrs BP (δ^{13}C: −30.6 ‰) and 671 ± 52 yrs BP (δ^{13}C: −28.7 ‰).

Core L25 (Figure 35D) was sampled from the upper terrace and reached 260 cm depth. Between 260 to 160 cm dark (2.5Y 4/1 over 2.5/1, 3/1.5 and 3.5/1 at the base), sandy (15–20% middle and 65–75% fine sand) sediment with increased C_{tot} (0.7–2.1), OM (1.3–3.6), C/N and N_{tot} was found. C/N varies between 48 to 72 and pH between 3.8 to 5.6. Taken together, these phenomena indicate another palaeosurface. Above 160 cm much lighter sediment (2.5Y 5/2 and 6/2 in 160–120 cm, 2.5Y 6/4, 5.5/4 and 5/4 in 80–20 cm) in a fining-upward sequence was found. Wood and organic sediment sampled from the lower (240–220 cm) and upper (180–160 cm) palaeosurface layers yielded ^{14}C-ages of 22,398 ± 316 yrs BP (δ^{13}C: −29.1 ‰) and 21,908 ± 302 yrs BP (δ^{13}C: −27.0 ‰), respectively.

Transect Nyabessan

At the Nyabessan site, and especially among a seasonally flooded swamp forest (Figure 36A) 15 cores were sampled and one pit was dug (exposure site), after during preliminary fieldwork in 2005 first 3 cores were taken from the northern river bench (L20–22) and 2 from the southern (L18–19, Figure 36A and C). While tropical rain forest dominates in some distance from the river, swamp forest highly increases towards the river bench. Especially across the southern bench the latter is very dense and documents a seasonal flooding of the site (Figure 36A). Here, initially L18 was cored, reaching a depth of 220 cm (Figure 36C). At the base (220–140 cm), dark (10YR 3/1 and 4/1), silty-clayey (30–40% silt, 30–50% clay) sediment layers with high C_{tot}, OM and N_{tot} were found: C_{tot} between 2.13 to 8.18%, OM 3.67–10.65% and N_{tot} 0.318–0.625%. Moreover, low pH (3.59–3.94) and increased C/N (21–33) as well as iron content (0.318–0.625%) occurred. Undecomposed leaf layers and other fossil remains (seeds, wood; Figure 36B) embedded in these clayey layers yielded ^{14}C-ages of 2337 ± ± 55 yrs BP (220–200 cm, δ^{13}C: −31.9 ‰) and 587 ± 64 yrs BP (160–140 cm, δ^{13}C: −29.1 ‰).

In an attempt to reach greater depth, another core (L19) was sampled a few metres closer to the river. It reached 420 cm and revealed very dark (10YR 2/1 and 3/1) organic sediments over sandy deposits at the base (420–400 cm: 18% middle sand and 50% fine sand). The former contained very high C_{tot} (up to 9.8%), OM (16.9%) and N_{tot} (0.300%); C/N (33) and pedogenic iron were increased (0.84%). Only pH was relatively high (4.05–4.26) in comparison to palaeosurfaces in other cores. The organic sediments were overlain by a coarsening-upward sequence above 280 cm. In terms of textures, silt and clay rapidly decreased and fine sand simultaneously increased. Wood sampled from 340–320 cm yielded a ^{14}C-age of 2189 ± 52 yrs BP (δ^{13}C: –29.5 ‰).

During extended and more detailed research, core C17 was taken near L18 and 19. Over a sandy base (290 cm) clayey to sandy organic sediment was uncovered until around 120 cm; partially embedded with very concentrated layers of fossil organic macro-remains (295–285 and 180–130 cm) and fossil leaf layers (280–278 and 261–258 cm; Figure 36B) of dark colour (10YR 3/1, 3/3 and 5Y 4/1, 5/1). These were

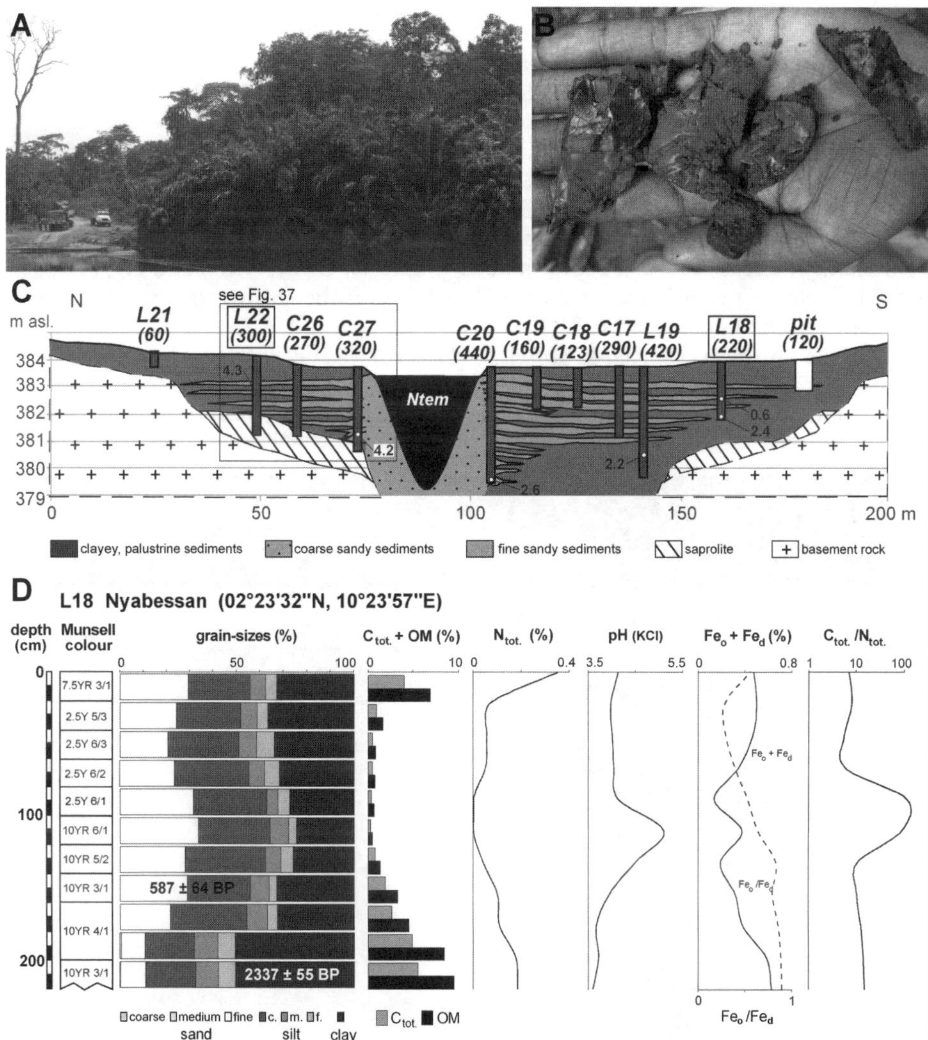

Figure 36. A Seasonally flooded swamp forest of the southern Nyabessan transect, where L18, L19 and C17–20 as well as an additional transect were sampled. **B** Fossil, undecomposed leaf layers (left) and other macro-remains (seeds, wood) from the Nyabessan palaeosurface (here L18, 220–215 cm; Photos: M. Sangen, January 2005). **C** Schematic overview of the Nyabessan stratigraphy and transect with corings (depths in cm and cal. ^{14}C-data in kyrs BP). **D** Sediment profile of L18 near Nyabessan with ^{14}C-data.

covered by bleached sediment layers having oxidation features (5Y 6/2 and 10YR 6/8) in 117–18 cm depth as well as a humic top layer (10YR 3/3).

In core C18, another 10 m closer to the river, an identical succession of sediment layers was found, although in a much sandier matrix. Again, organic sediments were sampled over a sandy base (160 cm). These were similarly overlain by bleached sediments having oxidation marks (5Y 6/2 and 10YR 6/8) and a humic top layer (10YR 3/3).

Core C19, further 10 m in direction of the river, showed the same profile. The final core of this transect, C20, was sampled 5 m from the river. From the base (440 cm) to the the surface uniformly organic sediments and macro-remains were found. The generally loamy-clayey sediments were repeatedly interrupted by sandy layers (especially in 430–380, 360–325, 180–150 and 120–100 cm). In 440–430 cm, 380–360 cm, and 325–200 cm very clayey organic layers having concentrated macro-remains were uncovered. Organic sediments from the base (440 cm) yielded a ^{14}C-age of 2479 ± 43 yrs BP (δ^{13}C: −28.8 ‰).

On the most southernmost end of this floodplain, a pit was dug to reveal the detailed stratigraphy of the alluvial sediments. At the base (10YR 5/1), organic sediments having low C_{tot} (0.40%), OM (0.69%), N_{tot} (0.002%), pH (2.70) and iron (0.12%) were identified. These values further decreased until 70 cm depth, while pH significantly increased (6.29). From there C_{tot} (2.97%), OM (5.12%) and N_{tot} (0.253%) strongly increase towards the surface (pH: 4.10), and the sediments become increasingly sandier and lighter (10YR 7/1).

On the northern river bank, core L20 (120 cm) was taken from a location outside the transect and ended like core L21 (60 cm) on a laterite crust. Core L22 (Figure 37B) reached a depth of 300 cm and laterite as well as saprolite (10YR 5.5/6, 5/6, 6/6 and

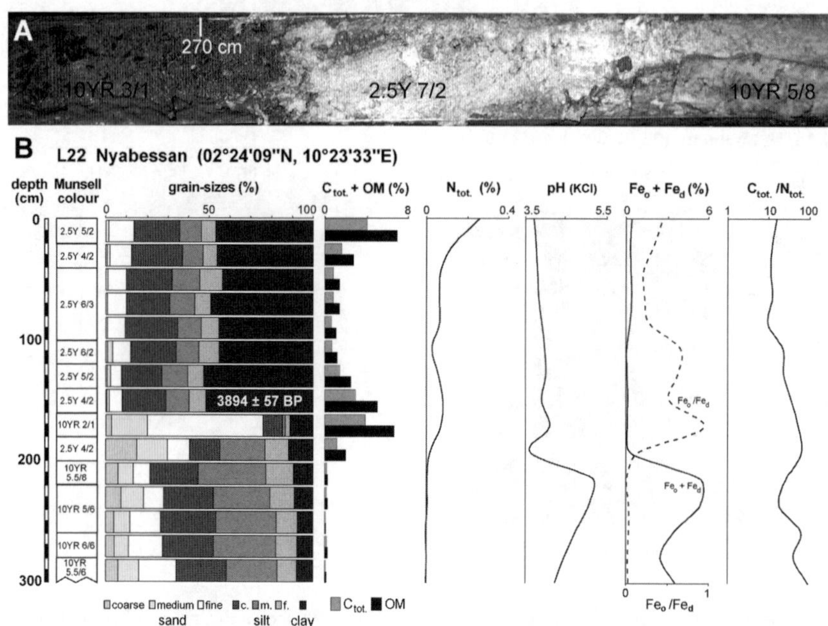

Figure 37. A Coring section around 270 cm from core C27 at Nyabessan, showing the transition from saprolite (right) to fossil organic sediment and macro-remains dated to 3829 ± 46 ^{14}C yrs BP (Photo: M. Sangen, February 2006). **B** Sediment profile of L22 near Nyabessan.

5.5/8) bellow 200 cm. These highly weathered layers were covered by, first sandy, and then from 160 cm onwards fine-grained (44–53% clay), dark (10YR 2/1, 2.5Y 4/2 and 5/2) organic sediments up to 120 cm. Here, C_{tot} (1.15–3.84%), OM (1.98–6.62%), N_{tot} (0.019–0.082%) and C/N (23–61) strongly increased, while pH (3.3–4) and iron (0.07–0.14%) rapidly decreased. Organic macro-remains (wood, leaves) from 140–160 cm were dated to a ^{14}C-age of 3894 ± 57 yrs BP (δ^{13}C: −28.1 ‰).

Between L22 and the Ntem channel, two further cores (C26 and 27) were sampled. In both cores, here again, fossil organic sediments over saprolite were found (Figure 36C). In C26 saprolite (2.5YR 6/8) was reached at 270 cm depth, covered thereafter by fossil organic sediments up to 105 cm depth. Between 240 and 190 cm, layers were highly concentrated (10YR 2/1) and contained undecomposed leaves at around 220 cm. Above this, the sediment becomes lighter (2.5Y 6/2) and indicates oxidation. At the base (320–270 cm) of C27 the same saprolite layer appeared (2.5YR 6/8). They are overlain by sandy to silty, fossil organic sediments (10YR 3/2 over 10YR 2/1) until 97 cm. Organic sediment from 270 cm (transition to saprolite; Figure 37A) yielded a ^{14}C-age of 3829 ± 46 yrs BP (δ^{13}C: −28.5 ‰).

Nyabibak

Nyabibak is located in the southeasternmost part of the research area. Beyond Nyabibak, the Ntem crosses a narrow gorge (basement) before it fans out into the interior delta. Before the gorge, core C02 (Figure 38) was taken on the northern river bench and reached 410 cm depth. At the base (410–320 cm), coarse-grained sediments were sampled (1.2–2.6% middle sand, 11–70% fine sand and 13–54% coarse

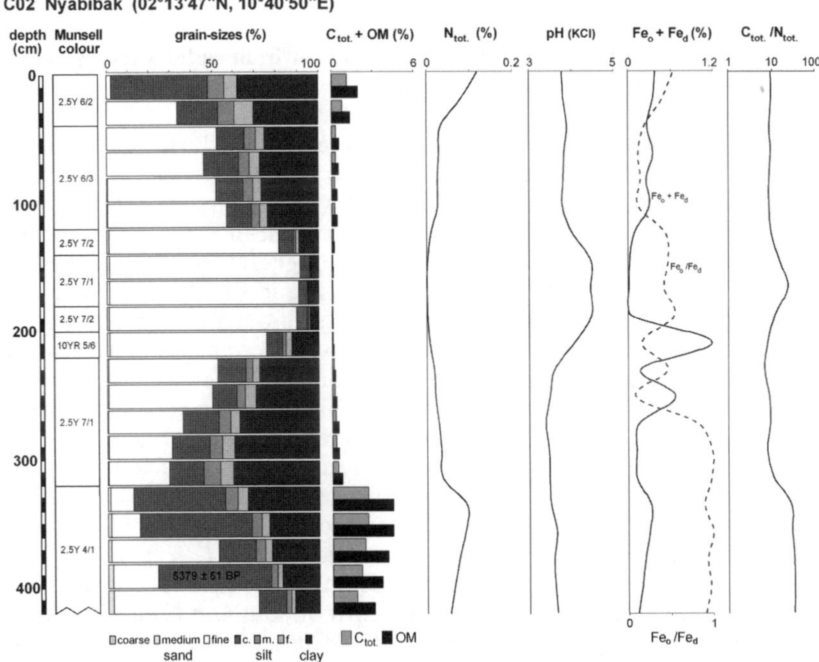

Figure 38. Sediment profile C02 Nyabibak with ^{14}C-age.

silt), which were followed by a fining-upward sequence. In 320–300 cm, fine-grained sediments already reach 55% (40% clay). At the base, again, fossil organic sediments of dark colour (2.5Y 4/1) were found. C_{tot} and OM reach values of 1.77–2.58% and 3.05–4.45% respectively. N_{tot} only reaches 0.055–0.094% (maximum of 0.120% in 20–0 cm). C/N was increased (27–33) and pH low (3.49–3.65). Organic macro-remains (wood) from 400–380 cm depth were dated to a ^{14}C-age of 5379 ± 51 yrs BP (δ^{13}C: −29.0 ‰).

Upper Ntem catchment near Ngoazik

Allthough much of the upper and middle Ntem catchment could not be visited and investigated because this area is inacessible, some corings (S01–04) could be made in Messi Messi (2°18'N, 11°29'E) and Ngoazik (2°18'N, 11°18'E) near Ambam and the border of Gabon. However, mostly sandy deposits were uncovered that were not useful for palaeoenvironmental research.

Lower Ntem near Campo Beach

The lower Ntem was investigated near its mouth in the Gulf of Guinea and the Atlantic Ocean, at the village of Campo Beach (2°20'N, 9°49'E). Between Ma'an and Campo Beach, is located the '*Parc National Campo Ma'an*' (261.443 ha), founded in 2000 (Akogo, 2002). The lower Ntem crosses this nature reserve, where research is not allowed, so corings could only be made near Campo Beach (Dipikar island). The two cores (L50 and 51) obtained at this site were not of sufficient interest to the overall research framework to bear mention in detail.

4.3 SYNTHESIS AND DISCUSSION

4.3.1 Interpretation of the fluvial and alluvial palaeoenvironmental archives in the context of Quaternary landscape evolution

In total, around 160 cores (up to 550 cm depth) were sampled during the dry seasons of 2004–2008 in 16 different research regions of the 6 greatest river catchments of southern Cameroon. From those river catchments three alluvial archives were described in detail. Cores were sampled first and foremost across meandering, anabranching and anastomosing river sections and, more precisely, along terraces, river benches (levees), seasonal and abandoned channels, islands, swamps and floodplains. By this means, various lithostratigraphic formations and form units were analysed with respect to their suitability as palaeoenvironmental sediment archives. Certain very intensively examined research areas (e.g., Akonolinga, Ntem interior delta) offered considerabe data on pedological, sedimentological and stratigraphical facts. Findings were summarized in the form of profiles and transects (using the graphical programme *Macromedia Free-Hand MX*), and these were used as the basis for interpretation of fluvial and alluvial forms. The quantity of findings and evidence allowed the results to be correlated, especially at sites where detailed transects were conducted. In many profiles, comparable sedimentary stratigraphic details and characteristic features (e.g., grain-size distributions) become evident, which also belong to identical time scales (^{14}C-data); from this information, a classification of the alluvial record of southern Cameroon was developed wherein sedimentary units were assigned to distinct Late Quaternary evolutionary stages/phases. Furthermore, a hypothetical reconstruction of processes, run-off and sedimentation conditions in the respective evolutionary phases and river sections is given. Additionally, a synthetic interpretation of landscape evolution will be presented based on combining the study area findings with preexisting published evidence for Central Africa. The interpretations and hypotheses are further supported by numerous ^{14}C (AMS)- and δ^{13}C-data.

Unit 1: Middle Late Pleistocene (Maluékien)—MIS 3 (~53–40 kyrs BP)

This sedimentary unit is present at 4 locations and represents the oldest dated palaeoenvironmentally interpretable record of this study (Table 2). It is characterized by very coarse-grained (sandy) sediments, which indicate increased stream power and discharge, as well as stronger fluvial-morphological activity. At both locations Meyos (L08) and Abong (L14) over 90% sandy fractions (5–30% coarse sand, 20–30% middle sand and 20–60% fine sand) were found in the sediment layers representing this phase. They are associated with high quantities of unsorted macro-remains (wood and leaf fragments), which form decimetres-thick layers and offer additional evidence for turbulent erosion and deposition processes. Two further locations across the Nyong (C28, Akonolinga) and Boumba (B03, Mankako) delivered similar findings, although with lesser proportions of coarse and middle sand.

Next to C28, an identical sediment profile, especially in terms of grain-size, was found at the base of C23. This core ended at 420 cm on basement (metamorphised gneiss) or laterite (maximum iron). Although that coarse-grained sediment layer did not contain any datable organic remains, it has nevertheless been dedicated to this same unit because it shows identical characteristics and pedological indices as the sediment in C28 and has thereby been assigned to the same sedimentation phase.

Unit 1 was always found at quite a distance (> 20 m) from the recent river bed. At these locations, it also marks the direct transition layer to the Precambrian basement (intrusiva in the surroundings of the Congo Craton and metmorphites to the north). In the center of the concentrated organic layers, which are embedded in these units, very high values (maximum in B03, L08) C_{tot} and OM, in addition to high C/N occur. The pH is always acid and pedogenic iron and nitrogen are low. The clear predominance of dithionite solvable iron confirms the advanced age of the sediments. All profiles show an abrupt change in grain-size distribution from a certain depth, which is correlated with a maximum of iron (boundary layers). Unfortunately, datable organic remains were never found in this transition zone, so it can not be exactly classified. Additionally, the occurrence of hiatus can not be excluded. However, these abrupt changes in sedimentation conditions clearly indicate modifications in the fluvial systems, which were triggered by intrinsic and extrinsic factors. Concerning the

Table 2. Dated sediment layers of unit 1.

core location	depth (cm)	OM	^{14}C (AMS) BP	cal. Age (cal BP)	δ^{13}C (‰)	lab.-no.
L08 Meyo Ntem	300–320	wood	45,596 ± 4899	50,851 ± 5663	−31.4	Erl–8252
L14 Abong	340–360	wood	48,230 ± 6411	53,096 ± 6876	−29.6	Erl–8254
C28 Akonolinga	420	organic sediment	42,940 ± 1500	46,680 ± 1817	−19.6	LTL2114A
B03 Mankako	300–320	organic sediment	> 45,000	> 50,000	−29.6	LTL3177A

landscape history, these archives evidence a very high dynamic because even intrinsic (river internal) driven modifications at the respective sites (e.g., channel migration/ wandering) always depend on extrinsic impulses (tectonics, climate). From the palaeo-hydrological view, this unit can be associated with braided to anabranching, aggrading and unsteadily draining fluvial systems, which delivered unstable flushes of sandy deposits and were associated with frequent channel migrations and floodplain reorganizations. The absence of pebble layers at all investigated sites is remarkable. Only downstream of the Ntem interior delta (below the waterfalls of Memvé'élé) were thick pebble layers identified, in what was probably a neo-tectonically induced v-shaped valley (see Eisenberg, 2008 and contribution J. Eisenberg in this volume). These could have been eroded from the interior delta during older (> 48 kyrs BP) flooding events. The core locations L08 and L14 most likely testify to positions of older palaeochannels inside the recent, very dynamic, anastomosing Ntem fluvial system.

Unit 2: Younger Late Pleistocene (Njilien)—MIS 3 (~40–27 kyrs BP)

Unit 2 (Table 3) shows a high correlation with unit 1 (80–90% sand), this said, the records from the Ntem interior delta, at least, can be differentiated by their distinctly finer texture (70–80% fine sand). At the sites Nnémeyong (L25) and Aya'Amang (L37), both located in the interior delta, these units are associated with palaeosurfaces, where maximum C_{tot} (1–2%) and OM (2–4%) as well as increased C/N appeared. However, pH, pedogenic iron and nitrogen are very low. These readibly delimitable palaeosurfaces, which are also strikingly coloured, are overlain by an up to 60 cm thick sand layer, in which pH rapidly increases to maxima around 5. Above these sand layers, distinguishable stratification boundaries emerge by maxima of pedogenic iron, which mark a changeover to abruptly fining textures. The findings from these palaeosurfaces suggest that at these sites, at least temporary, backswamps had formed.

Sediments of cores NY11 and NY13 across the Tetar, a Nyong tributary, are also assigned to this unit, although distinguished by higher proportions of middle (20–70%) and coarse sand (2–12%). Nonetheless, pedological and sediment colour analysis show comparable tendencies across the profiles with distinct fining-upward sequences in the upper profile. Like in unit 1, in unit 2, colour and pedological changes again provide clear evidence for erosional and pedogenic processes. Phases of relative formation rest,

Table 3. Dated sediment layers of unit 2.

core	location	depth (cm)	OM	^{14}C (AMS) BP	cal. age (cal BP)	δ^{13}C (‰)	lab.-no.
L25	Nnémeyong	160–180	wood	21,908 ± 302	26,239 ± 583	−27.0	Erl-8268
		220–240	wood	22,398 ± 316	26,978 ± 631	−29.1	Erl-8269
L37	Aya Amang	320–340	wood	30,675 ± 770	34,962 ± 702	−30.8	Erl-8264
		360–380	wood, leaves	31,253 ± 240	35,220 ± 377	−27.1	LTL3183A
C23	Akonolinga	150–160	organic sediment	28,358 ± 300	32,788 ± 411	−20.1	LTL2113A
NY11	Akonolinga	280	organic sediment	31,904 ± 300	35,961 ± 482	−33.5	LTL3179A
NY13	Akonolinga	180	organic sediment	18,858 ± 100	22,702 ± 248	−18.1	LTL3180A

having clay and iron transfer and soil formation (decomposition of C_{tot} and OM), are clearly distinguishable from phases of intense run-off and accumulation. For the findings of the Tetar sediment archive, a distinct and abrupt shift from sandy to clayey (80–90%) deposits has been dated to after 18 kyrs BP (in NY11, 180–160 cm). In the immediate vicinity of this location, in profile C23, another increase of sandy and silty facies was documented beyond 300 cm depth. Therefore, the base of C23 seems to correlate with the formation of unit 2. The otherwise high clay content (70–80%) decreases to 35% in 200–180 cm. In about 160–150 cm depth, a ^{14}C-age of 28 kyrs BP was found in this sediment layer, which continues until 120 cm. Similar to unit 1, unit 2 was found at distal to the recent river channel located sites, and in the case of L25 and L37, in older, upper terraces.

Unit 3: Outgoing Late Pleistocene, Last Glacial Maximum (LGM) and Pleistocene Holocene transition (Léopoldvillien)—MIS 2 (~27–12 kyrs BP)

The time period represented by this sediment unit (Table 4) forms a direct transition to unit 2 at the site Akonolinga (Tetar sediment archive). This transition is associated with an abrupt change of sandy-silty to clayey (> 80%) alluvial deposits, which manifest an alteration of fluvial-morphological processes from turbulent flows (bedform, channel deposits) to more seasonal and calm conditions (stillwater conditions; floodplain, overbank deposits). This process of change becomes most visible in grain-size profiles C23, C28, NY11 and NY13.

Across a braided river section of the Sanaga (upper catchment near Bélabo), this sedimentary modification was also identified in the profile of an abandoned channel (core C13, 2007). Here, successively increasing clayey facies (> 70%) over a coarse-grained sandy base (400–360 cm) indicate a fluvial-morphological alteration after 16 kyrs BP (400–380 cm), which gradually led to the silting-up of the abandoned channel.

Two further significant findings were derived from the Ntem interior delta (Meyos site, C13 and L17). In the profile of C13, between 263 to 120 cm depth, silty-clayey sediments of a palaeosurface were cored, which were dated to ^{14}C-ages of between 18 to 14 kyrs BP. Another 100 m to the south (Ntem 3 channel) in core L17, a further palaeosurface was found between 400–280 cm. The presence of the palaeosurface was further corroborated by additional evidence from other corings at this location (transect Meyos, Figure 26B). Between 400 to 300 cm, fine-grained sandy, dark organic sediments having abundant macro-remains yielded ^{14}C-ages of 14.5 kyrs BP. In 300–280 cm an abrupt change of grain-size distribution (30–40% silt and 50–60% clay) marks an incisive transformation of fluvial-morphological and hydrological conditions at this site, which was dated to around 14 kyrs BP. And although the findings correlate closely with the Akonolinga results, pedological data clearly indicate palustrine conditions at Meyos. Additionally, the sediments at Akonolinga consist of much more coarser-grained sandy textures, indicating a different, much more turbulent run-off and sedimentation condition, with probably stronger fluvial activity. More comparable palustrine alluvia were uncovered in an abandoned channel near Mengba (C31). Also at this site, too, there is a strong decrease of coarse-grained textures at about 13.3 kyrs BP and an increase of silty (20–30%) and clayey (70–80%) alluvia around 11 kyrs BP.

In contrast to units 1 and 2, these deposits have generally finer textures and were cored in recent or former floodplains (wetlands), which topographically set themselves apart from the rest of the respective alluvial ridge. A detailed interpretation concerning the formation of these archives remains speculative due to limited data and results.

Table 4. Dated sediment layers of unit 3.

core	location	depth (cm)	OM	^{14}C (AMS) BP	cal. age (cal BP)	$\delta^{13}C$ (‰)	lab.-no.
C13, 2006	Meyos	120–140	wood	$14{,}020 \pm 106$	$17{,}271 \pm 229$	−28.1	Erl-9567
		140–160	wood	$17{,}570 \pm 141$	$20{,}981 \pm 336$	−26.5	Erl-9568
		160–180	wood	$18{,}719 \pm 161$	$22{,}389 \pm 377$	−26.5	Erl-9569
		180–200	organic sediment	$17{,}197 \pm 132$	$20{,}647 \pm 322$	−28.8	Erl-9570
		250–263	organic sediment	$18{,}372 \pm 164$	$22{,}011 \pm 341$	−27.6	Erl-9571
C13, 2007	Bélabo	390–400	organic sediment	$16{,}117 \pm 200$	$19{,}294 \pm 325$	−28.2	LTL2107A
NY13	Akonolinga	180	organic sediment	$18{,}858 \pm 100$	$22{,}702 \pm 248$	−18.1	LTL3180A
L17	Meyos	280–300	wood, leaves	$14{,}263 \pm 126$	$17{,}467 \pm 265$	−27.3	Erl-8253
		280–300	wood, leaves	$13{,}690 \pm 100$	$16{,}781 \pm 251$	−24.2	LTL2103A
		380–400	wood, leaves	$14{,}516 \pm 80$	$17{,}658 \pm 248$	−23.1	LTL3181A
C31	Mengba	220	organic sediment	$11{,}107 \pm 90$	$13{,}011 \pm 138$	−28.9	LTL3171A
		260–280	organic sediment	$13{,}357 \pm 60$	$16{,}295 \pm 418$	—	LTL2118A

However, it seems likely that a combination of climatic and river internal fluvial morphological impacts and changes were responsible for these substantial alterations. Perhaps most significantly, among all sites there are ubiquitous, abrupt grain-size distribution shifts, which accumulate between 18 and 11 kyrs BP. For most locations, these are interpreted as channel migrations or wandering or phase-wise increased crevasse and cravasse-splays with corresponding fining-upward sequences. The alluvial archives at Meyos, though, point to the existence of former palustrine backswamps.

Unit 4: Younger Dryas and Early to Middle Holocene (Kibangien A)—MIS 1 (~12–4 kyrs BP)

This sedimentary unit (Table 5) is characterized by sandy textures with proportions of up to 70–90% fine and middle sand, which are covered by more or less thick and, greater to lesser developed, fining-upward sequences. This unit was predominantly found in proximal areas of the recent river bed and, with few exceptions (profiles C02, L22 and L38), only reached around 150–200 cm depth. Many locations are situated in the Ntem interior delta (C02, Nyabibak; L36 and 38, Aya'Amang; C11, Aloum II; L46 and 49, Nkongmeyos; C27 and L22, Nyabessan). Additional sites are located along the lower Nyong (B06, Donenda) and the upper Sanaga (C19, Sakoudi). With the exeption of Aloum II (C11), these deposits, again, appear as palaeosurfaces with thicknesses of 60 to 100 cm. Characteristic features are repeatedly dark colours (2.5Y 3/1, 4/1; 10YR 2/1, 3/1), increased C_{tot} (1–2%) and OM (4–8%), low pH (~3) and high C/N (~20–100). ^{14}C-data of macro-remains (wood, leaves) and organic sediment from the palaeosurfaces yielded ^{14}C-ages between $10{,}871 \pm 99$ and 3894 ± 50 yrs BP and thereby confirm Holocene deposition phases of these archives. The sandy textures provide evidence for turbulent run-off and sedimentation conditions. Furthermore, the fact that at only two locations relative maximum ages of this unit were found supports the hypothesis of increased fluvial-morphological activity during this phase. It includes the Pleistocene-Holocene transition and the African Humid Period (AHP, ~9–6 kyrs BP; cf. DeMenocal et al., 2000), which was the last very humid phase recognized so far. This climatic reversal, which initially induced a significant run-off, sediment transport and budget enhancement, led the equilibrium-striving fluvial systems to react with increased fluvial-morphological activity. Simultaneously with generally forced accumulation processes (aggradation, lateral and vertical accretion), there was a considerable rearrangement and reorganisation of the previous floodplain morphology, accompanied by remobilization of older sediments (cut and fill sequences), construction of new river networks and hiatus formation (e.g., profiles C11 and L46). As a result, across many sites significant amounts of sediment could have been eroded and redeposited. Not until the middle of the AHP, did a notable stabilization of climatic and hydrological conditions set in, which correlates with the maximum expansion of tropical rain forest. In profiles L36 and L38 from Aya'Amang, a constant sedimentation phase could be verified for the period 5–4 kyrs BP. Under a climatic and hydrological regime very similar to modern day, the river systems and their sedimentary structures were able to consolidate. This interpretation is additionally supported by the presence of dense vegetation in the study area (especially in the Ntem interior delta). Floodplain sedimentation during this phase occurred during seasonal flooding, with occasional deposition from crevasse splays or avulsions. The profiles B06 (Donenda, Nyong), C19 (Sakoudi, Sanaga) and L22 (Nyabessan, Ntem interior delta) are interpreted as crevasse splay deposits. They document initial deposition of coarse sands at around 4 kyrs BP followed by fining-upward sequences with a strong increase of fine-grained facies (up to 50% clay).

Table 5. Dated sediment layers of unit 4.

core	location	depth (cm)	OM	^{14}C (AMS) BP	cal. age (BP)	δ^{13}C (‰)	lab.-no.
B06	Donenda	180–200	wood	4093 ± 47	4656 ± 117	−29.7	Erl-8945
C02	Nyabibak	380–400	wood	5379 ± 51	6162 ± 95	−29.0	Erl-9573
C06	Dizangué	360–365	organic sediment	9321 ± 90	10,515 ± 132	−27.2	LTL2106A
Probeb.	Aloum II	78–88	organic sediment	8402 ± 67	9408 ± 75	−30.6	Erl-9574
C11	Aloum II	120–140	wood	6979 ± 58	7822 ± 76	−30.1	Erl-9575
C19	Sakoudi	213	organic sediment	3894 ± 50	4327 ± 72	−29.2	LTL2111A
C27	Nyabessan	270–280	organic sediment	3829 ± 46	4257 ± 87	−28.5	Erl-9577
L22	Nyabessan	140–160	wood	3894 ± 57	4324 ± 79	−28.1	Erl-8273
L36	Aya Amang	140–160	leaves	4341 ± 60	4945 ± 75	−27.2	Erl-8263
		200–220	wood, leaves	4514 ± 45	5176 ± 94	−27.3	LTL3182A
L38	Aya Amang	220–240	wood	5306 ± 64	6096 ± 85	−31.4	Erl-8265
		280–300	wood, leaves	5282 ± 45	6079 ± 79	−26.8	LTL3184A
L46	Nkongmeyos	80–100	wood	8291 ± 89	9277 ± 127	−27.5	Erl-8599
		120–140	wood	10,871 ± 99	12,846 ± 101	−28.3	Erl-8258
L49	Nkongmeyos	180–200	organic sediment	10,775 ± 144	12,744 ± 155	−30.1	Erl-8257

*Unit 5: Late Holocene and First Millennium BC Crisis (Kibangien B)—MIS 1
(~4–2.5 kyrs BP)*

At many sites, palaeosurfaces occurring among unit 5 (Table 6) sedimentary layers
are covered by deposits, which manifest at first reduction (grey colours) and later oxidation (red colours and iron concretions). Additionally, most profiles also indicate
removal and soil formation processes.

The sediment units (profiles B02, B06, C02, C11, C13 [Meyos], C19, L36, L38)
of this phase show a clear increase of sandy facies (80–95%, with seldom 20–50% middle and barely coarse sand), which could indicate either repeated turbulent run-off
and sedimentation conditions or provision of coarser transport material as a result
of a more recent aridification of the study area. Besides, modified climatic (seasonal)
conditions, it is also possible that the landscape was destabilised through enhanced
anthropogenic activity.

Natural as well as anthropogenic induced transformations for the period around 4
kyrs BP have been substantiated for the study area. Archaeological studies have found
that there was an increase of anthropogenic activities, and somewhere around 2600 yrs
BP, there were the first signs of metallurgy in the tropical rain forest (e.g., Kadomura
and Hori, 1990; Schwartz, 1992; Schwartz and Lanfranchi, 1993; Oslisly, 2006; Meister,
2008). Moreover, various terrestrial and marine archives (e.g., Vincens et al., 1999; Gasse,
2000; Maley, 2002; Marchant and Hooghiemstra, 2004; Nguetsop et al., 2004; Weldeab
et al., 2005, 2007, 2007a; Gasse et al., 2008) provide evidence of climatic and palaeoenvironmental changes. The widespread occurrence of iron concentration layers in similar depths of this sedimentary unit (60–100 cm) strongly hints that there were distinctly
drier conditions (prolonged dry-out of floodplains in consequence of reduced rainfalls
and flooding; modified seasonality with longer dry season).

Unit 6: Late Holocene to (sub)recent, river bed near sediments (~2,5–0 kyrs BP)

The sediments of this youngest sedimentary unit (Table 7) are deposited directly across
recent river channels. They typically show coarse- to fine-grained sandy facies (5–40%
coarse, 5–70% middle and 10–90% fine sand) in different combinations. These sediments
were found across the Ntem interior delta (L02–05, Meyo Ntem; L24, Nnémeyong;
L27–28, Akom; L30, Tom; L32–35 and 44, Nkongmeyos; L39, 40 and 42, Anguiridjang), the Nyong (B01–03, Njock and Makak; B05, Lipombe II; B07, Dehane; C21,

Table 6. Dated sediment layers of unit 5.

core	location	depth (cm)	OM	^{14}C (AMS) BP	cal. age (cal BP)	δ^{13}C (‰)	lab.-no.
B02	Ouesso	360–376	organic sediment	2728 ± 60	2846 ± 59	−35.5	LTL3176A
B06	Donenda	180–200	wood	4093 ± 47	4656 ± 117	−29.7	Erl-8945
C13	Meyos	60–80	organic sediment	4081 ± 56	4639 ± 129	−29.9	Erl-10366
C19	Sakoudi	213	organic sediment	3894 ± 50	4327 ± 72	−29.2	LTL2111A
C27	Nyabessan	270–280	organic sediment	3829 ± 46	4257 ± 87	−28.5	Erl-9577
L22	Nyabessan	140–160	wood	3894 ± 57	4324 ± 79	−28.1	Erl-8273

Abong Mbang; exposure site and C26, Akonolinga; exposure site, C30 and NY05, Ayos), and also the Sanaga (C01, Dizangué; C07 and C09, Monatélé; C14–15, Bélabo; C16, Ngok Etélé; C17 Mbargué; C20, Mbaki II). In the interior delta, this unit principally occurs along Ntem 1, where the river bed is mainly shaped into metamorphic amphibolite and noritic gneiss. Here, macro-remains yielded maximum ages of 1.4 kyrs BP. An exception is profile L40, where this sandy unit covered a palaeosurface that yielded ^{14}C-ages between 2.4 to 2.2 kyrs BP. Very often (e.g., cores L02–05, L24, L28, L35, L42, L44) profiles are also associated with fining-upward sequences.

Among sites across the Sanaga, these fining-upward sequences are strongly developed, providing evidence for a relative stabilization of the fluvial-morphological conditions in the younger history of the fluvial system. Above coarsening-upward sequences (C01, C08–09, C16–17, C19), fining-upward sequences have developed, which testify to a successively growing and partially stagnating supply of fine-grained overbank sediments. This was corroborated at the Bélabo site (C14 and 15), where, over an organic and coarse-grained sandy base (C14: 1.7 kyrs BP in 335–340 cm; C15: 1.2 kyrs BP in 380–385 cm), clayey-silty sediments, over 3 m thick, were deposited.

Weak fining-upward sequences were also identified along the middle Nyong catchment (Eséka region, profiles B01 and 02). The sandy bases (200–225 cm in B01) indicate maximum ages of ca. 2.1 kyrs BP. In contrast, a profile at Lipombe II (B05) shows a very mixed sedimentary profile. It documents a fining-upward sequence until around 100 cm depth (wood: 707 ± 38 yrs BP) and afterwards an abrupt coarsening of the facies. A 340 cm deep profile from the lower Nyong (B07 Dehane), by contrast, first documents fining from 280 cm (organic sediment in 300–280 cm: 400 ± 40 yrs. BP), then coarsening until 180 cm, and then, again, fining below 80 cm followed by repeated coarsening. Comparable fluctuating conditions were recorded across the upper Nyong catchment (exposure sites at Akonolinga, 0.9 kyr BP, and Ayos, 1.5 kyrs BP). In the environs of the Ayos exposure site, abandoned and silted up channels show strongly developed, sandy coarsening-upward sequences covering dark (10YR 2/1) organic (27–41% OM) sediments having abundant macro-remains dated to ^{14}C-ages of about 700 yrs BP. At Mokounounou, along the Ngoko River (cf. Sangen, 2011; Sangen et al., 2010), the development of an oxbow (abandoned meander loop) was dated to the same time period. While the sediments in the river bed of the abandoned meander loop indicate coarse bed load and increased fluvial-morphological activity around 1.2 kyrs BP, increasing silty-clayey sediments manifest progressive abandonment around 0.9 kyr BP.

However, the near river bed sites across the Nyong represent the only findings indicating recent coarsening sedimentation processes/profiles apart from recently observed aggradation processes and sand bank generation (Dja, Ngoko, Nyong and Sanaga rivers). And because these phenomena are most pronounced across the upper Nyong catchment, with its wide and unforested floodplains, the fluvial systems in these areas are postulated to be highly sensitive to run-off and sedimentation changes caused by anthropogenic activities and natural external triggers. This hypothesis is substantiated by profiles indicating multiple grain-size changes (alternating coarsening- and fining-upward) as well as the occurrence of rapid and frequent channel migrations and abandonments (young ages in abandoned channels) in the floodplains of Akonolinga and Ayos.

Of special interest in this context is the sediment archive of the southern Nyabessan transect, which can be placed in this unit because of its age. Here, palustrine sediments with high clay content have been deposited under stillwater conditions. The high pollen content of these sediments make them very suitable for palynological studies and the first results from such studies have been published by Höhn et al. (2008)

Table 7. Dated sediment layers of unit 6.

core	location	depth (cm)	OM	^{14}C (AMS) BP	cal. age (cal BP)	δ^{13}C (‰)	lab.-no.
B01	Njock	200–225	organic sediment	2134 ± 41	2161 ± 100	−27.3	Erl-8941
B03	Makak	160–180	wood	−2105 ± 34	recent	−29.8	Erl-8944
		235–245	leaves etc.	−4116 ± 31	recent	−29.4	Erl-8946
B05	Lipombe II	100–120	wood	707 ± 38	637 ± 48	−28.2	Erl-8943
B07	Dehane	280–300	wood	400 ± 40	428 ± 72	−28.7	Erl-8942
C01	Dizangué	200–218	organic sediment	128.52 ± 0.90 pMC	recent	−19.6	LTL2104A
C14	Belabo	335–340	organic sediment	1766 ± 50	1693 ± 72	−29.7	LTL2108A
C15	Belabo	381	wood	1196 ± 45	1.130 ± 58	−28.7	LTL2109A
C16	Ngok Etélé	240–260	leaves etc.	14 ± 100	108 ± 125	−24.2	LTL2110A
C20	Nyabessan	430–440	organic sediment	2479 ± 43	2571 ± 106	−28.8	Erl-9576
C30	Ayos	280	leaves etc.	724 ± 45	679 ± 28	−25.8	LTL2115A
exposure	Akonolinga	230	organic sediment	902 ± 45	832 ± 61	−24.2	LTL2117A
exposure	Ayos	180	organic sediment	1508 ± 50	1419 ± 65	−18.0	LTL2116A
NY05	Ebabodo	280	leaves etc.	732 ± 30	684 ± 12	−22.0	LTL3178A
L02	Meyo Ntem	100–120	wood	1066 ± 53	995 ± 51	−27.0	Erl-8249
		160–180	wood	1104 ± 52	1024 ± 53	−29.2	Erl-8250
L05	Meyo Ntem	220–240	leaves etc.	908 ± 50	835 ± 61	−29.4	Erl-8251

L18	Nyabessan	140–160	leaves etc.	587 ± 64	596 ± 46	−29.1	Erl-8270
		200–220	leaves etc.	2337 ± 55	2389 ± 73	−31.9	Erl-8271
L19	Nyabessan	320–340	wood	2189 ± 52	2216 ± 78	−29.5	Erl-8272
L24	Nnémeyong	80–100	wood	441 ± 46	482 ± 39	−30.6	Erl-8266
		120–140	wood	671 ± 52	623 ± 46	−28.7	Erl-8267
L27	Akom	100–120	wood, leaves	443 ± 57	453 ± 72	−26.4	Erl-8261
		160–180	leaves	427 ± 52	440 ± 74	−27.6	Erl-8262
L30	Tom	140–160	wood	1381 ± 49	1309 ± 30	−28.2	Erl-8260
L32	Nkongmeyos	60–80	charcoal	217 ± 46	196 ± 103	−28.0	Erl-8255
C32	Nkongmeyos	145–150	organic sediment	954 ± 39	865 ± 50	−28.9	Erl-9572
L34	Nkongmeyos	100–120	wood	435 ± 51	448 ± 70	−27.4	Erl-8256
L40	Anguiridjang	280–300	organic sediment	2180 ± 50	2209 ± 81	−26.9	LTL3185A
		360–380	wood	2339 ± 52	2393 ± 65	−27.9	Erl-8259
Probeb.	Mokounounou	480–485	organic sediment	898 ± 50	829 ± 64	−30.0	LTL3170A
N01	Mokounounou	550	organic sediment	938 ± 45	856 ± 52	−28.0	LTL3169A
N02	Mokounounou	393	wood	1041 ± 45	973 ± 41	−28.4	LTL3172A
N04	Mokounounou	470	wood (*Raphia*)	892 ± 40	827 ± 62	−26.0	LTL3173A
N05	Mokounounou	465	organic sediment	1228 ± 35	1167 ± 64	−28.1	LTL3174A
N06	Mokounounou	545–550	organic sediment	1126 ± 45	1046 ± 59	−29.4	LTL3175A

and Ngomanda et al. (2009). Although not done, it would have been possible to further classify this unit into additional sub-units. One sub-unit would have been the very coarse-grained sedimentary facies, which can be associated to repeated humidification (Medieval Climate Optimum [MCO]). This sub-unit would include all samples dated between 1 and 1.7 kyrs BP.

Another potential sub-unit would have related to the noticeably reduced fluvial-morphological processes, which correlate with overlaying sedimentary units. These mainly correspond with the abandoned channels of the Nyong floodplain and the Ngoko abandoned meander loop. These sedimentary units were generally dated to 0.7–0.9 kyr BP.

A last possible sub-unit, having similar characteristics, would consist of sediments dated to ages of 0.4–0.6 kyr BP. These sediments, again, clearly indicate reduced run-off and sedimentation conditions, and significantly correlate with the northern hemispheric climate pessimum of the Little Ice Age (LIA).

Taken together, the cored alluvial sediments show some significant similarities. The first common characteristic is the occurrence of palaeosurfaces at the majority of investigated sites (see Figures 6, 17, 18, 20, 23, 26, 28, 29, 31, 32 and 35–38). Although most of them originate from different time phases, they are all characterized by extreme acid pH (2–4) and high proportions of organic material (maxima C_{tot} of 0.5–4% and OM between 1 to 8%), with highest contents at the locations of Meyos (L17, 9.9 to 17%) and Nyabessan (L19, 9.8 and 16.9%). Additionally, high N_{tot} (0.05–0.2%) and high C/N (20–100) are typical. Furthermore, high values of amorphous iron (Fe_o) occur in these layers, which, in the light of the high ages of the sediments, is very surprising. Typically, as a consequence of increasing maturity and pedogenic removal, the cristalline value (Fe_d) should strongly increase (Torrent et al., 1980; McFadden and Hendricks, 1985). Fitze (1973) postulates that one reason for delayed conversion of Fe_o to Fe_d might be the presence of backwater or/and organic connections. Considering that the palaeosurfaces were mostly found below the groundwater level and contained very high values of organic material (and macro-remains), this could explain the unusual proportions of pedogenic iron ($Fe_o > Fe_d$). The good conservation and low degree of decomposition is likely the result of being covered for a long time by groundwater under anaerobic conditions. One main cause for this conservation is the great likelihood that the groundwater level rose during the transition to the Holocene. At several profiles, where unusually high Fe_o values were detected, there was also a generally low total amount of pedogenic iron ($Fe_o + Fe_d$, mostly near the limit of proof 0.01%). This is why the results for the activity grade (Fe_o/Fe_d) are somewhat strange. The fossil, sandy and very organic layers, which partially emerge as residual palaeo-channels (L08, L14), former backswamps (L17, L37, L46) or other humid stillwater locations or even lacustrine structures (e.g., Nyabessan), could have been buried during rapid, far-reaching climatic and environmental changes and, thereby, exposed to totally modified conditions and processes. It must have taken some hundreds of years to develop these palaeoenvironmental archives. And although for the most sites in the interior delta, datable organic remains are missing in the transition layers to the palaeosurfaces and additional hiatus can not definitely be excluded, all palaeosurfaces are on account of their thickness and constitution (colour, soil chemical and physical parametres) very easy verifiable. At some sites (L17–18, L25, L36–37, L40, L46) the upper and lower transitional layers were dated in an effort to provide information on their period of formation:

L17:   14,516–14,263 yrs BP (400–280 cm) = ~253 yrs
L18:   2337–587 yrs BP (220–140 cm) = ~1750 yrs

L25:   22,398–21,908 yrs BP (240–160 cm) = ~490 yrs
L36:   4514–4341 yrs BP (220–140 cm) = ~173 yrs
L37:   31,253–30,675 yrs BP (380–320 cm) = ~578 yrs
L40:   2339–2180 yrs BP (380–280 cm) = ~159 yrs
L46:   10,871–8291 yrs BP (140–80 cm) = ~2580 yrs

If one excludes the sites having long formation periods, L18 and L46, average periods of formation between 200 and 600 yrs can be assumed.

A further characteristic feature is the occurrence of iron concentration layers in the mayority of studied profiles (see Figures 6, 7, 15–17, 19, 23, 26, 28, 29, 32, 33, 35 and 38). They normally appear between around 120 to 60 cm depth and reach average thicknesses of 5–20 cm. Absolute maxima of iron (Fe) indicate stratification boundary layers. At the Nkongmeyos Ntem 3 site (L44-L47), such an iron concentration layer occurs directly above a palaeosurface. In the noticeably older profiles of units 1 and 2 (e.g., L08, L14, L25) these iron concentration layers appear remarkably thick between 200 to 60 cm depth and, in some profiles (C28, L37), even two layers were found. Above and/or below the iron concentration layers, oxidation horizons with iron concretions have developed, which are followed by reduction horizons in progressing depth. The latter finally covers the palaeosurface in the majority of profiles. The profiles document removal and soil formation processes (i.e., with growing depth there are decreasing soil chemical parameters) in the upper part (1–2 m). Besides the aforementioned possibility that the iron concentration layers may have formed during temporary extremly dry periods (LGM, 'First Millennium BC Crisis'), it is also possible that iron transfer takes place during seasonal groundwater fluctuations. This could thereby have led to the formation of the upper (120–60 cm depth) iron concentration layer. For the formation of the deeper (240–220 cm depth) concentration layers, there are two likely explanations. On the one hand, the origin of genesis can be placed into the LGM, particularly in the light of advanced age of the respective sediments (C28, L08, L14, L37), a period for which a very distinct reduction in humidity has already been postulated for the study area. On the other hand, the layers could have formed in the same way as the upper layers by groundwater fluctuations and appropriate iron migrations during a time period when the groundwater level was much lower (LGM). They would then be residuals of very much older morphological processes.

Albeit that the studied deposits are alluvial sediments in open systems, which are still subject to ongoing erosion and sedimentation processes, in some way possible soil formation, systematics and dissemination will here be discussed using diagnostic horizons. For example, podzolation processes, as demonstrated in the Congo catchment area for Pleistocene ages (Schwartz et al., 1986; Giresse et al., 1991; Preuss, 1986, 1990), can only be assumed for certain of the older profiles (L08, L14, L37). This speculation is based on the presence of middle to fine-grained sandy sediment layers and high amounts of organic material in these profiles. Further supporting factors are offered by the assumed hypothermal climate phases during the Pleistocene and the quartz-rich gneiss and granite basement of the catchment areas. Even though above the hypothetical illuvial horizons also possibly grey eluvial horizons were discovered, humus layers were missing. And, even if one assumes they represent residual podzols, which could have been eroded by adjacent fluvial-morphological activity, B_{sh}-horizons typical of podzols and found by Preuss (1986a) in the Congo basin, could not be identified. The limited findings from existing profiles and exposure sites respecting podzols formation do not allow any further interpretations.

In contrast, much more obvious and omnipresent is the process of gleyization and the presence of similigley. In the majority of cores, oxidation (G_o)- and reduction horizons (G$_r$) are clearly distinguishable. Depending on age, location and development of these typical horizons, it is possible to differentiate between wet gley (L02, L04, L31, L32, L35, L44, L50), turfy moulder gley (upper halves of NY11 and NY13) and floodplain gley (B06–07, C16, L05, L26). Floodplain soils (vega), the youngest soils developing on alluvial sediments, can easily be distinguished by regularly occurring meadow (iron) ore (G$_{mo}$-horizon) of different thicknesses (10–20 cm) between depths of 120–60 cm in the upper parts of the profiles (C11, C27, C31, L24, L30, L36, L38, L40, L45, L49). Consequently, the lower parts of the older profiles can also be interpreted as relictic gley. The initial profile was modified in the course of a groundwater rise and a Holocene superimposition. This is quite similar to what is seen for the locations where floodplain marshes and black swamp developed (C13–14, L17, L37, L46). These processes led to the generation of sub-hydric soil types like gyttja and mudd.

4.3.2 Synthetic Late Quaternary palaeoenvironmental evolution of southern Cameroon in the context of terrestrial and marine Central African sediment archives

Putting the findings of this study into a much broader palaeoenvironmental context, further terrestrial, semi-terrestrial and marine sediment archives of the wider study region will be hereforth integrated into an overall synthesis. For the reconstruction of southern Cameroonian palaeoenvironments, findings from Central African lakes and deep-sea fans of the major rivers Niger, Congo, Sanaga and Nyong (marine cores MD03–2707 and GeoB 4905–04 on the Cameroon shelf) are of the greatest relevance. Apart from these sedimentary archives for western equatorial Africa, WAM dynamics and hemispheric and global process cycles and coherences must also be considered and related to past climate changes in Africa. With respect to recent and past environmental conditions, it is clear that the interaction and variation of atmospheric, oceanic, cryospheric and terrestrial dynamics play a vital role. Planetary radiations and fluxes are influenced by cyclic varying orbital parameters. Milanković-scale orbital fluctuations are of fundamental importance when examining palaeoenvironmental and palaeohydrological evolution of the study area because insolation changes essentially control the intensity of the WAM. Moreover, the position and extension of the ITCZ, to which the WAM is strongly linked, are additionally modified by complex global teleconnections, feedbacks and reorganizations (Figure 39).

For the period of the last glacial (Würm, ~120–11 kyrs BP) and interglacial (Eem, ~130–120 kyrs BP) valuable evidence for past environmental conditions comes from cores of pelagic and hemi-pelagic sediments of the Gulf of Guinea, and ice cores from both hemispheres; indeed, Antarctic ice cores even span several glacial cycles (Vostok ~420 kyrs BP, Petit et al., 1999; EPICA Dome C ~800 kyrs BP, Jouzel et al., 2007). They confirm that during the period when glacials as well as interglacials appeared, palaeoclimatic evolution was strongly influenced by orbital parameters. During the respective orbital cycles, complex feedbacks between atmospheric, oceanic and continental systems emerged. Additionally, the last glacial era was accompanied by sub-Milanković climate fluctuations (Dansgaard/Oeschger [D/O]-cycles, Heinrich events [HE]), which led to repeated short-term (10^1–10^3 yrs) global climatic fluctuations with corresponding reactions on each hemisphere. As a consequence of these global oscillations, during which changes in concentration of greenhouse gases (especially CO_2 and CH_4), extension of ice masses, albedo, oceanic/thermohaline and atmospheric circulation and ecosystems occurred (e.g., Vidal et al., 1999; Blunier and Brook, 2001; EPICA community members, 2006; Loulergue et al., 2008), the climatic, hydrologic and

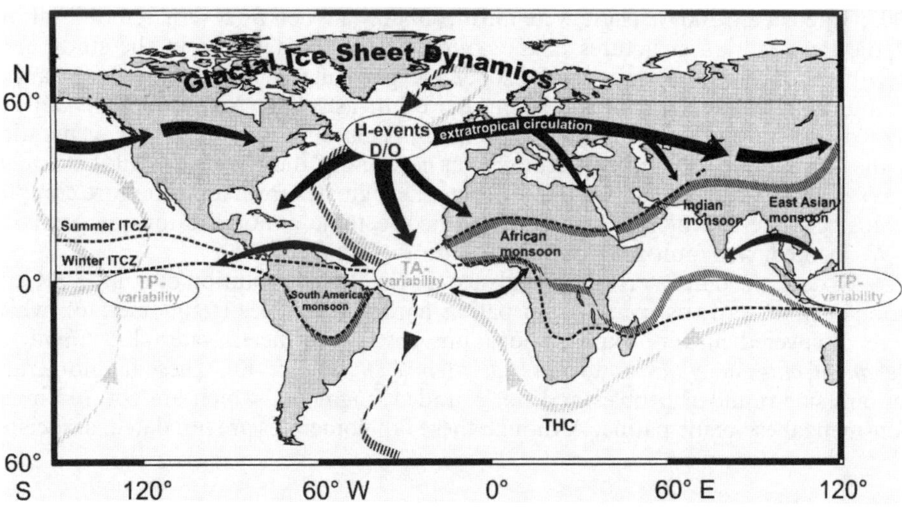

Figure 39. Overview of global teleconnections and climate influencing variables. Of special interest is the *TA-variability* (variability over the tropical Atlantic Ocean; Vidal and Arz 2004, p. 37).

ecological systems of Central Africa, too, experienced far-reaching modifications (e.g., Gasse, 2000; Maley, 2001; Barker et al., 2004; Elenga et al., 2004; Gasse et al., 2008). These proceeded in tight association with changes in the global oceanic (thermohaline) circulation (bipolar see-saw) as well as atmospheric flow and circulation conditions, which through teleconnections and feedbacks also modified the position of the meteorological equator and the intensity of the (African) monsoon (Figure 39).

It is still not entirely clear which phenomenon acted as a trigger for the abrupt sub-Milanković climate changes, although changes in thermohaline circulation and tropical ocean-atmosphere-dynamics seem to be the most reasonable causes (Blunier and Brook, 2001; Broecker, 2003; Vidal and Arz, 2004). Some much more small-scaled (decades to centuries) climate changes occurred during the Holocene (Bond et al., 2001; Mayewski et al., 2004; DeMenocal et al., 2000, 2001; Verschuren, 2004). In the Late Holocene, the influence of the Atlantic Niño on climate change and position as well as intensity divergences of the WAM strongly increased (e.g., Servain, 1991; Fontaine and Janicot, 1996; Paeth and Hense, 2006; Chang et al., 2006; Zhao et al., 2007; Paeth et al., 2008). Tropical regions exert a significant effect on climatic trends as they function as a major source and sink of atmospheric greenhouse gases (CO_2, CH_4, H_2O). The ^{14}C-ages of the southern Cameroonian alluvial sediment archives were calibrated using calpal-online (www.calpal-online.de, calibration curve CalPal2007_Hulu after Hughen et al., 2006).

The drainage system of the African continent has been affected by a very complex and dynamic evolution, especially since the break-up of Gondwana (Summerfield, 1985; Kuete, 1990; Goudie, 2005). Since the last orogenesis ('*Pan-Africain*', ~1000–600 Ma), the opening of the South Atlantic (~95 Ma) and associated evolutionary beginning of the Central African rift, primarily E-W and NE-SW oriented geological structures were generated (cf. Bessoles and Trompette, 1980; Ngako, 2006). During the Tertiary, further impulses followed, which were attended by plutonism and volcanism of the CVL (~70 Ma) as well as the initial peneplation processes between the coastal surface and the adjacent hinterland, i.e., the interior surface (Ngako et al.,

2003; Toteu et al., 2004). These were most probably aligned by several remobilisations of the Precambrian structures in the course of the Quaternary. In the study area, (neo-) tectonical and seismic active zones are particularly found across the Sanaga fault and the contact belt of the Nyong-series' thrust plate (*Nappe de Yaoundé*) and the Congo Craton (Kankeu and Runge, 2008). This is responsible for the generation of the morphologically very structured river network of the Nyong (middle catchment area) and the Sanaga. The Ntem interior delta originates from a much more complex geological genesis, which is manifested in neo-tectonic remobilizations of primarily E-W oriented, step fault-like Archaic structures (Eisenberg, 2008).

To date, the oldest terrestrial evidence for the fluvial evolution of southern Cameroonian river systems comes from patina hardened pebbles (fanglomerate), which were discovered in very exposed locations (bellow terraced waterfalls: *Chutes de Dehane, Chutes de Njock* and *Chutes de Memvé'élé*; Figure 40). These fanglomerates contain side-rounded pebbles and not rounded fragments, which are covered by an iron-manganese-oxide patina. Although these fanglomerates are not dated, associated

Figure 40. Basement and hardened, partially patina-coated pebbles bellow the waterfalls of Dehane (1), Njock (2, both lower Nyong catchment) and Memvé'élé (3), bordering the Ntem interior delta (Photos: M. Sangen, 2005).

genetic processes indicate extensive increased transport (large pebbles) and erosion (large, not rounded fragments) forces on the one hand, and significant weathering (sediment supply) and aridification (patina formation and hardening), on the other.

On the basis of relevant palaeoclimatic evidence that spans the greatest time scales (i.e., ice, deep-sea and shelf cores), the beginning of accumulation processes can be placed into the Eemian interglacial (~130 kyrs BP, OIS 5e). For this and several following phases (until around 70 kyrs BP, although of distinctly smaller amplitude) immense increases of discharges in the Niger (Dupont and Weinelt, 1996; Zabel et al., 2001), Congo (Schneider et al., 1994, 1997; Gingele et al., 1998) and Sanaga as well as Nyong (Adegbie et al., 2003; Weldeab et al., 2007, 2008) have been substantiated. Figure 41 shows that discharge peaks coincidence with warming in the Gulf of Guinea (SSTs).

They also correlate with the Eemian interglacial and the comparatively prolonged D/O-cycles 19–21, 23–24 (OIS 5 a and c) as well as insolation maxima at 15° N and increased temperatures over Antarctica. There seems to be a interrelationship between discharge intensity and duration of the D/O-cycles (slow cooling during interstadials 21 [OIS 5a] and 23 [OIS 5c]). The discharges generally tend to diminish towards the more arid (hyperthermal) *Maluékien*, which is accompanied by a strong impulse around 109–106 kyrs BP, and longer, but less accentuated ones around 104–94 (interstadial 23) and 86–70 kyrs BP (interstadials 21–19).

During OIS 5b and d, there are clearly reduced impulses, which in turn correlate with stadials and insolation minima (around 114 and 93 kyrs BP). The evidence that the Eemian interglacial witnessed the greatest warming event over Antarctica (cf. Jouzel et al., 2007), probably related to the highest humidity and discharge peaks in the Gulf of Guinea during the last eight glacial cycles (~800 kyrs), corroborates the hypothetical chronological classification of the pebble displacements. Patina coating and hardening could have formed during adjacent Late Pleistocene stadials and interstadials of the Würm glacial, when discharge dynamics were strongly reduced and arid and humid conditions often altered (cf. Dansgaard et al., 1993; Taylor et al., 1993; Vidal et al., 1999; Blunier and Brook, 2001). Further evidence is provided by studies on weathering intensities in the catchment areas of the Niger (Zabel et al., 2001) and Congo (Schneider et al., 1997), which reveal high sensitivity to humid phases and insolation maxima. Recently, and probably since the transition to the Holocene, the pebbles and fragments have been reworked and carved out of the matrix (see Figure 40/3 above).

However, the pebbles and fanglomerates were not discovered during coring in the alluvial ridges, therefore, palaeoenvironmental and palaeohydrological interpretations based on findings from this study can only start in OIS 3 and the *Maluékien*, at around 53 cal. kyrs BP (oldest sediment dated to ^{14}C-age of 48 kyrs BP, cf. Late Pleistocene unit 1). Although the alluvial sedimentary records are considerably younger then those found in, for example, South America or Australia, they show striking analogies with respect to Late Quaternary climatic and hydrological changes (OIS 1–3; cf. e.g., Latrubesse, 2003; Nanson et al., 2008). While OIS 5e, c, a and 1 mark basically stable, warm and humid conditions with maximum tropical rain forest expansion, OIS 6 and 2 were dry and cool climatic phases with spreading of savannas (*Gramineae*) in most parts of the Central African tropics. The periods 5b and d are characterised by an immense retreat of the tropical rain forest and an expansion of afromontane rain forest (*Podocarpus*) in lower (≥ 500 m) areas. The river systems eroded most of by aridifications disequilibrated landscape settings during the humid phases of OIS 5 (especially 5a and c) until the onset of the *Maluékien*. Favoured by the partial retreat of the rain forest and increased savannization, the Eemian clayey overbank floodplain deposits were probably eroded and replaced by sandy lower terraces. OIS 3 and 4 describe transition periods having much less tropical rain forest fragmentation.

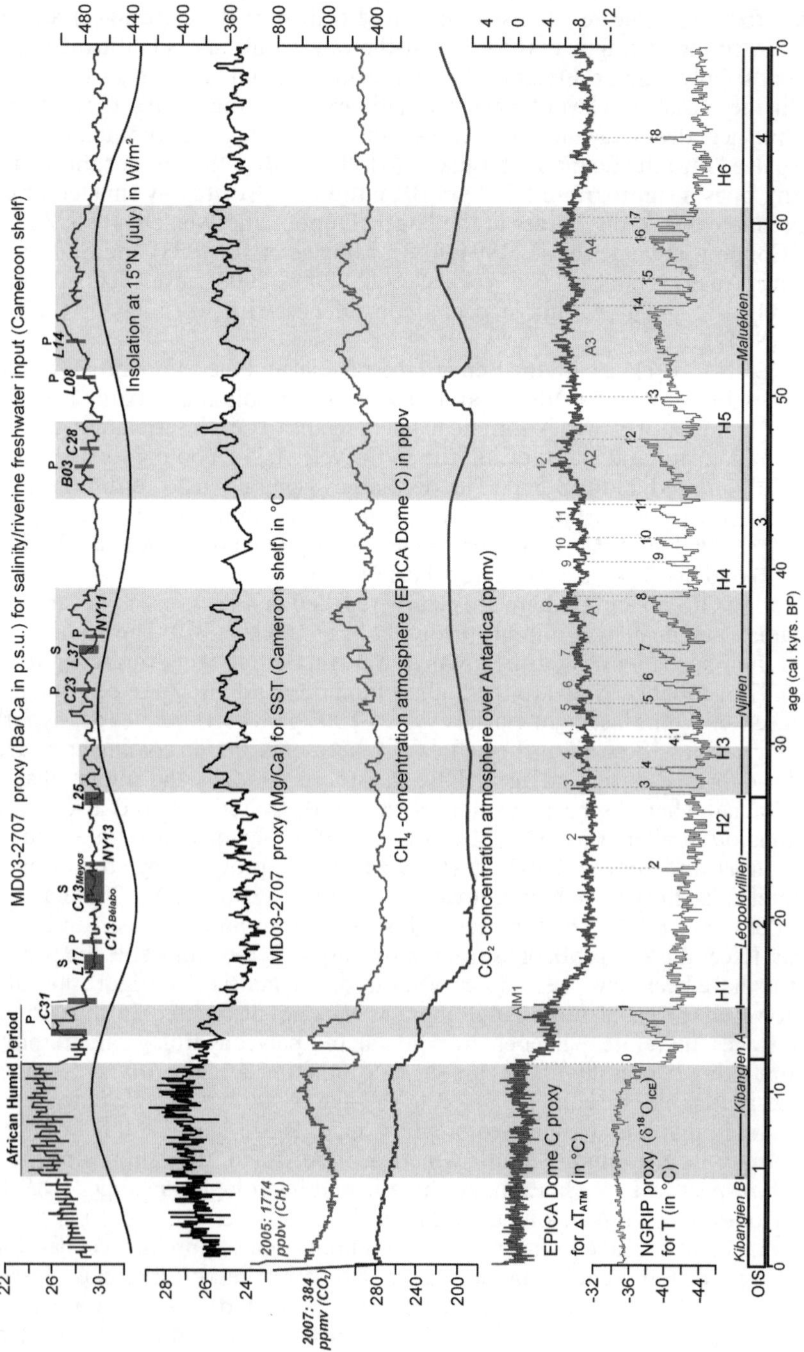

Figure 41. Correlation of salinity/riverine freshwater input and SSTs fluctuations across the Cameroon coast (Weldeab et al., 2007) with changes in greenhouse gas concentrations (CO₂ cf. Lüthi et al., 2008; CH₄ cf. Loulergue et al., 2008), temperatures over Antarctica (EPICA Dome C, Jouzel et al., 2007) and Greenland (NGRIP members, 2004), interstadials (no. 1–25), Heinrich events (H1–6; Bond et al., 1999), oxygen isotopic stages (OIS 1–5e) and insolation at 15°N (Berger and Loutre, 1991) for the last 130 cal. kyrs BP.

Although *Podocarpus* forest is again markedly reduced during these stages, *Poaceae* and *Cyperaceae* strongly increased, providing strong evidence for open savanna landscapes (Jahns, 1996; Dupont et al., 2000; Maley, 2001; Barker et al., 2004).

During several phases of OIS 4 (*Maluékien*) there was considerable rain forest fragmentation locally and landscape destabilization under hypothermal climate conditions and a displuvial rainfall regime; for these reasons, it is assumed that there were generally stronger erosion processes and fluvial systems having occasionally high erosive and fluvial-morphological potential. But overall, there must have been distinctly reduced and periodical discharges. In combination with the *Pré-Inchirien* regression (100 m under recent level), increased superficial flushing and incision processes occurred during short but intense rainfall and discharge events.

On the basis of terrestrial and marine evidence from lower latitudes, it is possible to speculate that in such a landscape setting, predominantly braided to anabranching river systems having coarse-grained sandy bed/sediment loads and river beds prevailed in equatorial Africa, even though there exist very few dated sedimentary layers for this period. Pokras (1987) and Gasse et al. (1989) on the basis of fluviatile (maximal insolation and, hence, intensified monsoon and river discharge) and eolian (minimal insolation and, thereby, increased NE trades [harmattan] and deflation) deposited diatoms in marine sediments, postulate the existence of considerably increased arid conditions during OIS 4. This means there must have been weakened monsoonal activity and a simultaneous increase of the NE trade (harmattan), and this has been corroborated by palynological studies of marine sediments from the Gulf of Guinea (especially cores GIK16856 and 16867). Giresse et al. (1991) proved the existence of *Maluékien*-age deposits and stone-lines in the Sangha catchment, south-east of Cameroon. Most significantly, there were major discharge reductions of the Congo, Niger, Sanaga and Nyong, too, which strong correlate with insolation minima and diminished monsoon intensity (see MD03–2707 discharge proxy in Figures 41 and 42). Lowest discharges in the Sanaga and Nyong appeared between ~70–62 cal. kyrs BP (OIS 4), and during the LGM (OIS 2). Congruently, phases of increased upwelling and marine palaeoproduction were evidenced during arid periods OIS 2, 4 and 6 in the Congo deep-sea fan (e.g., Schneider et al., 1994, 1997).

During the transition of OIS 4 to OIS 3, as well as in the course of OIS 3, the proxy data of the NGRIP ice core reveals an accumulation of relatively staggered 'short-term' interstadials (D/O 3–17), which correlate with preceding Antarctic warmings (Antarctic Isotopic Maxima [AIM] 3–12, EPICA community members, 2006), especially the longer D/O-cycles 8, 12, 14 and 16 (Antarctic warmings 1–4, Blunier and Brook, 2001; see Figure 42). Additionally, a high cohesion exists to Heinrich events 3–6 in the northern Atlantic (Vidal et al., 1999), after which the SSTs and discharges into the Gulf of Guinea rise, lagging in time by ca. 1000 yrs, followed, after another delay, by atmospheric CO_2-, and particularly CH_4-concentration of the atmosphere over Antarctica and Greenland. The high SST- and discharge-impulses show a certain dependence on the intensity and duration of the Antarctic warmings and D/O-cycles (A1–4 and D/O 8, 12, 14 and 16). In total, two phases of increased fluvial activity (61–44 and 39–27 cal. kyrs BP) can be distinguished, which occurred under increasing insolation (maxima around 58 and 32 cal. kyrs BP). This led to intensifications of the WAM and, thereby, to relatively increased rainfall, amplified erosion and sedimentation as well as fluvial-morphological activity. Also the *Njilien* (starting around 40 cal. kyrs BP) falls into this more humid and warmer (hyperthermal) phase having considerably moister climate conditions (Würm-interstadial, maximum around 35 cal. kyrs BP). Under what has been described as an isopluvial rainfall regime with broad and periodical showers, the tropical rain forest was able to partially re-expand, and the sea level rose to about 35–47 m under the present level (*Inchirien* transgression). Under semi-humid conditions, the superficial flushing predominated, particularly because rain forest recolonialisation proceeded in starts and stops.

De Ploey (1964, 1965, 1969) and Preuss (1986, 1990) both found evidence for increased fluvial activity in the western Congo region around 40–37 kyrs BP. Among other records for this are those from terrigenous deposits of Central African rivers' deep-sea fans. Even the Nile, whose major tributaries drain equatorial regions, shows increased discharges. For African lower latitudes, the oldest fluvial deposits also date back to this time period. The fact that the renewed discharge impulses were considerably reduced under a weakened monsoon (in contrast to OIS 5 a, c and e) has been attributed to glacial conditions. This fact could be the main reason for the preservation of alluvial sediments dated to this period. The sediment archives of unit 1 provide little but significant evidence for this dynamic phase (Figure 42). Albeit that in the absence of sound palaeohydrological research the following interpretation must remain speculative, the sediments of unit 1 seem to indicate increased fluvial activity in a fairly disequilibrated landscape. Clayey deposits, which should have been deposited during the Holocene and the quite comparable and climatically much more stable Eemian interglacial, were not found at any location and might therefore all have been eroded at this stage; they are only indirectly verifiable across the shelf. The oldest sediments from the Ntem interior delta (locations L08 and L14) can be tentatively classified as belonging to the most humid initial phase of OIS 3.

They represent former palaeochannels in a wide, neo-tectonically strongly formed alluvial ridge of the Ntem interior delta. Probably phase-wise debris flows and accretion of sandy sediments occurred along striking fault and obstruction lines (lineaments) inside the delta and this was combined with the construction of the main drainage pattern and swampy areas. Studies on opal-phytoliths and sponge spicules from the site Meyo Ntem (L08) in the Ntem interior delta proved that the site was very humid and had a partially lacustrine character (Sangen et al., 2011). In addition, at both sites (L08 and L14) very high amounts of macro-remains were found inside the sandy matrix. The $\delta^{13}C$ (−31.4 and −29.6 ‰) data signals a C_3-dominated vegetation. This strongly points to the likelyhood that in this study area, or at least across the riverine locations, tropical rain forest may largely have prevailed (even during arid and cool phases). This hypothesis of 'fluvial refuges' (Figure 43) is in agreement with earlier studies from equatorial Africa (e.g., Hamilton, 1976; Colyn et al., 1991; Maley, 2001).

A single piece of evidence from Mankako across the Boumba (SE Cameroon), likewise testifies to an increased fluvial impulse dated to ca. 50 cal. kyrs BP. However, the fine-grained sandy texture of the sediments, here, argues for less turbulent conditions. Again, the $\delta^{13}C$ (−29.6 ‰) indicates the presence of C_3-dominated vegetation.

A wide, but neo-tectonically less stressed, valley floodplain is hypothesised for the Nyong; close to Akonolinga further evidence exists for increased fluvial activity between 46 and 36 cal. kyrs BP (L23, 28 and NY11, 13). Here, over the basement, a lower terrace composed of coarse sandy deposits was banked up. The next overlaying and younger sandy layer is generally coarser, which substantiates once more increased fluvial activity towards the last insolation maximum. Further indicators from these archives are suggestive of reduced discharges, which must have lasted until around 22 cal. kyrs BP (C23, NY13). The overall results from Akonolinga allow the hypothesis that the main river bed successively migrated northwards towards its recent position during the documented sedimentation period. This can particularly be constrained by the strongly developed fining-upward sequences. However, the question as to whether the Tetar should be regarded as part of the proto-Nyong or as an independent tributary still remains unanswered. Most of the $\delta^{13}C$ (−18.1, −19.6, −20.1 ‰) found at

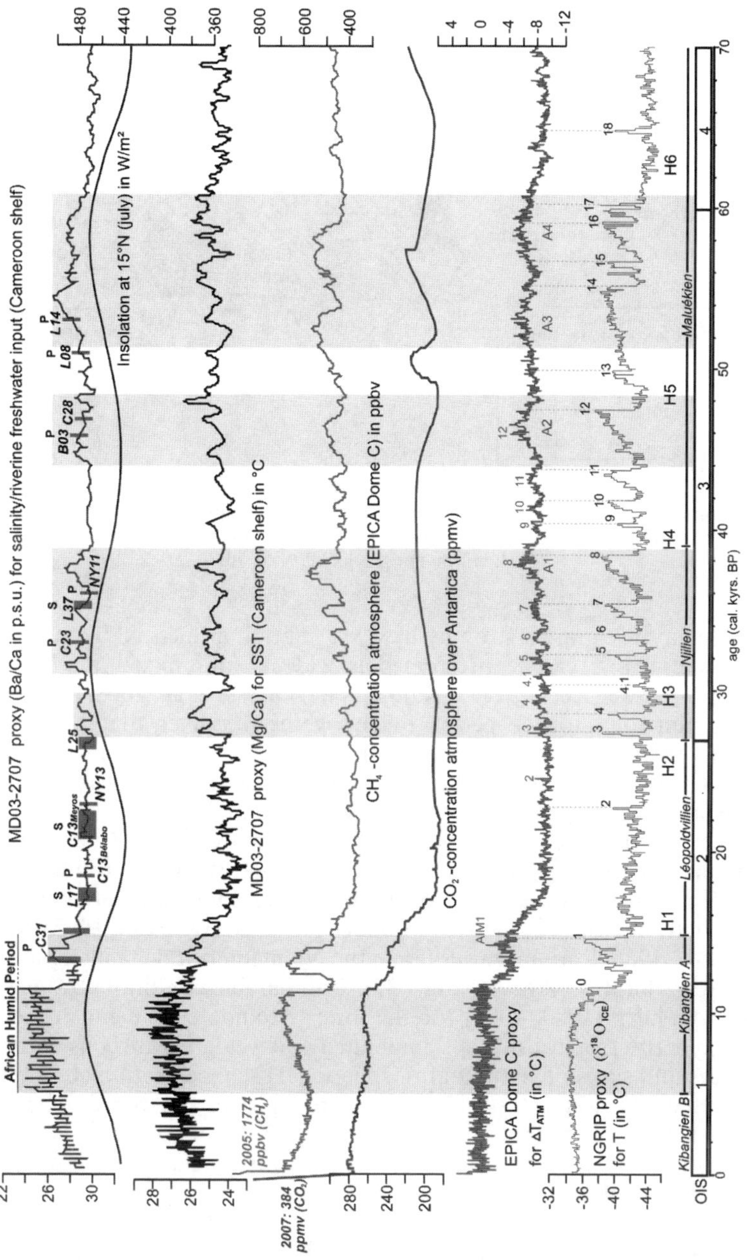

Figure 42. Correlation of salinity/riverine freshwater input and SSTs fluctuations across the Cameroon coast (Weldeab et al., 2007) with changes in greenhouse gas concentrations (CO_2 cf. Lüthi et al., 2008; CH_4 cf. Loulergue et al., 2008), temperatures over Antarctica (EPICA Dome C; Jouzel et al., 2007) and Greenland (NGRIP members, 2004), interstadials (no. 1–18), Antarctic warmings (A 1–4; Blunier and Brook, 2001), Antarctic isotope maxima (AIM 1–12; EPICA community members, 2006), Heinrich events (H1–6; Bond et al., 1999), oxygen isotopic stages (OIS 1–4), insolation at 15°N (Berger and Loutre, 1991), Central African climate stratigraphy following De Ploey (1965) and southern Cameroonian alluvial sediment archives (dark grey boxes with approx. time span; **P** palaeochannel, **S** swamp) for the last 70 cal. kyrs BP. Grey bars indicate more humid phases.

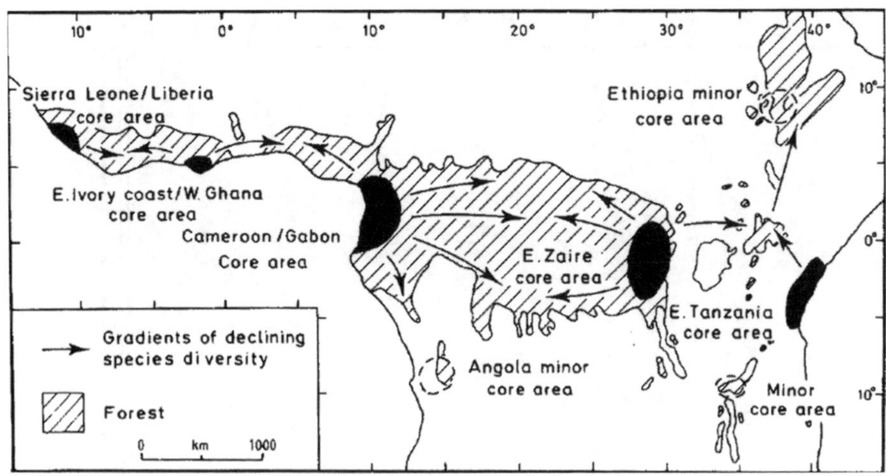

Figure 43. Position of possible refugial areas (black) of the tropical rain forest in West and Central Africa during arid phases (especially LGM) from different sources, and gradients of decreasing species diversity (Hamilton and Taylor, 1991, p. 70).

the Akonolinga site signify a strongly by C_4-vegetation influenced biome. The around −20 ‰ fluctuating $\delta^{13}C$ can be interpreted as evidence for a mixed rain forest-savanna vegetation (cf. Schwartz et al., 1995; Runge, 2002; Giresse et al., 2008) or, conversely, as evidence of numerous pioneer species. Taken together, it is likely that the environmental history and former landscape of the Akonolinga study region was a rain forest-savanna transition ecosystem having isolated gallery forests across the river system (Nyong) during the most arid stages. This is very close to present day characteristics.

The next younger sediments of the Akonolinga archives are assigned to unit 2. They demonstrate once again increasing humidity and fluvial-morphological activity in relation to another insolation maximum (~32 cal. kyrs BP) at the ending OIS 3. While regionally intense podzolation processes prevailed (De Ploey, 1965; Preuss, 1986; Schwartz et al., 1986; Kadomura, 1995), in the Ntem interior delta inundation swamps developed (e.g., location Aya'Amang L37). Comparable conditions are described by Thomas and Thorp (1980, 2003) for the lower Moinde catchment in Sierra Leone around 36 kyrs BP. Tropical swamps, inundated and wet lands are considered to be the main sources and sinks of natural atmospheric CH_4-concentration (Loulergue et al., 2008). Elevated CH_4-concentrations during humid phases can therefore be interpreted as additional evidence for the conservation of wet lands and swamps (see Figure 42). In the Ntem interior delta, the sandy, yet, rich in organic matter and macro-remains, sediment archive at Nnémeyong (L25) confirms the persistence of relatively humid conditions and C_3-dominated vegetation (−29.1 and −27.0 ‰) until around 26 cal. kyrs BP. Renewed intensification of the WAM prior to the transition to OIS 2 has been documented in several further West and Central African sediment archives. Equatorial lakes, e.g., Lake Chad, Lake Bosumtwi in Ghana and Lake Barombi Mbo in Cameroon, exhibited high levels (water stages). For these lakes, palaeoenvironmental data reaches back to around 32 kyrs BP, and provide evidence for the partial preservation of evergreen tropical rain forest (*Caesalpiniaceae*) at these sites from 32 to ca. 24 kyrs BP (Figure 44). Thereafter, towards the LGM, this forest type disappears from around Lake Bosumtwi, while in NW Cameroon, albeit strongly decimated, it outlives the cool and arid phase of the *Léopoldvillien* (Maley, 1991; Maley and Brenac, 1998). On the Batéké-plateau in Congo (Ngamakala swamp; Elenga et al., 1994) and at Lake Bambili

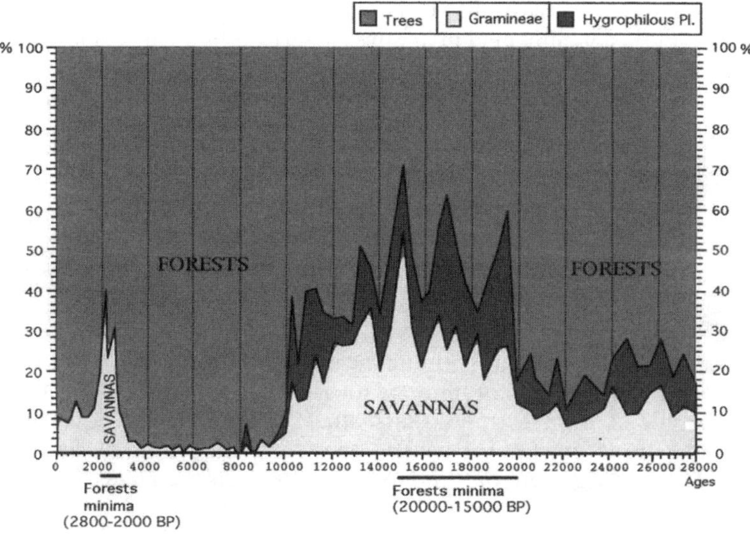

Figure 44. Simplified pollen diagram generated from Lake Barombi Mbo sediments (Maley, 2002).

(Stager and Anfang-Sutter, 1999) close to Lake Barombi Mbo, evidence for similar humid conditions until around 24 kyrs BP were found, although the strongly varying local boundary conditions suggest the need for critical and careful interpretation of the various lacustrine sediment archives.

Subsequently, with the approaching OIS 2 and the *Léopoldvillien* (~27–12 kyrs BP), western equatorial Africa was subjected to considerable and wide-spread aridification. Under much more arid and cooler conditions with strongly reduced rainfall (~30–50%) and land- and ocean temperatures (~3° C, which correlate with an insolation minimum during the LGM ~20 cal. kyrs BP), discharges clearly decreased and the landscape once again became disequilibrated. Tropical lowland rain forest was extensively displaced by spreading afromontane species (*Ilex, Olea, Podocarpus*) from higher regions and by savannas, and this forest type reached its last minimum expansion. In so-called 'core-areas' (refugian areas) the lowland rain forest could persist and later, expand again, from these sites (see Figure 43). There exist numerous theories on position and extent of these refugia, among them, a refugium is hypothised to be located along the coast of Cameroon and Gabon (cf. e.g., Hamilton, 1976, 1983; Littmann, 1987; Maley, 1987, 1989, 1991; Colyn et al., 1991; Hamilton and Taylor, 1991; Kadomura, 1995; Leal, 2004).

The Ntem interior delta might have been part of that refugium. In particular, recessional areas could have survived as gallery forests across fluvial systems, as demonstrated by the findings from Meyos (C13 and L17). Here, in a silty-clayey (C13) and fine-sandy matrix, inundation swamps developed between ~22–17 cal. kyrs BP, which testify to the preservation of a C_3-dominated vegetation ($\delta^{13}C$ from −29.9 to −23.1 ‰). Percussion core samples from this location having sediments suitable for palynological research, have unfortunately, not been evaluated yet. While here the tropical rain forest could survive as a fluvial refugium, the afromontane rain forest at Lake Barombi Mbo could, according to Maley (1987, 1989, 1996), have persisted because of sufficient humidity supplied via the atmosphere. However, this humidity was not provided by the WMA, but rather, through the regular formation of stratus clouds. These provided sufficient humidity, in the form of clouds and fog, to the higher laying lands, such as Lake Barombi Mbo. Likewise, at similar favourable sites near the coast, for example, as documented in archives described by Elenga et al. (1994), in the west of the Republic of Congo and coast near refugia across the Gulf of Guinea.

As a result of growing global ice sheets and cooling, in addition to appropriate boundary conditions and feedbacks, it is thought that the monsoon belt was compressed from both hemispheres by intensified temperature gradients and correlated trade wind currents. The very glacial conditions led to the expansion of effects on the Gulf of Guinea from polar anticyclones, polar fronts and cold waters of the Benguela current (e.g., Dupont et al., 1998, 2000; Shi et al., 2000, 2001; Stuut et al., 2002, 2004; Gersonde et al., 2005; Chase and Meadows, 2007). Thus, the SSTs cooled down and the affectiveness of the Walker- and Hadley-circulation diminished. This implied that WAM circulation and humidity transfer from the ocean to the continent were extensively reduced. For this phase, it is thought that the compressed or almost dissolved ITCZ was located in a distinctly more southern position, which is explained by an increased dynamic of the NE trades (harmattan) in the northern equatorial regions (deflation, dune displacements, intensification of upwelling and palaeoproduction in the Gulf of Guinea; e.g., Dupont and Weinelt, 1996; Jahns, 1996; Schneider et al., 1997; Dupont et al., 2000; Gasse et al., 2000, 2008; Marret et al., 2001; Lézine and Cazet, 2005). Due to these modifications, the global hydrological cycle was much weaker and seasonal dry periods were considerably prolonged. Water levels of Central African lakes as well as those of oceans reached their lowest (*Ogolien* regression with a drop of around 120 m in the Gulf of Guinea). For Cameroon, in particular, the levels of Barombi Mbo and Lake Chad experienced remarkable drops, the latter of which is largely fed by the main tributaries of the Sanaga from the Adamaoua plateau.

In this most arid phase of the last glacial cycle, wherein there was maximum rain forest retreat and widespread savannisation in equatorial Africa, mainly braiding to anabranching and fluvial-morphologically less active fluvial systems developed, which were accompanied by gallery forests. In a generally less vegetated and morphodynamically more instable landscape, during short but very intense storms (excessive precipitation), phasewise river benches were eroded and river patterns partially modified, but overall, surface formation predominated over valley formation (cf. e.g., Kadomura, 1995; Gasse, 2000; Thomas and Thorp, 2003; Runge, 1992, 1996, 2008). Because of modified morphodynamics, at many locations hillwash- and pebble layers, stone-lines and ferricretes (pedogenic crusts) were formed. Under these climatic conditions (increased seasonality), chemical weathering and saprolithisation were able to proceed more intensely (Runge, 2008). These incision and mass movement processes were favoured by strong deflation and considerable sea level lowering. This might also have been conducive to the transformation of inundated and wet lands into seasonal swamps, dry regions and valleys. Seasonal flooding events in combination with extensive reduction of vegetation types that held back erosion (especially rain forest) created widespread sandy sediment layers, although fluvial-morphological processes (transport, erosion and sedimentation) were most reduced during the LGM. In the Cameroonian study region, thin and middle sandy layers lacking embedded macro-remains (organic material) are typical for this sedimentation period. In several older sediment profiles, iron concentration layers can be found, which can be associated with this aridification (see profiles B03, C28, L08, L14, L25, L37 and NY11). Preuss (1990) also describes the increased deposition of sandy sediment layers by anastomosing rivers in the Congo basin for the period between 23 and 17 kyrs BP. In subtropical regions during this period a hiatus often occurs (e.g., Thomas and Thorp, 1980, 2003; Hall et al., 1985).

In the outgoing LGM and *Léopoldvillien*, and with the transition to the Holocene and the OIS 1, initially a successive humidification sets in, which correlates with a gradual Antarctic warming and insolation increase around 17.5 cal. kyrs BP (see Figure 42). Between ~17 to 13.5 cal. kyrs BP, climatic conditions in western equatorial Africa (< 8° N) became increasingly humid because of a gradual northward shift of the ITCZ and, thereby, increasing influence of the repeatedly intensified WAM. This trend was accompanied by an abrupt warming of the northern hemisphere around 14.5 kyrs BP ('Bølling–Allerød') and a subsequent cooling on both hemispheres ('Antarctic Cold Reversal—ACR'; Blunier and Brook, 2001), which lead to the Younger Dryas. The most reasonable causes for the Younger Dryas cooling are postulated to have been the rapid melting of the northern hemisphere ice masses and, particularly, the flow of Lake Agassiz (during glacial times) from the Laurentide ice sheet into the Atlantic, by which means the thermohaline circulation was interrupted (Broecker, 2003). The Younger Dryas was finally followed by the abrupt transition to the Holocene after 11 kyrs BP (humid hyperthermal phase of the *Kibangien A*, ~12–3.5 kyrs BP) and the African Humid Period (AHP) around 9 kyrs BP (Holocene climatic optimum).

During the AHP the Sahel and the southern Sahara experienced their last pronounced humid phase and during its climax the region flourished, while most tropical African lakes overflowed. This humidification is, in the first instance, conditioned by, what are by recent standards, clearly elevated insolation in the lower latitudes and a related maximum northward (~27° N) displaced position of the monsoon front (Kutzbach and Liu, 1997; Gasse, 2000; Leziné et al., 2005; Weldeab et al., 2007). The Holocene is also marked by several periods of rapid climate fluctuations, which are linked to cooling of the poles, aridification of the tropics, far reaching atmospheric circulation changes and extreme cultural upheavals (Bond et al., 1999; DeMenocal et al., 2000a, 2001; Nicholson, 2000; Alley et al., 2003; Mayewski et al., 2004; Denton and Broecker, 2008; Wanner et al., 2008). Like the dramatic climate fluctuations (both in terms of duration and amplitude) of the last glacial cycle, these global events occurred in intervals of 1500 ± 500 yrs centred (with partially regional varying magnitude) around 9–8, 6–5, 4.2–3.8, 3.5–2.5, 1.3–1 and 0.6–0.15 cal. kyrs BP (Mayewski et al., 2004). They show the same pattern as the former Pleistocene turnovers (D/O-cycles), but after ca. 8 cal. kyrs BP proceed under considerably reduced global glaciation. As a consequence of this, partially abrupt climate oscillations set in, which sometimes led to far reaching cultural upheavals, such as the downfall of prominent and highly developed cultures (e.g., Acadians in Mesopotamia, Ancient Egypt, Classical Maya in Mexico, Mochica in Peru and Tiwanacu in Bolivia).

The event around 6–5 cal. kyrs BP marks the end of the AHP and moreover a changeover to increased seasonality (prolonged dry season) and variability of precipitation in tropical Africa (Gasse, 2000). At around 10 cal. kyrs BP, the ITCZ was already displaced to around 14° N, and this favoured the onset of humidification of West African and Sahelian regions. Under notably increased humidity, discharges and fluvial-morphological activity swelled immensely during the onset and climax (AHP) of this warming. Hereupon, equatorial and West African fluvial systems reacted by dissection and transformation of the morphological structure of their river patterns and alluvial plains. These processes were exceptionally pronounced in subtropical and less vegetated regions (e.g., Niger inland delta: Gallais, 1967; Makaske, 1998, 2001; Lespez et al., 2008; Nile: Butzer, 1980; Sierra Leone: Thorp and Thomas, 1992; Ghana: Hall et al., 1985; Thorp and Thomas, 1992; Nigeria: Gumnior, 2005). In many cases, through the remobilisation of older, partially unconsolidated sediment layers, hiatus were created in the alluvial architecture (Thomas and Thorp, 1995, 2003; Thomas, 2000). Other regions affected by monsoon, such as South America (Latrubesse, 2003) and Australia (Nanson et al., 2008) exhibit a similar evolution. In the study area, these initially very abrupt climatic and fluvial-morphological transitions towards the Holocene, can be related to unit 3 and the oldest archives of unit 4. The sites Akonolinga (Tetar-tributary, NY13), Bélabo (C13) and Mengba (C31) document abrupt transformations in grain-size distribution (from sand to clay), which are interpreted as channel migrations and palaeomeanders, respectively (fining-upward sequences).

The very humid AHP, which is characterised by progressive stabilising wetter and warmer conditions, as a result of global deglaciation, can be substantiated by maximum rises of several lakes in equatorial Africa (e.g., Street and Grove, 1976, 1979; Gasse and Van Campo, 1994; Damnati, 2000; Gasse, 2000). Furthermore, the hemi-pelagic sediments of the Congo, Niger and Cameroonian rivers evince increased discharges. The sea level rose by around 90 m between 12,000 and 9000 BP (Holocene transgression). As a consequence of the re-expansion of the tropical rain forest until its maximum dissemination and the return to an isopluvial to pluvial rainfall regime (with a similar seasonality as today), the landscape progressively stabilized, accompanied by a decline in erosive and fluvial-morphological processes. In subtropical (West African) regions as well as in the Nile valley, increased aggradation, redeposition and incision processes took place in the floodplains, which can be chronologically defined through increased alluviation phases at 15–12.5, 11–8.2 and 4.8–3.5 cal. kyrs BP (Rossignol-Strick, 1983; Thomas and Thorp, 2003). As a result, the coastal fluvial systems show progressively meandering behaviour, while around the Niger inland delta braiding to anabranching fluvial systems dominate (Makaske, 2001; Lespez et al., 2008). In agreement with what was found by Kadomura (1995) for central Cameroon, Preuss (1986, 1990) for the Congo basin and Knox (1995) for several other tropical regions, southern Cameroonian fluvial systems experienced a general metamorphosis in fluvial behaviour towards meandering and anastomosing rivers having sandy beds and fine-grained sediment load. The Holocene is represented by the deposition of thick (2–4 m), for the most part, fine-grained alluvia. For the Nyong near Akonolinga, Ayos and Mengba, an initial transformation towards a meandering river network and channels probably began as early as 22 cal. kyrs BP (site NY13). In the course of the adjacent Holocene stabilisation of the study area relatively intact upper catchment areas developed having vegetated river banks. Consequently, with reinforced seasonal flooding events following increased aggradation during the Pleistocene-Holocene transition, fluvial systems reacted with a tendency towards anastomosing behaviour, augmented avulsions, stream bifurcation and channel migrations. Anastomosing flow properties in the Ntem interior delta are primarily favoured by its being in an intra-cratonic graben structure (cf. McCarthy, 1993) and the prior neo-tectonical stress (Schumm et al., 2000; Latrubesse and Rancy, 2000; Eisenberg, 2008). In the lower catchments of these rivers, these transformations were aided by rising discharges and sediment loads, and displacement of the erosion level due to a rise in sea level. To this fluvial disequilibrium, the anastomosing Ntem river section in the interior delta responded with increased mobility, frequent avulsions and channel migrations. The sites Nkongmeyos (L46 and L49) embody sandy sediments dated to 13 cal. kyrs BP, which progressively refine towards Holocene ages (up to 30% clay) and afterwards coarsen again. A similar tendency is seen by the sediments from Dizangué near Lac Ossa (lower Sanaga catchment) for the period around 10 cal. kyrs BP. The sediment units of the initiating AHP, and δ^{13}C (from −30.1 to −27.2 ‰) of these archives indicate survival, and probably, also an expansion of C_3-dominated vegetation.

There exist abundant sedimentary archives in the wider study region (especially Gabon, Cameroon and Republic of Congo) covering the subsequent Middle and Late Holocene (mainly from the ECOFIT programme), which clearly evince small-scaled climate oscillations with seasonality changes. A first aridification occurred between 8.5 and 7.8 cal. kyrs BP. This event was also registered in ice cores from Antarctica and Greenland, and marks the most striking cooling during the AHP. A further distinct lesser reduction of humid conditions, although restricted to the northern sub-equatorial domain, took place between 7 and 6.6 cal. kyrs BP (Gasse and Van Campo, 1994; Gasse, 2000; Thomas and Thorp, 2003). Humid conditions definitely diminished with the outgoing AHP around 6 cal. kyrs BP, while aridification tendencies accelerated again from 5.5 cal. kyrs BP onwards (Figure 42). This reversal proceeded in concurrence with decreasing insolation and cooling SSTs. Consequently, the intensity of the WAM and accompanied humidity transfer from the ocean to the continent were reduced. Moreover, the monsoon belt was displaced to the south, thereby lowering rainfalls and discharges in Cameroon and reinforcing the influence of the harmattan on the continent and upwellings in the Gulf of Guinea. An adjacent, significant dry phase around 4–3 cal. kyrs BP is corroborated by low lake levels in western equatorial Africa and marks the transition to the generally arid *Kibangien B* (~3.5–1.7 cal. kyrs BP). Among other things, reduced hygric conditions are documented for the lakes of Sinnda (Vincens et al., 1994; Elenga et al., 2004), Bosumtwi (Russell et al., 2003) and Sélé (Salzmann and Hoelzmann, 2005), where the so-called 'Dahomey Gap' (a 200 km wide savanna corridor inside the West African rain forest belt) evolved. In the West African domain discharges and sedimentation strongly decrease between 7 and 4.5 cal. kyrs BP (Thorp and Thomas, 1992). In contrast, the sediment archives of the Cameroonian lakes Barombi Mbo and Ossa for this period indicate an expansion of tropical lowland and mountain rain forests. From sediments of the lakes Kitina on the Batéké plateau and Maridor in Gabon, a similar progression could be reconstructed. Further indications for aridification trends and considerable reduction of the monsoonal activity were derived from the eastern equatorial domain (cf. Gasse, 2000). In the Ntem interior delta, the sites Aloum II (C11), Aya'Amang (L36 and L38) and Nyabibak (C02) manifest increased hygric conditions and fluvial-morphological activity for the period between 7.8 and 4.9 cal. kyrs BP (predominantly fine to middle sandy and coarse silty [Nyabibak] sediments). Also, here, the sediments contain abundant macro-remains, indicating C_3-dominated vegetation habitats ($\delta^{13}C$ from −31.4 to −26.8 ‰).

For West Africa increased discharge and sedimentation dynamics have been postulated for the period since 4.5 cal. kyrs BP, which however were also associated with advancing anthropogenic activity (Thomas and Thorp, 2003). The next younger sedimentary layers (unit 5) in the study area manifest a comparable metamorphosis of discharge and sedimentation conditions. Most remarkable is the transition to considerably coarser, sandy facies units dated to around 4.6–4.2 cal. kyrs BP, which were found in the Sanaga (C19) and Nyong (B06) catchments as well as in the Ntem interior delta (C13, L22, L36, L38). This abrupt changeover in grain-size distribution was also identified in a great quantity of undated sediment profiles. On the basis of often significant affinities with regionally associated and dated reference profiles for this period, identical regional processes can be assumed. Thus, the evidence from the Mbargué site (C17), some 40 km downstream, shows a strong analogy to C19. A basal coarsening-upward sequence (~4.3 cal. kyrs BP in C19) is covered by a fining-upward sequence, which is interpreted as a transition from turbulent to calm conditions at this location. In the Nyong upper catchment area the findings from the tributary Mfoumou (C27 Messa and C29 Ekoko) show similar (undated) modifications. Most evidence, however, was discovered in the Ntem interior delta. Here sedimentary archives from the sites Meyo Ntem (L43), Meyos (C13 and 14), Meyos II (L09–12), Nkongmeyos (L44 and 45), Nnémeyong (L23 and 26), Nyabessan (C16) and Nyabibak (C02) substantiate a very similar development dated to around 4 cal. kyrs BP. Further indications for significantly modified hygric and fluvial-morphological conditions are at several locations

(C13, C14, C16, L22, L43, L44 and L45) found: embedded palaeosurfaces (palustrine, organic sediments and macro-remains layers) between coarsening- and fining-upward sequences. Similar to the aforementioned older palaeosurfaces of the units 1–3, these also exhibit a sandy matrix, which increasingly refines (up to 50% clay; see profile L22) towards the next superior layer boundary (surface of organic sediment layers). They additionally allocate repeated, externally induced fluvial modifications of the fluvial system's internal fluvial-morphological and sedimentological responses to large-scale changes in environmental conditions. Hence, again, location dependent transformations of wetlands and inundated areas into seasonal swamps and dry areas took place, which were induced by external and internal impulses. However, $\delta^{13}C$ data (from −29.9 to −27.2‰), which correlations strongly with findings from the lakes of Barombi Mbo and Ossa in Cameroon as well as some lakes in Gabon (Ngomanda, 2005; Giresse et al., 2008), confirms the existence of primarily C_3-dominated vegetation. The spreading of savannas, which for this period has been corroborated for the domain north of ~5° N and south of ~5° S, can therefore be excluded. Such 'savannisations' have been reported for the Adamaoua plateau, the western Republic of Congo and parts of the Central African Republic during the period 5–2 cal. kyrs BP (cf. Elenga et al., 2004; Runge, 2002; Neumer, 2007). Locally, like in parts of Cameroon and West Africa, this arid trend is replaced by repeated humid pulses around 3.5–2.5 cal. kyrs BP (cf. Vincens et al., 1999; Marret et al., 2006; Ngomanda et al., 2009; Giresse et al., 2008; Thomas and Thorp, 2003; Gumnior, 2005). In the study area, prevalent and supra-regional occurring iron concentration horizons in several profiles (B06, C02, C11, C13–14, C16, C17, C19, C21, C23, C27, C28, C31, L09–11, L16–17, L22, L28, L36–38, L45–47, L49) can be interpreted as relics of developing pedogenic crusts (ferricrete), which, since about ~5 cal. kyrs BP formed in the context of the start of a reduction of hygric and hydromorphic conditions. This association however remains speculative because of limited ^{14}C-data. These horizons, found in some profiles, mark groundwater levels during the dry season, which argues for seasonal groundwater oscillation linked iron displacements. Furthermore they could be residuals of a younger arid phase around ~2.8 cal. kyrs BP ('First Millennium BC Crisis').

The latter phase has, in the majority of sedimentary archives of the study area, been substantiated on the basis of multiple proxy data, especially pollen, microfossils, sapropels and clay minerals (e.g., DeMenocal, 2000; Gasse, 2000; Elenga et al., 2004; Nguetsop et al., 2004; Lézine et al., 2005; Weldeab et al., 2005, 2007; Gasse et al., 2008; Giresse et al., 2008). Nonetheless, some authors (e.g., Fontaine and Bigot, 1993; Moron et al., 1995; Maley, 1997, 2001) think that Cameroon, at least initially, maintained hygric conditions due to the preferential development of fog and nimbostratus clouds in the region. The ultimate and abrupt reversal to arid climate conditions in equatorial Africa took place between 3.5 and 2.5 kyrs BP (Marchant and Hooghiemstra, 2004; Elenga et al., 2004; Nguetsop et al., 2004).

There is a lot of evidence in the study area for this aridification, mainly in terms of vegetation changes and lake level drops (Maley, 1992; Reynaud-Farrera et al., 1996; Jolly et al., 1998; Maley and Brenac, 1998; Vincens et al., 1999; Gasse, 2000; Giresse et al., 2005, 2008; Marret et al., 2006; Gasse et al., 2008; Ngomanda et al., 2009). All over western equatorial Africa, considerable fragmentations of the tropical rain forest and savannisations have been proved for this period, culminating between 2.8 and 2.5 cal. kyrs BP ('*péjoration climatique*' or 'First Millennium BC Crisis'). In Cameroon, further evidence is provided by the sediment archives from lakes Barombi Mbo and, particularly, Ossa (see Figure 45). The Dahomey Gap repeatedly opens, and across the investigated lakes, pioneer species disperse. Laminar denudation and gullying are perhaps the most remarkable accessory phenomena in reaction to newly opened landscape (Kadomura,

1995). Besides repeated teleconnections in the context of a general cooling trend in the northern hemisphere (Bond et al., 1999; Mayewski et al., 2004), the seasonal meridional dipole between the SSTs in the northern and southern Atlantic (interaction between NAO and SO) is largely considered to be the main trigger for serious rainfall anomalies in the Sahel and the Gulf of Guinea; it is additionally responsible for displacing the ITCZ sometimes far to the south in boreal winter (Servain, 1991; Fontaine and Bigot, 1993; Fontaine and Janicot, 1996; Paeth and Stuck, 2004). Linked to this, are increased numbers of global ENSO-phenomena during that period (Figure 45). Recently, these are believed to be responsible for occasional general reductions of rainfalls in these regions (Fontaine and Janicot, 1996; Moron et al., 1995; Zhao et al., 2007). While strong yearly fluctuations are largely attributed to oscillating dipole conditions and associated ENSO in the North and South Atlantic, decadal variations are ascribed rather to a comparable, global inter-hemispheric dipole (Folland et al., 1986; Janicot et al., 1998; Camberlin et al., 2001), which is most extensive in times of a cooling northern hemisphere. And last, but not least, this period is characterised by a marked increase of anthropogenic impacts on the ecological balance of the landscape.

Evidence from alluvial sediment archives partially derives from the aforementioned last aridifcation records. Here, some sites (C02, 11, 13, 14, 16, 19) document rather turbulent fluvial discharge and sedimentation processes in a, once more, disequilibrated landscape setting and fluvial environment, while others (B06, C16, C23, C28, C31, L22, L36–38, L43–45) manifest relatively consistent to consolidating tendencies. Although not all sedimentary profiles have been dated, many marked and abrupt revulsions in these profiles since ca. 5 cal. kyrs BP seem to nearly correlate with the development of pedogenic crusts and swamp. There is a distinct decrease of discharges between 3.5 and 1.4 cal. kyrs BP, interrupted by a wetter phase between

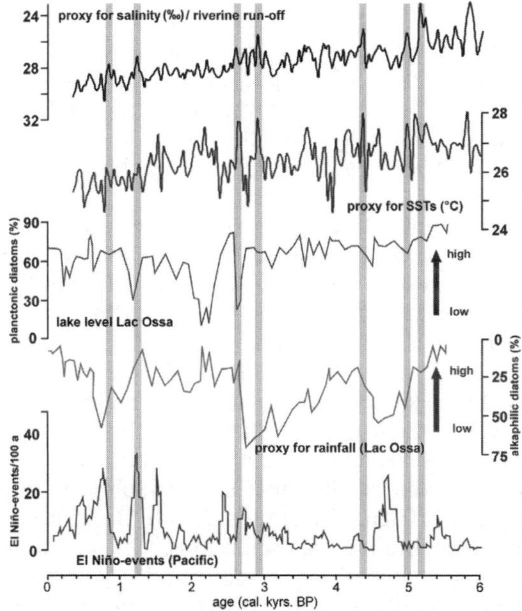

Figure 45. Correlation of proxy-data from core MD03–2707 (Weldeab et al., 2007) with data from Lac Ossa (Nguetsop et al., 2004) and proxies for El Niño-event intensity (Rodbell et al., 1999). Grey bars indicate more humid phases with increased fluvial-morphological activity.

3 and 2.7 cal. kyrs BP. The latter could be verified at the Ouesso site (B02) in the upper Boumba catchment (SE Cameroon), where a conspicuous accretion of sandy facies was identified for the period around 2.8 cal. kyrs BP. Additionally, the lowest so far $\delta^{13}C$ data (-35.5 ‰) was found. Further evidence for increasing fluvial activity and/or landscape opening was provided by the oldest dated sedimentary archives of unit 6. Here, the locations Njock (B01), in the middle Nyong catchment area, and Nyabessan (especially C20 and L18–19) as well as Anguiridjang (L40), in the Ntem interior delta, attest to an increase of sandy facies and the development of widespread (palustrine) palaeosurfaces between 2.6 and 2.1 cal. kyrs BP. Their sediments having very low pH and elevated iron contents indicate lowland moor conditions, which might have been initiated by strongly reduced hygric and hydro-morph conditions. The trend towards reduced humidity was also substantiated in the sediment archives of lakes Barombi Mbo, Ossa, Kitina and Nguène (Maley, 2002; Nguetsop et al., 2004; Ngomanda et al., 2009). The $\delta^{13}C$ of the alluvial sediments (from -31.9 to -26.9 ‰), however, once again point to predominantly C_3-dominated vegetation, without any indication for the presence of savanna (C_4-species). Also for this time period, there was a cooling over Antarctica followed by a warming, after which, SSTs and atmospheric CH_4-concentration rise. Palynological studies on a core sampled from the Nyabessan palaeosurface depicted a vegetation change at this site around 2.5–2.4 cal. kyrs BP (Ngomanda et al., 2009). Evergreen tropical rain forest (having many *Caesalpiniaceae*) and inundation forest (mainly *Raphia*) were considerably displaced by pioneer species and secondary rain forest. Given the facts that during this period SSTs across Cameroon significantly rise and at the same time discharges remain relatively unchanged, a seasonality change to a prolonged winter dry season seems most reasonable for this part of the study area (Nguetsop et al., 2004; Maley, 2004; Ngomanda et al., 2009). This brought about higher evapotranspiration and seasonally increased dryness, which induced a serious retreat of the highly sensitive evergreen tropical rain forest. This retreat and fragmentation was greatly aggravated locally by a seasonally boosted harmattan influence and frequently occurring convective thunderstorms ('squall-lines/*Lignes de grains*') in the outer-equatorial domain (Maley, 2004; Nguetsop et al., 2004). In addition, the strong increases in anthropogenic activity in the rain forest exerted collateral pressure on these sensitive ecosystems, especially around 2.6 cal. kyrs BP, as registered by metallurgical finds (Schwartz, 1992; Eggert et al., 2006; Höhn et al., 2008; Meister, 2008; Ngomanda et al., 2009).

Even though since about 2 cal. kyrs BP, more humid conditions have commenced, principally confirmed by renewed, but possibly locally delayed, expansion of the tropical rain forest in the *Subactuel* (~1.7 cal. kyrs BP until 1990s), after previous widespread regression (Maley, 2004), reduced discharges occur across the Cameroonian coast around 1.3, 1.1, 0.8 and 0.4 cal. kyrs BP. These are preceded by relatively elevated discharge peaks around 1.7 and especially 1.2 as well as 0.9 cal. kyrs BP (Figure 42). These successions correlate with globally registered Holocene climatic reversals, of which the MCO and the LIA are the most striking (Bond et al., 1999; Mayewski et al., 2004; Verschuren, 2004; Denton and Broecker, 2008; Wanner et al., 2008). The LIA is seen as the most severe Holocene northern hemispheric cooling (DeMenocal et al., 2000a; Verschuren, 2004). But both events also triggered anew equatorial African environmental oscillations, which regionally mainly resulted in modified seasonality (prolonged dry season) and vegetation changes. The SSTs in the Gulf of Guinea generally decreased from 1.3 cal. kyrs BP onwards, reaching minima at 0.8 and 0.5 cal. kyrs BP. This aridification was most pronounced across the northern West African domain and involved widespread rain forest regression, denudation, gully erosion and aggradation, and which was intensified by accelerating anthropogenic activity (e.g., DeMenocal et al., 2001; Nguetsop et al., 2004; Giresse et al., 2008). Moreover, a repeated, and considerable southward displacement of the ITCZ position during boreal winter is thought to have resulted from opposed dynamics south of the equator (Haug et al., 2001; Nguetsop et al., 2004; Broccoli et al., 2006). There is strong coherence between SSTs and rainfall anomalies and a strong dipole in the Atlantic and El Niño- and La Niña-events (Atlantic Niño). The latter strongly accumulated during the Late Holocene through atmospheric-oceanic teleconnections, which considerably affected the temperature and current as well as drift conditions in the Gulf of Guinea. Rodbell et al. (1999) found evidence for increasing intensity of El Niño- and La Niña-events across the Ecuadorian coast since 15 kyrs BP and for the establishment of the recent recurring periodicity of 2–8.5 yrs. around 5 kyrs BP. This was subsequently substantiated by results from the Peruvian coast (Sandweiss et al., 2001). Van Geel et al. (1999), Bond et al. (2001) and Denton and Broecker (2008), conversely, emphasize the notable relevance of fluctuating global radiation (solar radiation intensity and sun spot activity) as a driving force of global temperature oscillations. Analysis of the temporal variation of $\delta^{14}C$ and ^{10}Be has provided evidence for the occurrence of sun spot activity minima (Wolf-Minimum at 680, Spörer-Minimum at 500 and Maunder-Minimum at 300 cal. yrs. BP) during the MCO. Simultaneously, Neolithic human activity (exploitation and settling in the tropical rain forest) greatly affected ecosystems in the study area, as evidenced by major archaeological and archaeobotanical results of the DFG-Research Unit 510 (e.g., Eggert et al., 2006; Meister, 2008).

During the periods of the MCO and adjacent LIA the southern Cameroonian fluvial systems experienced oscillating discharge pulses, with relative progressions around 1.2 and 0.9 cal. kyrs BP and regressions at 1.1 and 0.7 cal. kyrs BP. This succession of two humid and arid stages resulted in a phase of minimum riverine freshwater input across the Cameroonian coast around 0.4 cal. kyrs BP. Sediment archives of this period are found closest to river beds. Alluvia from Bélabo (C14–15, upper Sanaga), Akonolinga, Ayos (exposure sites, Nyong), Benana (no data, Nyong), Meyo Ntem (L02 and L05) and Tom (L30), in the Ntem interior delta, in addition to a multitude of undated sediment profiles, corroborate a successive increase of fine- to coarse-grained deposits since ca. 1.9 cal. kyrs BP. For this period, other authors postulate increased erosion processes and the deposition of a '*basse terrasse*' in many parts of Nigeria, Cameroon, Gabon and the Republic of Congo (cf. Preuss, 1986; Maley and Brenac, 1998a; Ngos III et al., 2003). In southern Cameroon, this period is represented by sedimentary unit 6; it is associated locally with the formation of organic and macro-remain-rich layers (palaeosurfaces) in a sandy matrix, which were dated to 1.1- 0.8 cal. kyrs BP, and was probably interrupted by another discharge peak culminating around 0.9 cal. kyrs BP. These palaeosurfaces are very similar to those dated to the Late Pleistocene and Early Holocene, although less distinct in terms of dimension and thickness. The associated $\delta^{13}C$ (from −30.6 to −25.8 ‰) values attest the presence of C_3-dominated vegetation at all locations, with the exception of the Nyong floodplain (Akonolinga, Ayos and Ebabodo).

Most significantly, the aggradation of a partially flooded seasonal oxbow channel (palaeomeander) of the Ngoko River (SE Cameroon), where recently Lake Mokounounou has formed, seems to fall into this same period. The silting-up, with fine- to coarse-grained alluvial sediments, was dated to 1.2–1 cal. kyrs BP (cores N02–03 and 05; cf. Sangen et al., 2010; Sangen, 2011).

The next phase of widespread transformations across southern Cameroonian floodplains is restricted to the upper Nyong floodplain, and culminates around 0.7 cal. kyrs BP. The Ayos (C30 and NY08) and Ebabodo (NY05) sites display another phase of intense aggradation, manifested by coarse-grained sandy stratigraphies. It is followed by the deposition of extremely organic, swampy to marshy, sediment archives in many parts of the study area. At Lipombe II near Eséka (middle Nyong catchment) middle- to coarse-grained sandy layers having organic bags were dated to the same sedimentation phase. In the Ntem interior delta, more specifically at Nnémeyong (L24), Akom (L27), Tom (L31) and Nkongmeyos (L34), middle to fine-grained sandy sediments were cored close to the river bed (lower terrace, dry season) and dated to slightly younger ages (~0.65–0.6 cal. kyrs BP). They were covered by sandy organic sediments having abundant macro-remains, dated to 0.45–0.4 cal. kyrs BP. Similar sedimentary archives were found at Akom (L28), Nkongmeyos (L35) and Nyabessan (L18). In fact, all these results were found across the most northern of the anastomosing Ntem channels (Ntem 1). Also, here, macro-remains embedded in a sandy to silty matrix were dated to ages around 0.6 cal. kyrs BP. In the lower Nyong catchment (Dehane, B07), similar sediments were dated to 0.4 cal. kyrs BP. Again, all $\delta^{13}C$ data (from −30.6 to −26.4 ‰) indicate C_3-dominated vegetation at all sites. While these palaeosurfaces were forming, the northern West African lake levels (especially Bosumtwi and Chad) were low, conversely, Barombi Mbo and Ossa levels rose. Humid conditions also prevailed in Gabon (Ngomanda, 2005), which could again imply a southward displacement of the ITCZ. In contrast to the western domain of the African monsoon, east African lakes had high lake levels (Gasse, 2000; Verschuren, 2004). This E-W-anomaly of monsoon related humidity could have been induced by the strong ENSO dynamics during that period, which modulates circulation mechanisms by means of atmospheric/tropospheric-oceanic teleconnections (Rodbell et al., 1999; Chang et al., 2006). Crowley (2000) and Wanner et al. (2008) explain the considerable cooling during this 'mini-stadial' by, among other things, strong volcanic activity in tropical regions. Lastly, similar to the aforementioned periods 4.8–4.6, 2.8–2.6 and especially 1.6–1.2 cal. kyrs BP (maximum), El Niño-events occured very frequently between 0.8 und 0.5 cal. kyrs BP (see Figure 45).

After 0.4 cal. kyrs BP and until the *Actuel*, as well as in recent times, the sediment archives manifest antithetic developments. They often show generally fining-upward sequences, which, though, are covered in the upper 20–40 cm by coarsening-upward sequences. The only dated sediment layer covering this period was sampled from Nkongmeyos (L32) and shows strongly fluctuating coarse-grained sandy facies. This data fits well with a trend of strongly fluctuating discharges, which has been documented since the beginning of run-off recording (early 1950s) in the study area. Observations made during fieldworks indicate that there has been a tendency towards increased aggradation in southern Cameroonian rivers, manifested by increased sandy sediment load and frequently occurring sand banks. These processes may be strongly amplified by increasing human activities in Cameroonian ecosystems (e.g., settlement and cultivation expansion, logging, shifting cultivation, mining and illegal hunting).

4.4 CONCLUSIONS

The methods chosen for use in this study (cf. Chapter 4.1.2) have generally proven to be very satisfactory, especially in combination with the remote sensing data (LAND-SAT ETM+- and ASTER-data). With few exceptions (i.e., middle Sanaga catchment near Monatélé, upper Nyong near Abong Mbang, Mfoumou tributary, upper Ntem at Ngoazik and lower Ntem at Campo Beach), the identified river sections contained very valuable alluvial sediment archives for palaeoenvironmental reconstructions. In particular, the anastomosing Ntem river section of the Ntem interior delta

(Chapter 4.3.3.1.) yielded chronologically quite old and informative sedimentary records. The ASTER-data helped locate abandoned channels, which were potentially promising reserves of proxy data archives (abandoned channels of the Nyong at Akonolinga, Chapter 4.2.3.3.; abandoned channels between Endom and Mbalmayo, Chapter 4.2.3.4. and the palaeomeander of the Ngoko near Mokounounou). Additionally, remote sensing data provided excellent visualisation of the morphological structure (physiognomy) of the examined river sections, and, hence, useful information on fluvial-morphological dynamics and formation of sedimentary archives.

There were, of course, uncertainties surrounding the selection of suitable coring sites, especially regarding preliminary data respecting palaeoenvironmentally relevant time frames. Although it turned out that sedimentary archives ages increased with distance from recent river beds, it still remained very difficult to find those archives of most interest to the DFG-Research Unit 510's research goals (i.e., 'First Millennium BC Crisis'). In the end, the southern Cameroonian alluvia proved to contain very useful palaeoenvironmental information, covering multiple sediment-stratigraphic and chrono-stratigraphic periods, which reach as far back as the Late Pleistocene (i.e., 53 cal. kyrs BP).

Findings from this study, together with pre-existing evidence from the western African monsoonal domain (Chapters 4.1.3 and 4.3.2), contribute to the deciphering of past environmental conditions in a hithertofore largely misunderstood region. Furthermore, this research shows that palaeoenvironmentally utilisable sedimentary archives are preserved in the widespread alluvial plains of tropical African fluvial systems. It disproves the long-held notion that intense chemical weathering in tropical regions has led to the loss of datable organic material in sedimentary archives. It demonstrates that abundant localities exist where datable and palaeoenvironmentally valuable alluvial sedimentary records have been preserved; moreover, these constitute a heretofore inadequately exploited terrestrial palaeoenvironmental archive. In spite of the difficulties and uncertainties inherent in the interpretation of tropical alluvial sediments (e.g., hiatus, interbedding with other strata, etc.) and the fact that sedimentary profiles only allow one-dimensional insights into the stratigraphies, overall, this research produced numerous significant findings. Principally based on grain-size distributions, proxy data from palaeosurfaces, related interpretations of multiple fluvial systems' alluvial ridges, (including over 160 sedimentary profiles and many transects, which generated a very high density of site-specific data), realistic hypotheses have been drawn on the genesis of the respective sedimentary archives and corresponding processes as well as environmental conditions. Notwithstanding the often divergent positions inside the fluvial system as a whole, geological substrate, spatial archetypes and environmental conditions, irrespective of location, identical and temporally correlatable fluvial-morphological and sedimentological development tendencies are distinguishable, which indicate supra-regional modifications. As a result of local variation in external controlling factors, fluvial-morphologically different fluvial systems and corresponding fluvial and alluvial sedimentary archives have evolved; these are most complex in the Ntem interior delta, which has a neo-tectonic origin (Figure 46). Here, partially temporal and supra-regional synchronic developed alternations in discharge, erosion and sedimentation conditions can be substantiated, which are manifested in the formation of swamps, palaeosurfaces and abandoned channels. Palaeosurfaces up to 120 cm thick provide some of the most significant palaeoenvironmental archives, covering Holocene, LGM-age and even Late Pleistocene periods. These mostly clayey laminated sediment layers harbouring abundant partially undecomposed macro-remains and various other proxy-data (e.g., phytoliths, sponge spicules, diatoms, etc.), provided some of the most precious information for the reconstruction of past fluvial

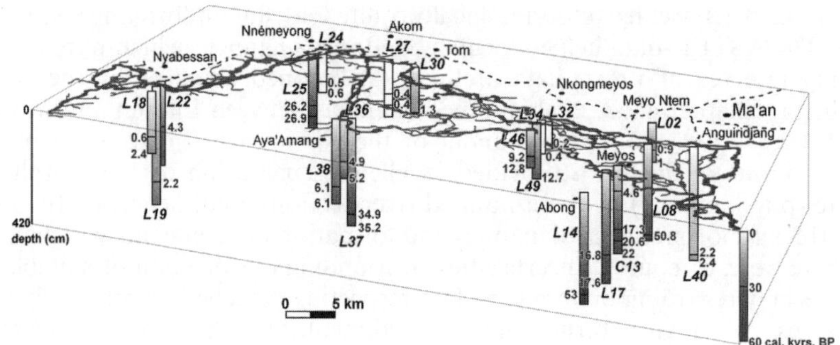

Figure 46. Overview of sedimentary archives found in the Ntem interior delta with ages in cal. kyrs BP (modified from Sangen, 2008, p. 95).

landscapes and environmental conditions. At several locations inside the Ntem 1 delta (Aya'Amang, L37; Nyabessan, L18 and Meyos, L17), these archives provided robust detailed data, from which solid conclusions for the periods around 34–35 (Aya'Amang), 22–17 (Meyos) and 4.3–2.3 cal. kyrs BP (Nyabessan) could be drawn.

Most strikingly, the macro-remains, charcoal and organic sediment embedded in these fossil organic layers and palaeosurfaces yielded around 70 radiocarbon dates, which enabled the drawing up of a chronological classification of the alluvial sedimentary archives. It was a surprise that the discovery of fossil, organic horizons, mudds and swamps was not restricted to single fluvial systems or alluvial ridges, but could be regularly found across all southern Cameroonian fluvial systems (Boumba, Ngoko, Ntem, Nyong and Sanaga). Similarly, among all sedimentary profiles significant changes in grain-size distribution occur (alternating coarse- and fine-grained sequences, coarsening- and fining-upward sequences), which point to modifications of discharge, transport and sedimentation. These indications are, for the most part, corroborated by clearly distinguishable changes in soil colour or pedochemical parameters. Furthermore, iron concentration layers and otherwise recognizable boundary layers were identified. Iron concretions, patches and even horizons (ferricretes) testify to considerable iron oxide dynamics. These last are also useful for examining alluvial weathering rates and, in this way, can provide indirect clues to sediment age. Conversely, these iron concentrations, which sometimes reach laterite-like dimensions (e.g., Ntem interior delta, Akonolinga), might reflect high evaporation rates and the formation of pedogenic crusts during arid phases. Last, but not least, episodes of increased or decreased fluvial activity can be defined and discussed in the context of widespread landscape and environmental dynamics of the Late Quaternary in southern Cameroon.

Most evidence is provided for periods of the Late Pleistocene and the Late Pleistocene-Holocene transition. Corresponding sediments indicate phases of unsteady und turbulent fluvial activity and sedimentation, which are mostly marked by fluxes of sandy deposits having abundant macro-remains and organic material. It could be demonstrated that these discharge and sedimentation pulses, which took place mainly during the *Maluékien* and *Njilien*, occurred in concert with insolation maxima; moreover, the proposed hypotheses correlate well with findings from other Western and Central African hemi-pelagic and terrestrial sediment archives. Correlations with lacustrine sediment archives, which, because they are closed systems, are very prone to misinterpretation, turned out to be somewhat problematic. Numerous palynological studies allow very detailed reconstructions of former ecosystems, but are very location dependent. An additional refinement of the study would be to learn, whether the $\delta^{13}C$-signals from the alluvial archives indicate autochthonous or allochthonous sources. Apart from this incertitude, the studied alluvia provide further evidence for the existence of so-called 'fluvial rain forest refugees' across certain fluvial systems during arid and cool phases (especially LGM). This said, it remains difficult to assign specific vegetation types with great accuracy. With the exception of locations in the Nyong floodplain, almost all sites show preservation of C_3-dominated vegetation. For the Nyong sites near Akonolinga, it seems that there was a more transitional vegetation composition, although the $\delta^{13}C$ data might also indicate the presence of pioneer species (J. Maley, personal communication). For the Holocene, which is generally characterized by a stabilisation of internal and external controlling forces, unequivocal climatic perturbations can be proved for the Middle to Late Holocene; in addition for the latter and the adjacent *Subactuel*, it is probable that accelerating anthropogenic impacts on the ecosystems also played a major role. Some evidence was found by the DFG-Research-Unit 510 that these processes were initiated around 2.6 cal. kyrs BP (e.g., Eggert et al., 2006; Meister, 2008). This trend has been seriously amplified during recent years, particularly in terms of industrial logging (cf. e.g., Semazzi and Song, 2001; Betti, 2003; Laporte et al., 2007), which will doubtless have far-reaching consequences in the near future (Hulme et al., 2001; Justice et al., 2001; Delire et al., 2008; Molua, 2009). Though the tropical rain forest either appears to be expanding in comparison to savanna (progression of the rain forest into savanna areas; cf. Youta Happi, 1998), which might, among other things, be induced by the increasing (anthropogenic induced) CO_2-concentration in the atmosphere ('CO_2-fertilization'), unprecedented CO_2- (384 ppmv, 30% higher than during the last 420,000 yrs.) and CH_4-concentrations (1774 ppbv, 250% higher than the last 650,000 yrs.) are of great concern. Possible side effects of drastically increasing greenhouse gases are usually discussed in the context of global warming (cf. IPCC, 2007), and the long-term effects are unclear. However, in light of the present state of knowledge, anthropogenic intensification of natural greenhouse gas effects, notably through tropical rain forest logging, industrial emissions, fossil fuel burning and land use changes, will have far-reaching consequences. Several studies highlight the increasing variability of Central African rainfalls and discharges in recent years (Nicholson, 1980, 1986, 2000; Moron et al., 1995; Servat et al., 1996; Jury, 2003; Lienou et al., 2008). With respect to global warming and its effects on equatorial Africa, it seems that the most striking source of climate variability comes from the strong increase in intensity and frequency of ENSO-phenomena and its, as yet, unclear relationship to the Atlantic Niño (Diaz et al., 2001; Paeth and Hense, 2006; Redelsperger et al., 2006; Zhang et al., 2006; Paeth et al., 2008). Of paramount importance in this respect are recent results on greenhouse gas concentration variations (Loulergue et al., 2008; Lüthi et al., 2008) and the increasing evidence that tropical regions and climate system play a key role on global climate trends. Problems related to recent natural hazards and impacts in Cameroon will be in detail discussed in the following chapter of this book. The future development of tropical fluvial systems will be very strongly linked to this problematic. For these reasons, the reactivation of hydrological gauging and monitoring stations in Central Africa is of extreme importance ('World Hydrological Cycle Observations System— WHYCOS', cf. WMO, 2005) as is the urgency of further research on fluvial systems in tropical Africa, which is still very limited.

REFERENCES

Abrantes, F. 2003, A 340,000 year continental climate record from tropical Africa—news from opal phytoliths from the equatorial Atlantic. *Earth and Planetary Science Letters*, **209**, pp. 165–179.

Achoundong, G. 2006, Vegetation du Cameroun. In *Atlas du Cameroun*, edited by Yahmed, D.B. (Paris: Les Éditions J.A.), pp. 64–65.

Adegbie, A.T., Schneider, R.R., Röhl, U. and Wefer, G. 2003, Glacial millennial-scale fluctuations in central African precipitation recorded in terrigenous sediment supply and freshwater signals offshore Cameroon. *Palaeogeography, Palaeoclimatology, Palaeoecology*, **197**, pp. 323–333.

Akogo, G. 2002, *Etude de cas d'aménagement forestier exemplaire en Afrique centrale. La zone de Campo-Ma'an, Cameroun.* Document FM/10F, FAO, Rome, pp. 1–31.

Anhuf, D., Ledru, M.-P., Behling, H., Da Cruz Jr., F.W., Cordeiro, R.C., Van Der Hammen, T., Karmann, I., Marengo, J.A., De Oliveira, P.E., Pessenda, L., Siffedine, A., Albuquerque, A.L. and Da Silva Dias, P.L. 2006, Paleo-environmental change in Amazonian and African rainforest during the LGM. *Palaeogeography, Palaeoclimatology, Palaeoecology*, **239**, pp. 510–527.

Barker, P.A., Talbot, M.R., Sreet-Perrott, F.A., Marret, F., Scourse, J. and Odada, E.O. 2004, Late Quaternary climate variability in Intertropical Africa. In *Past Climate Variability through Europe and Africa*, edited by Battarbee, R.W., Gasse, F. and Stickley, C.E. (Dordrecht: Springer), pp. 117–138.

Bernard, E.A. 1962, *Théorie astronomique des pluviaux et interpluviaux du Quaternaire africain.* Memoires de l'Académie Royale des Sciences d'Outre-mer 2, 1–232.

Bessoles, B. and Trompette, R. 1980, *Géologie de l'Afrique. La chaine panafricaine «* Zone mobile d'Afrique Centrale (partie sud) et zone mobile Soudanaise . Orléans: Mémoires du Bureau de recherches géologiques et minières, Èditions B.R.G.M.

Betti, J.L. 2003, *Impact of forest logging in the Dja biosphere reserve, Cameroon.* Ministry of Environment and Forestry/PSRF, Yaoundé, Cameroon, 1–13.

Blunier, T. and Brook, E.J. 2001, Timing of millennial-scale climate change in Antarctica and Greenland during the last glacial period. *Science*, **109**, pp. 109–112.

Bond, G.C., Kromer, B., Beer, J., Muscheler, R., Evans, M., Showers, W., Hoffmann, S., Lotti-Bond, R., Hajdas, I. and Bonani, G. 2001, Persistent solar influence on north Atlantic climate during the Holocene. *Science*, **294**, pp. 2130–2136.

Bond, G.C., Showers, W., Elliot, M., Evans, M., Lotti, R., Hajdas, I., Bonani, G. and Johnson, S. 1999, The North Atlantic's 1–2 kyr climate rhythm: relation to Heinrich events, Dansgaard/Oeschger cycles and the little ice age. In *Mechanisms of Global Change at Millenial Time Scales*, Geophysical Monograph 112, edited by Clark, P.U., Webb, R.S. and Keigwin, L.D. (Washington D.C.: American Geophysical Union), pp. 59–76.

Broccoli, A.J., Dahl, K.A. and Stouffer, R.J. 2006, Response of the ITCZ to Northern hemisphere cooling. *Geophysical Research Letters*, **33**, L01702, doi:10.1029/2005GL024546.

Broecker, W.S. 2003, Paleocean circulation during the last deglaciation: A bipolar seesaw? *Paleoceanography*, **13**, pp. 119–121.

Butzer, K.W. 1980, The Holocene lake plain of north Rudolf, East Africa. *Physical Geography*, **1**, pp. 42–58.

Camberlin, P., Janicot, S. and Poccard, I. 2001, Seasonality and atmospheric dynamics of the teleconnection between African rainfall and tropical sea-surface temperature: Atlantic vs. ENSO. *International Journal of Climatology*, **21**, pp. 973–1005.

Chase, B.M. and Meadows, M.E. 2007, Late Quaternary dynamics of southern Africa's winter rainfall zone. *Earth-Science Reviews*, **84**, pp. 103–138.

Chang, P., Fang, Y., Saravanan, R. Ji, L. and Seidel, H. 2006, The cause of the fragile relationship between the Pacific El Niño and the Atlantic Niño. *Nature*, **443**, pp. 324–328.

Colyn, M., Gauthier-Hion, A. and Verheyen, W. 1991, A re-appraisal of palaeoenvironmental history in Central Africa: evidence for a major fluvial refuge in the Zaïre basin. *Journal of Biogeography*, **18**, pp. 403–407.

Crowley, T.J. 2000, Causes of climate change over the past 1000 years. *Science*, **289**, pp. 270–277.

Damnati, B. 2000, Holocene lake records in the Northern Hemisphere of Africa. *Journal of African Earth Sciences*, **31/2**, pp. 253–262.

Dansgaard, W., Johnsen, S.J., Clausen, H.B., Dahl-Jensen, H.S., Gundestrup, N.S., Hammer, C.U., Hvidberg, C.S., Steffensen, J.S., Sveinbjörnsdottir, A.E., Jouzel, J. and Bond, G. 1993, Evidence for general instability of past climate from a 250-kyr ice-core record, *Nature*, **364**, pp. 218–220.

DeMenocal, P., Ortiz, J., Guilderson, T., Adkins, J., Sarnthein, M., Baker, L. and Yarusinsky, M. 2000, Abrupt onset and termination of the African Humid Period: rapid climate responses to gradual insolation forcing. *Quaternary Science Reviews*, **19**, pp. 347–361.

DeMenocal, P., Ortiz, J., Guilderson, T. and Sarnthein, M. 2000a, Coherent high- and low-latitude climate variability during the Holocene Warm Period. *Science*, **288**, pp. 2198–2202.

DeMenocal, P.B. 2001, Cultural responses to climate change during the Late Holocene. *Science*, **292**, pp. 667–673.

Denton, G.H. and Broecker, W.S. 2008, Wobbly ocean conveyor circulation during the Holocene? *Quaternary Science Reviews*, **27**, pp. 1939–1950.

De Ploey, J. 1964, Cartographie geomorphologique et morphogenese aux environs du Stanley-Pool (Congo). *Acta Geographica Lovaniensia*, **3**, pp. 431–441.

De Ploey, J. 1965, Position géomorphologique, génèse et chronologie de certains dépots superficiels du Congo occidental. *Quaternaria,* **7**, pp. 131–154.

De Ploey, J. 1969, Report on the Quaternary of the Western Congo. *Palaeoecology of Africa*, **4**, pp. 65–70.

Diaz, H.F., Hoerling, M.P. and Eischeid, J.K. 2001, ENSO variability, teleconnections and climate change. *International Journal of Climatology*, **21**, pp. 1845–1862.

Dupont, L.M. and Weinelt, M. 1996, Vegetation history of the savanna corridor between the Guinean and Congolian rain forest during the last 150,000 years. *Vegetation History and Archaeobotany*, **5**, pp. 273–292.

Dupont, L.M., Jahns, S., Marret, F. and Ning, S. 2000, Vegetation changes in equatorial West Africa: time slices for the last 150 ka. *Palaeogeography, Palaeoclimatology, Palaeoecology*, **155**, pp. 95–122.

Dupont, L.M., Marret, F. and Winn, K. 1998, Land–sea correlation by means of terrestrial and marine palynomorphs from the equatorial East Atlantic: phasing of SE trade winds and the ocean productivity. *Palaeogeography, Palaeoclimatology, Palaeoecology*, **142**, pp. 51–84.

Eggert, M.K.H., Höhn, A., Kahlheber, S., Meister, C., Neumann, K., and Schweizer, A. 2006, Pits, Graves and Grains: Archaeological and Archaeobotanical Research in Southern Cameroon. *Journal of African Archaeology*, **4**, pp. 273–298.

Eisenberg, J. 2008, A palaeoecological approach to neotectonics: The geomorphic evolution of the Ntem River in and below its interior delta, SW Cameroon. *Palaeoecology of Africa*, **28**, pp. 259–271.

Elenga, H., Schwartz, D. and Vincens, A. 1994, Pollen evidence of late Quaternary vegetation and inferred climate changes in Congo. *Palaeogeography, Palaeoclimatology, Palaeoecology*, **109**, pp. 345–356.

Elenga, H., Maley, J., Vincens, A. and Farrera, I. 2004, Palaeoenvironments, palaeoclimates and landscape development in Atlantic equatorial Africa: A review of key sites covering the last 25 kyrs In: Battarbee, R.W., Gasse, F. and Stickley, C.E. (Eds.): *Past climate variability through Europe and Africa.* Springer, pp. 181–198.

Eno Belinga, S.-M. 1984, *Géologie du Cameroun.* Librairie Universitaire, Université de Yaoundé, 1–307.

Fitze, P. 1973, Erste Ergebnisse neuerer Untersuchungen des Klettgauer Lößes. *Geographica Helvetica*, **2/28**, pp. 96–102.

Folland, C.K., Palmer, T.N. and Parker, D.E. 1986, Sahel rainfall and worldwide temperatures, 1901–1985. *Nature*, **320**, pp. 602–607.

Fontaine, B. and Bigot, S. 1993, West African rainfall deficits and sea surface temperatures. *International Journal of Climatology*, **13**, pp. 271–285.

Fontaine, B. and Janicot, S. 1996, Sea surface temperature fields associated with West African rainfall anomaly types. *Journal of Climate*, **9**, pp. 2935–2940.

Gallais, J. 1967, *Le delta intérieur du Niger et ses bordures; étude morphologique*. Mémoires et documents, nouvelle série 3, Centre National de la Recherche Scientifique, Paris.

Gasse, F. 2000, Hydrological changes in the African tropics since the Last Glacial Maximum. *Quaternary Science Reviews*, **19**, pp. 189–211.

Gasse, F., Lédéé, V., Massault, M. and Fontes, J.C. 1989, Water level fluctuations of Lake Tanganyika in phase with oceanic changes during the last glaciation and deglaciation. *Nature*, **342**, pp. 57–59.

Gasse, F. and Van Campo, E. 1994, Abrupt post-glacial climate events in West Asia and North Africamonsoon domains. *Earth and Planetary Science Letters*, **126**, pp. 435–456.

Gasse, F., Chalié, F., Vincens, A., Williams, M.A.J. and Williamson, D. 2008, Climatic patterns in equatorial and southern Africa from 30,000 to 10,000 years ago reconstructed from terrestrial and near-shore proxy data. *Quaternary Science Reviews*, **27**, pp. 2316–2340.

Gersonde, R., Crosta, X., Abelmann, A. and Armand, L. 2005, Sea-surface temperature and sea ice distribution of the Southern Ocean at the EPILOG Last Glacial Maximum—a circum-Antarctic view based on siliceous microfossil records. *Quaternary Science Reviews*, **24**, pp. 869–896.

Gingele, F.X., Müller, P.M. and Schneider, R.R. 1998, Orbital forcing of freshwater input in the Zaire Fan area—clay mineral evidence from the last 200 kyr. *Palaeogeography, Palaeoclimatology, Palaeoecology* **138**, pp. 17–26.

Giresse, P. 1978, Le contrôle climatique de la sédimentation marine et continentale en Afrique centrale atlantique à la fin du Quaternaire—Problèmes de corrélation. *Palaeogeography, Palaeoclimatology, Palaeoecology*, **23**, pp. 57–77.

Giresse, P., Kinga-Mouzeo, Schwartz, D. 1991, Breaks in the sedimentary and environmental equilibrium in the Congo Basin and incidences on the Oceanic sedimentation during the Quaternary. *Journal of African Earth Sciences*, **12**, pp. 229–236.

Giresse, P., Maley, J., and Brenac, P. 1994, Late Quaternary palaeoenvironments in the Lake Barombi Mbo (West Cameroon) deduced from pollen and carbon isotopes of organic matter. *Palaeogeography, Palaeoclimatology, Palaeoecology*, **107**, pp. 65–78.

Giresse, P., Maley, J. and Kossoni, A. 2005, Sedimentary environmental changes and millennial climatic variability in a tropical shallow lake (Lake Ossa, Cameroon) during the Holocene. *Palaeogeography, Palaeoclimatology, Palaeoecology*, **218**, pp. 257–285.

Giresse, P., Mvoubou, M., Maley, J. and Ngomanda, A. 2008, Late-Holocene equatorial environments inferred from deposition processes, carbon isotopes of organic matter, and pollen in three shallow lakes of Gabon, west-central Africa. *Journal of Paleolimnology*, doi: 10.1007/s10933–008–9231–5.

Goudie, A.S. 2005, The drainage of Africa since the Cretaceous. *Geomorphology*, **67**, pp. 437–456.

Gumnior, M. 2005, *Zur spätquartären Flussgeschichte NE-Nigerias. Morphologische, lithostratigraphische und pedologische Untersuchungen im Sedimentationsbereich der Tschadsee-Tributäre Komadugu Yobe und Komadugu Gana*. Doctoral thesis, University Frankfurt, 1–277.

Hall, A.M., Thomas, M.F. and Thorp, M.B. 1985, Late Quaternary alluvial placer development in the humid tropics: the case of the Birim Diamond Placer, Ghana. *Journal of the Geological Society*, **142**, pp. 777–787.

Hamilton, A.C. 1976, The significance of patterns of distribution shown by forest plants and animals in tropical Africa for the reconstruction of Upper Pleistocene environments: a review. *Palaeoecology of Africa*, **9**, pp. 63–97.

Hamilton, A.C. 1983, *Environmental history of East Africa—A study of the Quaternary*. Academic Press, New York, London, 1–328.

Hamilton, A.C. and Taylor, D. 1991, History of climate and forests in tropical Africa during the last 8 million years. *Climate Change*, **19**, pp. 65–78.

Haug, G.H., Hughen, K.A., Sigman, D.M., Peterson, L.C. and Röhl, U. 2001, Southward migration of the intertropical convergence zone through the Holocene. *Science*, **293**, pp. 1304–1308.

Hay, W.W. 1998, Detrital sediment fluxes from continents to oceans. *Chemical Geology*, **145**, pp. 287–323.

Heinrich, H. 1988, Origin and consequences of cyclic ice rafting in the northeast Atlantic Ocean during the past 130,000 years. *Quaternary Research*, **29**, pp. 142–152.

Höhn, A., Kahlheber, S., Neumann, K. and Schweizer, A. 2008, Settling the rain forest: The environment of farming communities in southern Cameroon during the First Millennium BC. *Palaeoecology of Africa*, **28**, pp. 29–41.

Holbrook, J. and Schumm, S.A. 1999, Geomorphic and sedimentary response of rivers to tectonic deformation: a brief review and critique of a tool for recognizing subtle epeirogenic deformation in modern and ancient settings. *Tectonophysics*, **305**, pp. 287–306.

Hughen, K., Southon, J., Lehman, S., Bertrand, C. and Turnbull, J. 2006, Marine-derived ^{14}C calibration and activity record for the past 50,000 years updated from the Cariaco Basin. *Quaternary Science Reviews*, **25**, pp. 3216–3227.

Hulme, M., Doherty, R., Ngara, T., New, M. and Lister, D. 2001, African climate change: 1900–2100. *Climate Research*, **17**, pp. 145–168.

IPCC, 2007: Climate Change 2007: Synthesis Report. Contribution of Working Groups I, II and III to the Fourth Assessment. Report of the Intergovernmental Panel on Climate Change [Core Writing Team, Pachauri, R.K and Reisinger, A. (Eds.)]. IPCC, Geneva, Switzerland, 1–104.

Jahns, S. 1996, Vegetation history and climate changes in West Equatorial Africa during the Late Pleistocene and Holocene, based on a marine pollen diagram from the Congo fan. *Vegetation History and Archaeobotany*, **5**, pp. 207–213.

Janicot, S., Harzallah, A., Fontaine, B. and Moron, V. 1998, West African monsoon dynamics and eastern equatorial Atlantic and Pacific SST anomalies (1970–88). *Journal of Climate*, **11**, pp. 1874–1882.

Jolly, D., Harrison, S.P., Damnati, B. and Bonnefille, R. 1998, Simulated climate and biomes of Africa during the Late Quaternary: comparison with pollen and lake status data. *Quaternary Science Reviews*, **17**, pp. 629–657.

Jouzel, J., Masson-Delmotte, V., Cattani, O., Dreyfus, G., Falourd, S., Hoffmann, G., Minster, B., Nouet, J., Barnola, J.M., Chappellaz, J., Fischer, H., Gallet, J.C., Johnsen, S., Leuenberger, M., Loulergue, L., Luethi, D., Oerter, H., Parrenin, F., Raisbeck, G., Raynaud, D., Schilt, A., Schwander, J., Selmo, E., Souchez, R., Spahni, R., Stauffer, B., Steffensen, J.P., Stenni, B., Stocker, T.F., Tison, J.L., Werner, M. and Wolff, E.W. 2007, Orbital and millennial Antarctic climate variability over the past 800,000 years. *Science*, **317/5839**, pp. 793–797.

Justice, C., Wilkie, D., Zhang, Q., Brunner, J. and Donoghue, C. 2001, Central African forests, carbon and climate change. *Climate Research*, **17**, pp. 229–246.

Kadomura, H. 1995, Palaeoecological and palaeohydrological changes in the humid tropics during the last 20.000 years, with reference to Equatorial Africa. In *Global Continental Palaeohydrology*, edited by Gregory, K.J., Starkel, L. and Baker, V.R. (Chichester: Wiley and Sons), pp. 177–202.

Kadomura, H. 2000, Late Holocene and historical forest degradation and savannization in Equatorial Africa. *Regensburger Geographische Schriften*, **33**, pp. 76–98.

Kadomura, H. and Hori, N. 1990, Environmental implications of slope deposits in humid tropical Africa: evidence from southern Cameroon and western Kenya. *Geographical Reports, Tokyo Metropolitan University*, **25**, pp. 213–236.

Kankeu, B. and Runge, J. 2008, Late neoproterozoic palaeogeography of Central Africa: Relations with Holocene geological and geomorphological setting. *Palaeoecology of Africa*, **28**, pp. 245–257.

Knox, J.C. 1995, Fluvial systems since 20,000 years BP. In *Global Continental Palaeohydrology*, edited by Gregory, K.J., Starkel, L. and Baker, V.R. (Chichester: Wiley and Sons), pp. 87–108.

Kuete, M. 1990, *Géomorphologie du plateau sud-Camerounais à l'ouest du 13°E*. Thèse de Doctorat d'Etat, Université de Yaoundé 1.

Kutzbach, J. and Liu, Z. (1997): Response of the African monsoon to orbital forcing and ocean feedbacks in the Middle Holocene. *Science*, **278**, pp. 440–443.

Laporte, N.T., Stabach, J.A., Grosch, R., Lin, T.S. and Goetz, S.J. 2007, Expansion of industrial logging in Central Africa. *Science*, **316**, p. 1451.

Latrubesse, E.M. 2003, The Late-Quaternary palaeohydrology of large South American fluvial systems. In *Palaeohydrology: Understanding Global Change*, edited by Gregory, K.J. and Benito, G. (New York: Wiley and Sons), pp. 193–212.

Latrubesse, E.M. 2008, Patterns of anabranching channels: The ultimate end-member adjustment of mega rivers. *Geomorphology*, doi: 10.1016/j.geomorph.2008.05.035.

Latrubesse, E.M. and Rancy, A. 2000, Neotectonic influence on tropical rivers of southwestern Amazon during the Late Quaternary: the Moa and Ipixuna river basins, Brazil. *Quaternary International*, **72**, pp. 67–72.

Latrubesse, E.M. and Franzinelli, E. 2002, The Holocene alluvial plain of the middle Amazon River, Brazil. *Geomorphology*, **44**, pp. 241–257.

Latrubesse, E.M. and Franzinelli, E. 2005, The late Quaternary evolution of the Negro River, Amazon, Brazil: Implications for island and floodplain formation in large anabranching tropical systems. *Geomorphology*, **70**, pp. 372–397.

Latrubesse, E.M., Stevaux, J.C. and Sinha, R. (2005), Tropical Rivers. *Geomorphology*, **70**, pp. 187–206.

Leal, M.E. 2004, *The African rainforest during the Last Glacial Maximum, an archipelago of forests in a sea of grass*. Doctoral thesis, Wageningen University, 1–96.

Lespez, L., Rasse, M. Le Drezen, Y., Tribolo, C., Huysecom, E. and Ballouche, A. 2008, Signal climatique et hydrosystèmes continentaux entre 50 et 4 ka en Afrique de l'Ouest: l'exemple de la vallée du Yamé (Pays dogon, Mali). *Géomorphologie: reliefs, processus, environnement*, **3**, pp. 169–186.

Lewin, J. and Macklin, M.G. (2003), Preservation potential for Late Quaternary river alluvium. *Journal of Quaternary Science*, **18/2**, pp. 107–120.

Lézine, A.-M. and Cazet, J.P. 2005, High-resolution pollen record from core KW31, Gulf of Guinea, documents the history of the lowland forests of West Equatorial Africa since 40,000 yr ago. *Quaternary Research*, **64**, pp. 432–443.

Lézine, A.-M., Duplessy, J.C. and Cazet, J.P. 2005, West African monsoon variability during the last deglaciation and the Holocene: Evidence from fresh water algae, pollen and isotope data from core KW31, Gulf of Guinea. *Palaeogeography, Palaeoclimatology, Palaeoecology*, **219**, pp. 225–237.

Lienou, G., Mahé, G., Paturel, J.E., Servat, E., Sighomnou, D., Ekodeck, G.E., Dezetter, A. and Dieulin, C. 2008, Evolution des régimes hydrologiques en région équatoriale camerounaise: un impact de la variabilité climatique en Afrique équatoriale? *Hydrological Sciences Journal,* **53**(4), pp. 789–801.

Littmann, T. 1987, Klimaänderungen in Afrika während der Würm-Eiszeit. *Geoökodynamik*, **8**, pp. 245–257.

Loulergue, L., Schilt, A., Spahni, R., Masson-Delmotte, V., Blunier, T., Lemieux, B., Barnola, J.-M., Raynaud, D., Stocker, T.F. and Chappellaz, J. 2008, Orbital and millennial-scale features of atmospheric CH_4 over the past 800,000 years. *Nature*, **453**, doi:10.1038/nature06950.

Lüthi, D., Le Floch, M., Bereiter, B., Blunier, T., Barnola, J.M., Siegenthaler, U., Raynaud, D., Jouzel, J., Fischer, H., Kawamura, K. and Stocker, T.F. 2008, High-resolution carbon dioxide concentration record 650,000–800,000 years before present. *Nature*, **453/7193**, pp. 379–382.

Makaske, B. 1998, *Anastomosing Rivers. Forms, processes and sediments*. Nederlandse Geografische Studies 249, Faculteit Ruimtelijke Wetenschappen, University Utrecht. 1–298.

Makaske, B. 2001, Anastomosing rivers: a review of their classification, origin and sedimentary products. *Earth-Science Reviews*, **53**, pp. 149–196.

Maley, J. 1981, Etudes palynologiques dans le bassin du Tchad et paléoclimatologie de l'Afrique nord-tropicale de 30,000 ans à l'epoque actuelle. *Travaux et Documents de l'ORSTOM*, **129**, Paris, 1–586.

Maley, J. 1987, Fragmentation de la forét dense humide africaine et extension des biotopes montagnards au Quaternaire récent: nouvelles données polliniques et chronologiques. Implications paléoclimatiques et biogéographiques. *Palaeoecology of Africa*, **18**, pp. 307–334.

Maley, J. 1989, Late Quaternary climatic changes in the African rain forest: Forest refugia and the major role of sea surface temperature variation. In *Paleoclimatology and Paleometeorology: Modern and Past Patterns of Global Atmospheric Transport*, NATO ASI, Series C: Mathematical and Physical Sciences, **282**, edited by Leinen, M. and Sarnthein, M., pp. 585–616.

Maley, J. 1991, The African rain forest vegetation and palaeoenvironments during Late Quaternary. *Climatic Change*, **19**, pp. 79–98.

Maley, J. 1992, Mise en évidence d'une péjoration climatique entre ca. 2 500 et 2 000 ans B.P. en Afrique tropicale humide. *Bulletin de la Société géologique de France*, **163/3**, pp. 363–365.

Maley, J. 1996, The African rain forest—main characteristics of changes in vegetation and climate from the Upper Cretaceous to the Quaternary. In: Alexander, I.J., Swaine, R.D. and Watling, R., (Eds.): *Essays on the Ecology of the Guinea-Congo Rain Forest.* Proceedings of the Royal Society of Edinburgh, **104B**, pp. 31–73.

Maley, J. 1997, Middle to Late Holocene Changes in Tropical Africa and other Continents: Paleomonsoon and Sea Surface Temperature Variations. In *Third Millenium BC Climate Change and Old World Collapse*, edited by Dalfes, H.N., Kukla, G. and Weiss, H. (Berlin: Springer), pp. 611–639.

Maley, J. 2001, The Impact of Arid Phases on the African Rain Forest Through Geological History. In *African Rain Forest Ecology and Conservation, An interdisciplinary perspective,* edited by Weber, W., White, L., Vedder, A. and Naughton Treves, L. (Yale: University Press), pp. 68–87.

Maley, J. 2002, A Catastrophic Destruction of African Forests about 2,500 Years Ago Still Exerts a Major Influence on Present Vegetation Formations. *IDS Bulletin*, **33/I**, pp. 13–30.

Maley, J. 2004, Les variations de la végétation et des paléoenvironnements du domaine forestier africain au cours du Quaternaire récent. In *Évolution de la Végétation Depuis Deux Millions d'années*, edited by Sémah, A.-M. and Renault-Miskovsky, J.L. (Paris: Artcom/Errance), pp. 143–178.

Maley, J. and Brenac, P. 1987, Analyses polliniques préliminaires du quaternaire récent de l'ouest Cameroun: mise en évidence de refuges forestiers et discussion des problèmes paléoclimatiques. Mémoires et Travaux de l'E.P.H.E. **17**, pp. 129–42.

Maley, J. and Brenac, P. 1998, Vegetation dynamics, Palaeoenvironments and Climatic changes in the Forests of West Cameroon during the last 28,000 years BP. *Review of Palaeobotany and Palynology*, **99**, pp. 157–187.

Maley, J. and Brenac, P. 1998a, Les variations de la végétation et des paléoenvironnements du Sud Cameroun au cours des derniers millénaires. Étude de l'expansion du Palmier à huile. In *Géosciences au Cameroun*. Collection GEOCAM, edited by Vicat, J.P. and Bilong, P. (Yaoundé: Université Yaoundé I), pp. 85–97.

Marchant, R. and Hooghiemstra, H. 2004, Rapid environmental change in African and South American tropics around 4000 years before present: a review. *Earth-Science Reviews*, **66**, pp. 217–260.

Marret, F., Scourse, J.D., Versteegh, G., Jansen, F.J.H. and Schneider, R. 2001, Integrated marine and terrestrial evidence for abrupt Congo River palaeodischarge fluctuations during the last deglaciation. *Journal of Quaternary Science*, 16, pp. 761–766.

Marret, F., Maley, J. and Scourse, J. 2006, Climatic instability in west equatorial Africa during the Mid- and Late Holocene. *Quaternary International*, **150**, pp. 71–81.

Maurizot, P. 2000, Carte géologique du sud-ouest Cameroun. Échelle 1:500 000. BRGM, Orléans Cedex 2.

Mayewski, P.A., Rohling, E.E., Stager, J.C., Karlén, W., Maasch, K.A., Meeker, L.D., Meyerson, E.A., Gasse, F., van Kreveld, S., Holmgren, K., Lee-Thorp, J., Rosqvist, G., Rack, F., Staubwasser, M., Schneider, R.R. and Steigl, E.J. 2004, Holocene Climat Variability. *Quaternary Research*, **62**, pp. 243–255.

McCarthy, T.S. 1993, The great inland deltas of Africa. *Journal of African Earth Sciences*, **17/3**: pp. 275–291.

McFadden, L.D. and Hendricks, M.D. 1985, Changes in the content and composition of pedogenic iron oxyhydroxides in a chronosequence of soils in southern California. *Quaternary Research*, **23**, pp. 189–204.

Mehra, O.P. and Jackson, M.L. 1960, Iron oxide removal from soils and clays by dithionite-citrate system buffered with sodium bicarbonate. *Clays and Clay Minerals Proceedings*, **7**, pp. 317–377.

Meister, C. 2008, Recent archaeological investigations in the tropical rain forest of south-west Cameroon. *Palaeoecology of Africa*, **28**, pp. 43–57.

Meiwes, K.-J., König, N., Khanna, P.K., Prenzel, J. and Ulrich, B. 1984, Chemische Untersuchungsverfahren für Mineralboden, Auflagehumus und Wurzeln zur Charakterisierung und Bewertung der Versauerung in Waldböden. *Berichte des Forschungszentrums Waldökosysteme/Waldsterben*, **7**, p. 67.

Molua, E.L. 2009, An empirical assessment of the impact of climate change on smallholder agriculture in Cameroon. *Global and Planetary Change*, doi: 10.1016/j.gloplacha.2009.02.006.

Moron, V., Bigot, S. and Roucou, P. 1995, Rainfall variability in Sub-equatorial America and Africa and relationships with the main sea-surface temperature modes (1951–1990). *International Journal of Climatology*, **15**, pp. 1297–1322.

Nanson, G.C., East, T.J. and Roberts, R.G. 1993, Quaternary stratigraphy, geochronology and evolution of the Magela Creek catchment in the monsoon tropics of northern Australia. *Sedimentary Geology*, **83**, pp. 277–302.

Nanson, G.C. and Price, D.M. 1998, Quaternary change in the Lake Eyre basin of Australia: an introduction. *Palaeogeography, Palaeoclimatology, Palaeoecology*, **144**, pp. 235–237.

Nanson, G.C., Price, D.M., Jones, B.G., Maroulis, J.C., Coleman, M., Bowman, H., Cohen, T.J., Pietsch, T.J. and Larsen, J.R. 2008, Alluvial evidence for major climate and flow regime changes during the middle and late Quaternary in eastern central Australia. *Geomorphology*, **101**, pp. 109–129.

Neumer, M. 2007, *Reliefformen, fluviale Morphodynamik und Sedimente in den wechselfeuchten Tropen Zentralafrikas: Indikatoren für die subrezente und rezente Landschaftsentwicklung*. Doctoral thesis, University Frankfurt, 1–304.

Ngako, V. 2006, Carte de Géologie, ressources minières. In *Atlas du Cameroun*, edited by Yahmed, D.B. (Paris: Les Éditions J.A.), pp. 60–61.

Ngako, V., Affaton, P., Nnange, J.M. and Njanko, T. 2003, Pan-African tectonic evolution in central and southern Cameroon: transpression and transtension during sinistral shear movements. *Journal of African Earth Sciences*, **36**, pp. 207–214.

Ngomanda, A. 2005, *Dynamique des écosystèmes forestiers du Gabon au cours des cinq derniers millénaires*. Thèse de Doctorat d'Etat, Université de Montpellier II.

Ngomanda, A., Neumann, K., Schweizer, A. and Maley, J. 2009, Seasonality change and the third millennium BP rainforest crisis in Central Africa: a high resolution pollen profile from Nyabessan, southern Cameroon. *Quaternary Research*, **71/3**, pp. 307–318.

Ngos III, S., Giresse, P. and Maley, J. 2003, Palaeoenvironments of Lake Assom near Tibati (south Adamawa, Cameroon). What happened in Tibati around 1700 years BP? *Journal of African Earth Sciences*, **37**, pp. 35–45.

Nguetsop, V.F., Servant-Vildary, S. and Servant, M. 2004, Late Holocene climatic changes in West Africa, a high resolution diatom record from equatorial Cameroon. *Quaternary Science Reviews*, **23**, pp. 591–609.

Nicholson, S.E. 1980, The nature of rainfall fluctuations in subtropical West Africa. *Monthly Weather Review*, **108**, pp. 473–487.

Nicholson, S.E. 1986, The spatial coherence of African rainfall anomalies— interhemispheric teleconnections. *Journal of Climate and Applied Meteorology*, **25/10**, pp. 1365–1381.

Nicholson, S.E. 2000, The nature of rainfall variability over Africa on time scales of decades to millennia. *Global and Planetary Change*, **26**, pp. 137–158.

North Greenland Ice Core Project (NGRIP) Members, 2004, High-resolution record of Northern Hemisphere climate extending into the last interglacial period. *Nature*, **431**, pp. 147–151.

Nott, J. and Price, D. 1999, Waterfalls, floods and climate change: evidence from tropical Australia. *Earth and Planetary Science Letters*, **171**, pp. 267–276.

Olivry, J.C. 1986, *Fleuves et rivières du Cameroun*. ORSTOM, Paris, 1–733.

Oslisly, R. 2006, Les traditions culturelles de l'Holocene sur le littoral du Cameroun entre Kribi et Campo. In *Grundlegungen. Beiträge zur europäischen und afrikanischen Archäologie für Manfred K.H. Eggert*, edited by Wotzka, H.-P. (Tübingen: Francke Attempto Verlag), pp. 303–317.

Paeth, H. and Hense, A. 2006, On the linear response of tropical African climate to SST changes deduced from regional climate model simulations. *Theoretical and Applied Climatology*, **83**, pp. 1–19.

Paeth, H. and Stuck, J. 2004, The West African dipole in rainfall and its forcing mechanisms in global and regional climate models. *Mausam*, **55**, pp. 561–582.

Paeth, H., Scholten, A., Friederichs, P. and Hense, A. 2008, Uncertainties in climate change prediction: El Niño-Southern Oscillation and monsoons. *Global and Planetary Change*, **60**, pp. 265–288.

Pastouret, L., Chamley, H., Delibrias, G., Duplessy, J.C. and Theide, J. 1978, Late Quaternary climatic changes in Western Tropical Africa deduced from deep-sea sedimentation off the Niger Delta. *Oceanologica Acta*, **1**, pp. 217–232.

Petit, J.R., Jouzel, J., Raynaud, D., Barkov, N.I., Barnola, J.M., Basile, I., Bender, M., Chappellaz, J., Davis, J., Delaygue, G., Delmotte, M., Kotlyakov, V.M., Legrand, M., Lipenkov, V., Lorius, C., Pépin, L., Ritz, C., Saltzman, E. and Stievenard, M. 1999, Climate and atmospheric history of the past 420,000 years from the Vostok Ice Core, Antarctica. *Nature*, **399**, pp. 429–436.

Pokras, E.M. 1987, Diatom record of Late Quaternary climatic change in the eastern equatorial Atlantic and tropical Africa. *Paleoceanography*, **2/3**, pp. 273–286.

Prell, W.L. and Kutzbach, J.E. 1987, Monsoon variability over the past 150,000 years. *Journal of Geophysical Research*, **92**, pp. 8411–8425.

Prentice, I.C., Cramer, W., Harrison, S.P., Leeman, R., Monserud, R.A. and Solomon, A.M. 1992, A global Biome model based on plant physiology and dominance, soil properties and climate. *Journal of Biogeography*, **19**, pp. 117–134.

Preuss, J. 1986, Jungpleistozäne Klimaänderungen im Kongo-Zaire-Becken. *Geowissenschaften in unserer Zeit*, **4/6**, pp. 177–187.

Preuss, J. 1986a, Die Klimaentwicklung in den äquatorialen Breiten Afrikas im Jungpleistozän. Versuch eines Überblicks im Zusammenhang mit Geländearbeiten in Zaire. In *Geographische Forschung in Marburg –Eine Dokumentation aktueller Arbeitsrichtungen*, Marburger Geographische Schriften, **100**, edited by Andres, W., Buchhofer, E. and Mertins, G., pp. 132–148.

Preuss, J. 1990, L'évolution des paysages du bassin intérieur du Zaire pendant les quarante derniers millénaires. In *Paysages quaternaires de l'Afrique centrale atlantique*, edited by Lanfranchi, R. and Schartz, D. (Paris: ORSTOM), pp. 260–270.

Redelsperger, J.-L., Thorncroft, C.D., Diedhiou, A., Lebel, T., Parker, D.J. and Polcher, J. 2006, African Monsoon Multidisciplinary Analysis. An international research project and field campaign. *Bulletin of the American Meteorological Society*, **87/12**, pp. 1739–1746.

Reynaud-Farrera, I., Maley, J. and Wirrmann, D. 1996, Végétation et climat dans les forêts du Sud-Ouest Cameroun depuis 4770 ans B.P.: analyse pollinique des sédiments du Lac Ossa. *Comptes Rendus de l'Académie des Sciences, Séries IIa*, **322**, pp. 749–755.

Rodbell, D.T., Seltzer, G.O., Anderson, D.M., Abbott, M.B., Enfield, D.B. and Newman, J.H. 1999, An ~15.000-year record of El Niño-driven alluviation in southwestern Ecuador. *Science*, **283**, pp. 516–520.

Rodier, J. 1964, *Régimes hydrologiques de l'Afrique noire à l'ouest du Congo*. O.R.S.T.O.M., Paris, 1–137.

Rossignol-Strick, M. 1983, African monsoons, an immediate climate response to orbital insolation. *Nature*, **304**, pp. 46–49.

Runge, J. 1992, Geomorphological observations concerning palaeoenvironmental conditions in eastern Zaire. *Zeitschrift für Geomorphologie N. F., Suppl. Bd.* **91**, pp. 109–122.

Runge, J. 1996, Palaeoenvironmental interpretation of geomorphological and pedological studies in the rain forest "core areas" of eastern Zaire (Central Africa). *South African Geographical Journal*, **78**, pp. 91–97.

Runge, J. 2001, Landschaftsgenese und Paläoklima in Zentralafrika. Physiogeographische Untersuchungen zur Landschaftsentwicklung und klimagesteuerten quartären

Vegetations- und Geomorphodynamik in Kongo/Zaire (Kivu, Kasai, Oberkongo) und der Zentralafrikanischen Republik (Mbomou). *Relief Boden Paläoklima*, **17**, pp. 1–294.

Runge, J. 2002, Holocene landscape history and palaeohydrology evidenced by stable carbon isotope ($\delta^{13}C$) analysis of alluvial sediments in the Mbari valley (5°N/23°E), Central African Republic. *Catena*, **48**, pp. 67–87.

Runge, J. (Ed.), 2008, *Dynamics of forest ecosystems in Central Africa during the Holocene. Past-Present-Future*. Taylor and Francis/Balkema, Leiden, 1–306.

Runge, J. 2008a, Of deserts and forests: insights into Central African palaeoenvironments since the Last Glacial Maximum. *Palaeoecology of Africa*, **28**, pp. 15–27.

Runge, J., Eisenberg, J. and Sangen, M. 2005, Ökologischer Wandel und kulturelle Umbrüche in West- und Zentralafrika—Prospektionsreise nach Südwestkamerun vom 05.03.-03.04.2004 im Rahmen der DFG-Forschergruppe 510: Teilprojekt "Regenwald-Savannen-Kontakt (ReSaKo)". *Geoökodynamik*, **26**, pp. 135–154.

Runge, J., Eisenberg, J. and Sangen, M. 2006, Geomorphic evolution of the Ntem alluvial basin and physiogeographic evidence for Holocene environmental changes in the rain forest of SW Cameroon (Central Africa)—preliminary results. *Zeitschrift für Geomorphologie*, **145**, pp. 63–79.

Russell, J., Talbot, M.R. and Haskell, B.J. 2003, Mid-Holocene climate change in Lake Bosumtwi, Ghana. *Quaternary Research*, **60/2**, pp. 133–141.

Salzmann, U. and Hoelzmann, P. 2005, The Dahomey Gap: an abrupt climatically induced rain forest fragmentation in West Africa during the late Holocene. *The Holocene*, **15/2**, pp. 190–199.

Sandweiss, D.H., Maasch, K.A., Burger, R.L., Richardson, J.B., Rollins, H.B. and Clement, A.C. 2001, Variations in Holocene El Niño frequencies: climate records and cultural consequences in ancient Peru. *Geology*, **29**, pp. 603–606.

Sangen, M. 2007, Physiogeographische Untersuchungen zur holozänen Umweltgeschichte an Alluvionen des Ntem-Binnendeltas im tropischen Regenwald SW-Kameruns. *Zentralblatt für Geologie und Paläontologie*, **1/4**, pp. 113–128.

Sangen, M. 2008, New evidence on palaeoenvironmental conditions in SW Cameroon since the Late Pleistocene derived from alluvial sediments of the Ntem River. *Palaeoecology of Africa*, **28**, pp. 79–101.

Sangen, M. 2009, Physiogeographische Untersuchungen zur pleistozänen und holozänen Umweltgeschichte an Alluvionen des Ntem-Binnendeltas und alluvialer Sedimente der Flüsse Boumba, Ngoko, Nyong und Sanaga in Süd-Kamerun. Doctoral thesis, University Frankfurt, Frankfurt, 1–319.

Sangen, M. 2011, New results on palaeoenvironmental conditions in equatorial Africa derived from alluvial sediments of Cameroonian rivers. *Proceedings of the Geologists' Association*, **122**, pp. 212–223.

Sangen, M, Eisenberg, J., Kankeu, B., Runge, J. and Tchindjang, M. 2010, Preliminary results on palaeoenvironmental research carried out in the framework of the second phase of the ReSaKo-project in the upper catchments areas of the Nyong and Sanaga Rivers in Cameroon. *Palaeoecology of Africa*, **30**, pp. 165–188.

Sangen, M., Neumann, K. and Eisenberg, J. 2011, Climate induced fluvial dynamics in tropical Africa around the last glacial maximum? *Quaternary Research*, **76**, pp. 417–429.

Schneider, R.R., Müller, P.J. and Wefer, G. 1994, Late Quaternary paleoproductivity changes off the Congo deduced from stable carbon isotopes of planktonic foraminifera. *Palaeogeography, Palaeoclimatology, Palaeoecology*, **110**, pp. 255–274.

Schneider, R.R., Price, B., Müller, P.J., Kroon, D. and Alexander, I. 1997, Monsoon related variations in Zaire (Congo) sediment load and influence of fluvial silicate supply on marine productivity in the east equatorial Atlantic during the last 200,000 years. *Paleoceanography*, **12/3**, pp. 463–481.

Schumm, S.A. 1977, *The Fluvial System*. Wiley, New York, 1–338.

Schumm, S.A., Dumont, J. and Holbrook, J. (Eds.), 2000, *Active Tectonics and Alluvial Rivers*. Cambridge University Press, 1–276.

Schwartz, D. 1992, Assèchement climatique vers 3000 B.P. et expansion Bantu en Afrique centrale atlantique: quelques réflexions. *Bulletin de la Société géologique de France*, **3**, pp. 353–361.

Schwartz, D. and Lanfranchi, R. 1993, Les cadres paléoenvironnementaux de l'évolution humaine en Afrique centrale atlantique. *L'Anthropologie*, **97**, pp. 17–50.

Schwartz, D., De Foresta, H., Mariotti, A., Balesdent, J., Massimba, J.P. and Girardin, C. 1995, Present dynamics of the savanna-forest boundary in the Congolese Mayombe: a pedological, botanical and isotopic (^{13}C and ^{14}C) study. *Oecologia*, **106**, pp. 516–524.

Schwartz, D., Mariotti, A., Lanfranchi, R. and Guillet, B. 1986, $^{13}C/^{12}C$ ratios of soil organic matter as indicators of vegetation changes in the Congo. *Geoderma*, **39**, pp. 97–103.

Segalen, P. 1967, Les sols et géomorphologie du Cameroun. *Cahiers ORSTOM, Série Pédologie*, **5/2**, pp. 137–187.

Semazzi, F.H.M. and Song, Y. 2001, A GCM study of climate change induced by deforestation in Africa. *Climate Research*, **17**, pp. 169–182.

Servain J. 1991, Simple climatic indices for the tropical Atlantic ocean and some applications. *Journal of Geophysics Research*, **96**, pp. 15137–15146.

Servant, M. and Servant-Vildary, S. (Eds.), 2000, Dynamique à long terme des écosystèmes forestiers intertropicaux. Publications issues du Symposium international «*Dynamique à long terme des écosystèmes forestiers intertropicaux*», *Paris, 20–22 mars* 1996. Paris, UNESCO, 1–434.

Servat, É., Paturel, J.E. and Lubès, H. 1996, La sécheresse gagne l'Afrique tropicale. *La Recherche*, **290**, pp. 24–25.

Shi, N., Dupont, L.M., Beug, H.-J. and Schneider, R. 2000, Correlation between vegetation in southwestern Africa and oceanic upwelling in the past 21,000 years. *Quaternary Research*, **54**, pp. 72–80.

Shi, N., Schneider, R., Beug, H.-J. and Dupont, L.M. 2001, Southeast trade wind variations during the last 135 kyr: evidence from pollen spectra in eastern South Atlantic sediments. *Earth and Planetary Science Letters*, **187**, pp. 311–321.

Stager, J.C. and Anfang-Sutter, R. 1999, Preliminary evidence of environmental changes at Lake Bambili (Cameroon, West-Africa) since 24,000 BP. *Journal of Paleolimnology*, **22**, pp. 319–330.

Stevaux, J.C. 1994, The Upper Paraná River (Brazil); geomorphology, sedimentology and paleoclimatology. *Quaternary International*, **21**, pp. 143–161.

Street, F.A. and Grove, A.T. 1976, Environmental and climatic implications of Late Quaternary lake-level fluctuations in Africa. *Nature*, **261**, pp. 385–390.

Street, F.A. and Grove, A.T. 1979, Global maps of lake level fluctuations since 30,000 yr B.P. *Quaternary Research*, **12**, pp. 83–118.

Stuut, J.B.W., Crosta, X., Van der Borg, K. and Schneider, R. 2004, Relationship between Antarctic sea ice and southwest African climate during the late Quaternary. *Geology*, **32**, pp. 909–912.

Stuut, J.B.W., Prins, M.A., Schneider, R.R., Weltje, G.J., Jansen, J.H.F. and Postma, G. 2002, A 300-kyr record of aridity and wind strength in southwestern Africa: inferences from grain-size distributions of sediments of Walvis Ridge, SE Atlantic. *Marine Geology*, **180**, pp. 221–233.

Suchel, J.-B. 1987, *Les climats du Cameroun*. These Doctoral, Université de Bordeaux III, 1–1186.

Summerfield, M.A. 1985, Tectonic background to long-term landform development in tropical Africa. In *Environmental Change and Tropical Geomorphology*, edited by Douglas, I. and Spencer, T. (London, Boston: Allen and Unwin), pp. 281–294.

Talbot, M.R., Livingstone, D.A., Palmer, P.G., Maley, J., Maleck, J.M., Delibrias, G. and Gulliksen, S. 1984, Preliminary results from sediment cores from Lake Bosumtwi, Ghana. *Palaeoecology of Africa*, **16**, pp. 173–192.

Tchouto Mbatchou, G. P. 2004, Plant diversity in a tropical rain forest. Implications for biodiversity and conservation in Cameroon. Doctoral thesis, Wageningen University.

Thomas, M.F. 2000, Late Quaternary environmental changes and the alluvial record in humid tropical environments. *Quaternary International*, **72**, pp. 23–36.

Thomas, M.F. 2003, Late Quaternary sediment fluxes from tropical watersheds. *Sedimentary Geology*, **162**, pp. 63–81.

Thomas, M.F. 2004, Landscape sensitivity to rapid environmental change—a Quaternary perspective with examples from tropical areas. *Catena*, **55**, pp. 107–124.

Thomas, M.F. 2008, Understanding the impacts of Late Quaternary climate change in the tropical and sub-tropical regions. *Geomorphology*, **101**, pp. 146–158.

Thomas, M.F. and Thorp, M.B. 1980, Some aspects of geomorphological interpretation of Quaternary alluvial sediments in Sierra Leone. *Zeitschrift für Geomorphologie*, N.F. **36**, pp. 140–161.

Thomas, M.F. and Thorp, M.B. 1995, Geomorphic response to rapid climatic and hydrologic change during the late Pleistocene and early Holocene in the humid and sub-humid tropics. *Quaternary Science Reviews*, **14**, pp. 193–207.

Thomas, M.F. and Thorp, M.B. 2003, Palaeohydrological reconstruction for tropical Africa since the Last Glacial Maximum—evidence and problems. In: Benito, G. and Gregory, K.J. (Eds.): *Palaeohydrology: Understanding Global Change*. Wiley, Chichester, pp. 167–192.

Thorp, M.B. and Thomas, M.F. 1992, The timing of alluvial sedimentation and floodplain formation in the lowland humid tropics of Ghana, Sierra Leone and western Kalimantan (Indonesian Borneo). *Geomorphology*, **4**, pp. 409–422.

Torrent, J., Schwertmann, U. and Schulze, D.G. 1980, Iron oxide mineralogy of some soils of two river terrace sequences in Spain. *Geoderma*, **23**, pp. 191–208.

Toteu, S.F., Penaye, J. and Djomani, Y.P. 2004, Geodynamic evolution of the Pan-African belt in central Africa with special reference to Cameroon. *Canadian Journal of Earth Sciences*, **41**, pp. 73–85.

Van Geel, B., Raspopov, O.M., Renssen, H., Van Der Plicht, J., Dergachev, V.A. and Meijer, H.A.J. 1999, The role of solar forcing upon climate change. *Quaternary Science Reviews*, **18**, pp. 331–338.

Van Zinderen Bakker, E.M. 1967, Upper Pleistocene and Holocene stratigraphy and ecology on the basis of vegetation changes in sub-Saharan Africa. In *Background to Evolution in Africa*, edited by Bishop, W.W. and Eas, D.C. (Chicago: University Chicago Press), pp. 125–147.

Van Zinderen Bakker, E.M. 1969, *Palaeoecology of Africa and of the surrounding islands and Antarctica*, Volume 4, covering the years 1966–1968. A. A. Balkema, Cape Town, 1–274.

Van Zinderen Bakker, E.M. 1975, The origin and paleoenvironment of the Namib Desert biome. *Journal of Biogeography*, **2**, pp. 65–73.

Vandenberghe, J. 2003, Climate forcing of fluvial system development: an evolution of ideas. *Quaternary Science Reviews*, **22**, pp. 2053–2060.

Verschuren, D. 2004, Decadal and century-scale climate variability in tropical Africa during the past 2000 years. In: Battarbee, R.W., Gasse, F. and Stickley, C.E. (Eds.): *Past climate variability through Europe and Africa*. Springer, pp. 139–158.

Vidal, L. and Arz, H. 2004, Oceanic climate variability at millennial time-scales: Models of climate connections. In *Past climate variability through Europe and Africa*, edited by Battarbee, R.W., Gasse, F. and Stickley, C.E. (Dordrecht: Springer), pp. 31–44.

Vidal, L., Schneider, R.R., Marchal, O., Bickert, T., Stocker, T.F. and Wefer, G. 1999, Link between the North and South Atlantic during the Heinrich events of the last glacial period. *Climate Dynamics*, **15**, pp. 909–919.

Vincens, A., Buchet, G., Elenga, H., Fournier, M., Martin, L., de Namur, C., Schwartz, D., Servant, M. and Wirrmann, D. 1994, Histoire et Changement majeur de la végétation du lac Sinnda (vallée du Niari, Sud-Congo) consécutif à l'assèchement climatique holocène supérieur: apport de la palynologie. *C. R. Acad. Sci. Paris, Serie 2*, 318, pp. 1521–1526.

Vincens, A., Schwartz, D., Bertaux, J., Elenga, H. and de Namur, C. 1998, Late Holocene climatic changes in western Equatorial Africa inferred from pollen lake Sinnda, Southern Congo. *Quaternary Research*, **50**, pp. 34–45.

Vincens, A., Schwartz, D., Elenga, H., Reynaud-Farrera, I., Alexandre, A., Bertaux, J., Mariotti, A., Martin, L., Meunier, J.-D., Nguetsop, F., Servant, M., Servant-Vildary, S. and Wirrmann, D. 1999, Forest response to climate changes in Atlantic Equatorial Africa during the last 4000 years BP and inheritance on the modern landscapes. *Journal of Biogeography*, **26**, pp. 879–885.

Wang, S., Chen, Z. and Smith, D.G. 2005, Anastomosing river system along the subsiding middle Yangtze River basin, southern China. *Catena*, **60**, pp. 147–163.

Wanner, H., Beer, J., Bütikofera, J., Crowley, T.J., Cubasch, U., Flückiger, J., Goosse, H., Grosjean, M., Joos, F., Kaplan, J.O., Küttel, M., Müller, S.A., Prentice, I.C., Solomina, O., Stocker, T.F., Tarasov, P., Wagner, M. and Widmann, M. 2008, Mid- to Late Holocene climate change: an overview. *Quaternary Science Reviews*, **27**, pp. 1791–1828.

Weldeab, S. 2008, Control of West African monsoon precipitation: Insights from the past. *PAGES News*, **16/1**, pp. 16–18.

Weldeab, S., Frank, M., Stichel, T., Schneider, R., Haley, B. and Sangen, M. 2011, Decoupling between past ITCZ movement and rainfall intensity over the West African monsoon area. *Geophysical Research Letters*, **38**, doi:10.1029/2011GL047805.

Weldeab, S., Lea, D.W., Schneider, R.R. and Andersen, N. 2007, 155,000 years of West African monsoon ocean thermal evolution. *Science*, **316**, pp. 1303–1307.

Weldeab, S., Lea, D.W., Schneider, R.R. and Andersen, N. 2007a, Centennial scale climate instabilities in a wet early Holocene West African monsoon. *Geophysical Research Letters*, **34**, L24702, doi: 10.1029/2007GL031898.

Weldeab, S., Schneider, R.R., Kölling, M. and Wefer, G. 2005, Holocene African droughts relate to eastern equatorial Atlantic cooling. *Geology*, **33/12**, pp. 981–984.

WMO, 2005, Hydrological information systems for integrated water resources management. WHYCOS Guidelines. WMO/TD-No. 1282, 1–48.

Youta Happi, J. 1998, Arbres contre graminées: la lente invasion de la savane par la forêt au Centre-Cameroun. Thèse de Doctorat d'Etat, Université de Paris-Sorbonne, 1–238.

Zabel, M., Schneider, R., Wagner, T., Adegbie, A.T., De Vries, U. and Kolonic, S. 2001, Late Quaternary Climate Changes in Central Africa as inferred from Terrigenous Input in the Niger Fan. *Quaternary Research*, **56**, pp. 207–217.

Zhang, C., Woodworth, P. and Gu, G. 2006, The seasonal cycle in the lower troposphere over West Africa from sounding observations. *Quaterly Journal of the Royal Meteorological Society*, **132**, pp. 2559–2582.

Zhao, Y., Braconnot, P., Harrison, S.P., Yiou, P. and Marti, O. 2007, Simulated changes in the relationship between tropical ocean temperatures and the western African monsoon during the mid-Holocene. *Climate Dynamics*, **28**, pp. 533–551.

CHAPTER 5

Challenges of climate change, landscape dynamics and environmental risks in Cameroon

Mesmin Tchindjang
*Department of Geography, Faculty of Art, Letters and Social Sciences
University of Yaoundé I, Yaoundé, Cameroon*

Joseph Armathée Amougou
Ministry of Environment and Protection of the Nature, Cameroon

Samuel Aimé Abossolo
*Department of Geography, Faculty of Art, Letters and Social Sciences
Université de Yaoundé I, Yaoundé, Cameroon*

Stanislas Bessoh Bell
*Department of Plant Biology, Faculty of Sciences University
of Yaoundé I, Yaoundé, Cameroon*

ABSTRACT: Climate change since the Quaternary appears to be related to solar cycles as well as anthropogenic activities that have accelerated the phenomenon. Studies on variability and climate change have interested the world community after several large-scale climate events. Of these, one can identify the 1972–1973 and 1983–1984 droughts felt in Cameroon and which hit most of the countries of tropical Africa. Apart from these droughts, the global warming findings of the World Meteorological Organization (WMO) and those of the Intergovernmental Panel on Climate Change (IPCC), show a temperature increase of about 0.7° C since the beginning of the last century. This paper is based on basic research begun in 2008 focusing on climate change in Cameroon and its related risks. This work attests to these changes at different scales. Indeed, the drought over the last forty years in Sahelian countries has extended to the humid countries bordering the Gulf of Guinea (including Central Africa and Cameroon), with serious consequences (e.g., decreasing of rainfall, lowering of groundwater levels, drop and deficit of stream flows) to the whole region.The Sanaga River basin was chosen, because it has the advantage of belonging to three climatic zones and, thus, it is better able to illustrate the phenomenon, and gives an idea of its impact on availability of water resources for hydroelectric production. The large rainfall deficit has caused a significant decrease in flows and the mortgage of the country's electricity provision. Characterization of rainfall variability using statistical methods and the evaluation of its impacts on groundwater resources are the first major objectives of this study. Indeed, starting from the background of the Cameroonian geological setting, this paper makes the link between its configuration and ongoing challenges of climate change and subsequent modifications of the landscape. Special mention is made to related environmental hazards. The challenge to this landscape dynamic involves changes such as deforestation, increased flooding and erosive potential of rivers, processes that induce a new landscape dynamic with acceleration of geomorphic processes. Therefore, significant environmental risks may appear as a consequence. Deforestation is linked to agricultural activities, mining or logging, and has brought an increase in erosion and degradation of natural conditions. In addition to mass movements, changes in fluvial processes are an inevitable consequence because the increased sediment budget on the deforested watershed reduces or affects the erosive power of rivers by increasing accumulation. Sedimentation recorded at the Edéa dam on the Sanaga River, are proof of such a situation. Morever, the occurrence of increased flooding appears to be the main consequence of deforestation and changes in water processes. Mass movements are the most visible face of environmental risk arising from these activities. By means of a synthesis of landform dynamics in Cameroon, this paper seeks to analyze the effect climate change will have under a deforestation regime; with a particular emphasis on the causes (logging, mining and oil extraction, development of large dams), consequences (including environmental hazards like mass movements and flooding, increasing of sediment budget) and challenges (mining governance and afforestation or reforestation) of deforestation.

5.1 INTRODUCTION

With its wide latitudinal (2–13°N over 1500 km) and meridional (9–16°E over 800 km) extension, Cameroon by its triangular shape, holds a unique position in Africa. Its location above the Congo Craton and opening to the Atlantic Ocean gives it a special, morphologically pivotal, position on the African continent. Cameroon has many mountains and basins, and among these last, are included, in the north, the endorheic basin of Lake Chad and the Doba Trough; in the west, the Cretaceous Benue Trough; in the south-east the Congo basin; in the south-west, the Mamfe basin in the Douala coastal basin. Indeed, all the geodynamic events that marked the African continent (since the break-up of Gondwana to the Present) are recorded in Cameroon. Globally, it summarizes the main morphological features of Africa. This country also intersects the main morpho- and bio-climatic tropical areas, and the dynamics of landscapes follow these contours. Thus, the basins and pediplains, bristling with inselbergs and dominated by massive concave shaped slopes observed in the Sahel region of the North, contrast with large polyconvex shaped features, marked by forest degradation, in the southern Cameroon plateau overhanging the Atlantic coastal plain. Between these two entities, mountain ranges emerged built up by the endogenous processes (volcanism). These landscapes are limited by steep tectonic and lithological escarpments. They are essentially cut in the Precambrian basement, and were the result of a long development marked by planation surfaces, derived from an alternating process of backwearing and downwearing. In fact, the geological diversity explains the morphological structure of the country and, therefore, understanding of Cameroonian landscape dynamics has come from a thorough grasp of these features by pedologists, geologists and geophysicists. For soil scientists, correlations in soil studies conducted in West Africa by Grandin (1976) and Boulangé (1984), produced the first hypothesis developed for their study are based, in part, on the presence and age of a bauxite cuirass.

From geologists like Déruelle *et al.* (1983, 2007) and Tchoua (1974), studying Cameroon mountain volcanism, came the hypothesis of hot spots and plumes. In addition, new geophysical studies from Poudjom Djomani *et al.* (1992), Boukéké (1994), Burke and Whiteman (1972), and Burke and Gunnell (2008) attempt to establish the development of Cameroonian mountains and features in the Adamawa plateau.This plateau represents a batholith, which was formed during the Pan African thermo-tectonic stage (550–600 Ma). This axial massive consisting of a Precambrian basement, which was raised from the Upper Cretaceous to Tertiary, is consider as the starting point for development of the Cameroon highlands, in particular, and the configuration of the country features, in general.

A "domes" hypothesis has been invoked by these geophysicists to justify its occurrence. Since the establishment of the Adamawa Dome, three major tectonic structures have been associated with this plateau: the Cameroon Volcanic Line (CVL, which constitutes a series of volcanic and anorogenic events and built from the Cretaceous to Quaternary along a N30°E fault line); the Benue Trough, presented like an aborted rift (Dumont, 1986, 1987); and the Central African shear zone, which gathered major accidents and is oriented along N70°E, extending from the Gulf of Guinea to Sudan (Cornacchia and Dars, 1983; Ngako *et al.*, 1991; Ngako, 1999; Moreau *et al.*, 1987). This structure and the Sanaga fault, which was set up before the opening of the Atlantic Ocean, extend to Brazil by the Pernambuco fault (Almeida and Black, 1967). These central African lineaments (Guiraud *et al.*, 1985), cross both continental crustand upper mantle to a depth of 190 km. Finally, the geological history of Cameroon shows alternating warm and humid periods until the late Tertiary. These changes affected the volumes in place in the Tertiary, leading to hardening. During this period and until the Quaternary, erosion of the structure carved features and shaped their planation, the result of backwearing and downwearing processes (compare contribution by Boniface Kankeu in this volume).

For palynologists and geomorphologists (Kadomura, 1984, 1995; Kadomura and Hori, 1990; Kadomura and Kiyonoga, 1994; Maley and Brenac, 1998, 1998; Maley, 2001; Giresse *et al.*, 1994, 2008; Maley and Brenac, 1987; Sangen, 2009), proxy data analysis has allowed to locate and outline major landscape changes since the Quaternary. These authors (see above) also detected past deforestation crises in Cameroon.The significance of the landscape coupled with palaeoclimatic reconstitution of its approximate rhythms, provides a better understanding of the current development of environments and palaeoenvironments around lakes and rivers.

In terms of external geodynamics, one can identify with Dewolf and Bourrié (2008), that "the surface deposits are correlated to landscape evolution, their identification in the field are logically linked to their relation to the forms and this, what ever the original shape and the scale envisaged". Therefore, processes and landforms depend on the types of rocks and formation of the volumes. Natural dynamics of slopes is, thus, amplified by the presence of tectonic or cyclic scarps, where gravity and widespread subsidence, erosion and slippage occur regularly. Anthropogenic processes finally come as an aggravating factor of the natural processes. If Quaternary palaeoclimatic oscillations set up palaeo-landforms and palaeoenvironments linked to the morphogenesis of the fluvial systems, the geological history retains the "memory" of the genesis of materials constituting the relictual forms that contributed to the finalization of features from the Precambrian to Present. Given this, the tectonic and neotectonic landscape will be affected by significant changes related to anthropogenic activities such as deforestation, which peaked around 3 kyrs BP. This exposed the rivers to denudation and accelerated morphological processes.

Also, since the Quaternary, climate change affecting the world appears to have been related to solar cycles as well as anthropogenic activities. To understand its effect on the landscape in Cameroon, several studies have been undertaken by Eisenberg (2009); and Sangen (2009, see contributions in this volume) or IRD (Institut de Recherche pour le Développement, e.g., palaeoenvironments along the Nyong and Sanaga, Liénou *et al.*, 2008; Sighomnou, 2004). These works attest to changes at different scales. The challenges associated with this situation concern the occurrence of environmental risks such as deforestation, which lead to the increased likelihood of stream erosion and flooding. All these events induce a new landscape dynamic with the acceleration of geomorphic processes. The particular geodynamic configuration of Cameroon exposes it to certain geological hazards such as earthquake and volcanism, in addition to other hazards associated with gravity, rainfall and anthropogenic activities. Four types of hazards have been identified in Cameroon in past tectonic and neotectonic areas.

- Seismic events would seem to be of great concern, but this hazard is only a minor threat because of its low magnitude <VI. Seismic events have been recorded since 1852 and are seemingly possible any where in the country.
- Volcanic hazard recorded since1815, is closely related to earthquakes and subsidence. This hazard highlights the role of Atlantic transformed faults and represents a considerable threat (vulnerability of over 1,000,000 persons today).
- Landslide hazard involves a combination of gravity, earthquake, subsidence, volcanism, and human activities (deforestation). The vulnerable population exposed to this hazard is more than 2,000,000 souls. Mass movements, e.g., bulk movements of soil and debris down slopes in response to gravity, are recorded in the collective memory since 1954.
- Flood risks are related to deforestation, gravity, and rain, among other things. This risk represents a constant threat for more than 3,000,000 persons.

Deforestation, which can trigger landslides and floods, has several causes including agricultural activities and mining. Either way, the result is the same: acceleration of erosion, modification of natural conditions, which lead to landscape changes. In addition, changes in fluvial processes are an inevitable consequence, for, the increase of sediment loads on the deforested watershed reduces or affects the erosive power of rivers by increasing accumulation. Our research results, presented in this paper, on sedimentation at the Edéa dam on the Sanaga river, demonstrate this.

Moreover, the occurrence of increased flooding appears, too, like the essential consequence of deforestation and the changes in water processes that accompany it. Mass movements are perceived as the most visible face of environmental risks arising from these activities.

This chapter attempts to analyze the challenges posed by climate change in Cameroon with respect to: (1) deforestation and its causes (logging, mining and extraction of gold and sapphire in Adamawa), (2) the development of large dams such as Edéa dam on the Sanaga River its consequences (including environmental hazards like mass movements, flooding, sediment budget) and (3) particular challenges (reforestation at Kaélé, mining and governance).

5.2 PERCEPTION, CONTEXT AND DEMONSTRATION OF CLIMATE CHANGE IN CAMEROON

Several large-scale climate events have led the world scientific community to conduct studies on climate variability and change. Among these, one can identify the particularly severe drought years of 1972–1973 and 1983–1984. These droughts were in the tropical Sahelian zone and extended to the wetlands of the Gulf of Guinea, with serious consequences (decreased rainfall, lower groundwater levels, drop in stream flows (Sircoulon, 1976; Servat *et al.*, 1999). There are also the findings of the World Meteorological Organization (WMO) and the Intergovernmental Panel on Climate Change (IPCC) reporting that during the twentieth century there was a warming of about 0.7 °C for almost the entire African continent.

At different times since creation of the continent, the tropical African climate has undergone alternating wet and dry periods, in response to global climate change.

According to Gasse (2006), a staged increase of aridity for nearly 4–3 Ma can be explained by astronomical forcing, and the development of high-latitude glacial cycles. At scales of 10^4–10^3 kyrs, these changes involved interactions between the orbital surface conditions and the oceans and vegetation. High frequency variations are associated with oscillations of the major modes of atmospheric circulation, solar activity, or major volcanic events. Of course, even if climate changes are linked to global climate variability, the fact remains that human activities have a great role to play in these changes (seasonal bush fires, deforestation, uncontrolled deforestation without sufficient reforestation and extensive slash and burn).

5.2.1 Climate change at the astronomical level in Tropical Africa

According to the IPCC, the whole of Africa appears to be highly vulnerable to climate change (IPCC, 2001a, b). This continent is home to a fast growing population (nearly 1 billion in 2010), and a very large diversity of ecosystems ranging from tropical evergreen forests to savannas, grasslands and desert. The diverse climate regimes associated with these ecosystems are subject to large inter-annual, inter-decadal and multi-decadal variations, particularly with regard to precipitation (Hulme *et al.*, 2001; Nicholson and Grist, 2001). Africa also has a very sparse and inconsistent observational network (IPCC, 2001a, b). Palaeogeological records show an increase in environmental aridity since about 3 Ma, punctuated by intervals of amplification of the phenomenon, synchronous with the stepwise development and amplification of high-latitude glacial cycles. In East Africa, the pollen data recorded by Bonnefillle (1983) and Bonnefille *et al.* (2004) highlight a marked cooling from 3.3 Ma to 2.5 Ma. A spectacular manifestation of climate change is the presence of a humid green Sahara during the lower and Middle Holocene. The data indicate a migration of tropical vegetation 5–10 degrees to the north of the tropical rainfall belt.

Apart from these changes in the very distant past, severe weather events reported for Africa at time scales less than 1000 years (Gasse and Van Campo, 1994; Gasse, 2000; Verschuren *et al.*, 2000; Tyson *et al.,* 2002) are often associated with imbalances in the climatic system affecting large parts of the World. Thus, for example, a very dry interval interrupted the great wet period of the lower and middle Holocene in the northern Tropics around 9–8 ka BP (Gasse and Van Campo, 1994; Gasse 2000; Gasse 2002; Gasse and Roberts, 2004); it is associated with a weakening of the summer monsoon in the Arabian Sea and Tibet and a cold interval in the northern hemisphere.

Whatever the time scale, wet/dry oscillations identified in tropical Africa during the Plio-Quaternary appear associated with profound changes in the global climate system or regional dynamic exchange among the atmosphere, ocean and biosphere. In central Africa including Cameroon, several authors (Mahé *et al.*, 1990; Mahé and Olivry, 1991; Liénou *et al.*, 2005; Liénou, 2007; Liénou *et al.*, 2008) have shown flood fluctuations and rainfall variations, which are most obvious when observed at the seasonal scale.

5.2.2 Appearances and demonstrations of climate change in Cameroon

In Cameroon as whole, despite its wide extension in latitude, human activities (deforestation, uncontrolled deforestation, extensive farming, slash and burn) have been instrumental in these climate changes. Some graphics show a decrease in precipitation over a period of between 30 to 50 years in dry regions (Garoua, Maroua), coastal areas (Douala, Kribi), mountain sites (Bamenda, Bafoussam) and in the hinterland (Bafia, Ebolowa) (Figure 1).

The overall trend is a decrease in rainfall. This is confirmed by the decrease in the number of days of rainfall per year observed in 85% of stations. Overall three tendencies emerged: excess rainfall (25% of stations) for some stations in dry regions (Kousseri, Yagoua, Kaélé Mokolo) and for a sub-dry mountain (Foumbot); a stable rainfall trend (15% of stations) found at some sites, including at Kribi (coastal), Bafoussam (a humid mountain located in the heart of the Bamileke plateau; see Tchindjang, 1996) and Edéa (littoral wetlands). A deficit rainfall trend characterizes the remaining stations (60%). However, it should be noted that interannual rainfall variation at different stations is characterized by alternating wet years, normal and dry years. But, changes in climate, i.e., low rainfall, coupled with population growth, which represents a pressure on water resources, forests and soils, lead to a constant environmental imbalance.

5.2.3 Impacts of climate change on water resources

Cameroonian rivers have experienced a serious rainfall shortage. By analyzing the spatial and temporal variability (monthly, seasonal and annual) of total rainfall, and flow and flow coefficients of three river basins in the southern forest (Kienke, Nyong and Ntem), Liénou *et al.* (2008) reveal that annual rainfall and flows decreased during

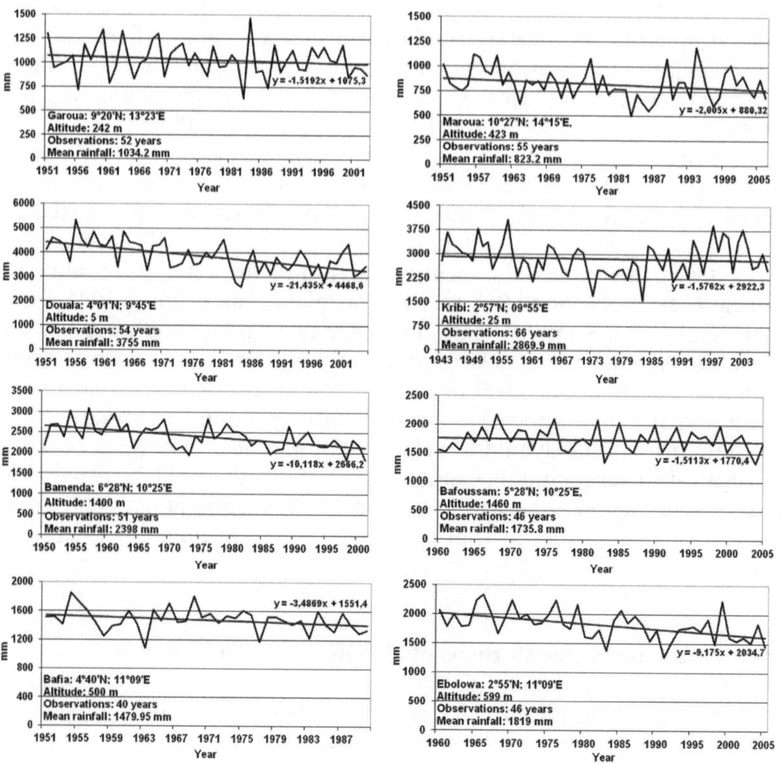

Figure 1. Graphs of rainfall from various Cameroonian stations (Douala Meteorological Service).

very dry years (1972–1973 and 1983–1984). The most significant climate change is in dry season rain regimes as well as the annual hydrological cycle. There is evidence for climate change in south-western Cameroon, and also in Gabon and the Republic of Congo.

According to Liénou *et al.* (2008) and actual observations within the Sanaga basin, the evolution of annual rainfall and rainy seasons shows a tendency to a very moderate decrease between the 1950s and 1960s, and high variability around the 1970s, which became more pronounced from the 1980s onwards.

According to Bricquet *et al.* (1996, 1997), Olivry (1987), Olivry *et al.* (1993); Mahé and Olivry (1991), the rainfall deficits, observed globally for over 25 years in Sub-Saharan Africa, both in the Sudano-Sahelian zone and in the humid tropics, have had a significant impact on water resources in these regions. The cumulative effect of many years of drought, besides an immediate response in lowered annual flow of rivers in the rainy season, calls into question the hydrological sustainability given this deficit.

In the other forested areas of the South Cameroonian plateau, Sigha-Nkamdjou *et al.* (2005), in a study covering 40 years (1950–1991), noted a rainfall deficit from 1975 to 1984 of around 5%. Over the same period, flow decreased about 10%. The resumption of more typical rainfall in 1985 was immediately accompanied by regular flow. We conclude that contrary to what occurs in the Sudano-Sahelian area, the memory effect of water stressed by Olivry *et al.* (1995) with respect to this regions' basins do not apply to those of equatorial Africa. River drying and associated low flows have changed very little in the forest-covered basins of the Dja and Boumba Rivers. Within the Nyong basin around Mbalmayo, human activities could explain the increase in depletion coefficients from 1983 onwards. In the context of the 'worsening' climate of tropical Africa, southern Cameroonian forest basins were less affected overall by the nearly three decades-long drought than the Sudano-Sahelian region.

The observed depletion of the strength of the annual flooding (volume, duration or maximum flow), the chronic ruptures of the maximum annual flood with contrasted frequency analysis, the different relationships between annual runoff and maximum flood for wet and dry periods, reflected significant changes in the hydrological regimes of West and Central African regions. Such reductions affect water availability in the watersheds and rivers, increase drought and mortgage water available to vegetation, and pose many environmental problems. Thus, climate change is detected by changes in rainfall and regimes of rivers. This has consequences on the environment, including the decline in plant and animal biodiversity with habitat fragmentation as well as the disruption of ecological balance. Most of these changes also affect agricultural production (Molua, 2003, 2006; Molua and Lambi, 2006), affect pastures, disrupt the agricultural calendar (forcing peasants to reorganize), mortgage food security, increase flooding, lead to erosion and land mass movement.

Erosion and flooding are the two main natural processes responsible for the loss of land. Erosion will grow with consequent deforestation, while floods are increasing in response to huge sediment build-up coupled with mangrove recession. On the basis of the IPCC flood model, which takes into account changes in sea level and micro-topography of mangroves, land loss is estimated at about 4950 ha (4.5% of the total area of mangroves in the estuary of Cameroon) for a rise of 20 cm sea level and about 33,000 ha (30% of the total mangrove area) for a maximum sea level rise of 90 cm. So, reduction in rainfall will be characterized by a sharp increase in temperature, an increase in bush fires, erosion, wind and rain erosion and exacerbation of aridity. This will have negative impacts on the vegetation and some species having narrow ecological requirements will become extinct. The water table will be affected, too, because water will be withdrawn. The worsening water deficit will decrease the area of cultivatable land thus, accentuating the process of land degradation and, hence, desertification.

5.3 DEFORESTATION IMPACTS

Tropical forests are particularly rich (CIFOR, 2009). Forest fragmentation increases the impact of deforestation and forest degradation on biodiversity, because it cuts migration routes and provides easy access to invasive species and humans, who may exploit or destroy the biome (UNDP, UNEP, World Bank and WRI, 2000). Forest cover in Africa was estimated at 650 million hectares, representing 17% of the world's forests (FAO, 2007, 2008). The main forest types are tropical dry forests of the Sahel and Eastern and Southern Africa, rainforests of West and Central Africa, forests and subtropical forest formations of southern Europe and North Africa, and coastal mangroves. There are a number of ecosystems that are particularly important for international biological diversity (Mittermeier *et al.*, 1998). Tree plantations make up only 1% of African forests, which provide goods and services (e.g., medicines, boost cash incomes, dietary needs). Deforestation, whether by commercial logging or clearing for agricultural purposes, is the main threat and represents an enormous loss of natural economic resources. The selective removal of certain plants also contributes to the deterioration of the quality and biological diversity of forests. Moreover, over exploitation of Non Timber Forest Products (NTFPs), including medicinal plants, compounds the problem.

Between 1990 to 2000, for the whole Africa, estimated total forest area (land with forest cover of 10% or more and larger than 0.5 ha) lost reached –0.74%, meaning a loss of more than 5 million hectares of forest per year (an area roughly that of Togo). The deforestation rate in Africa is the highest in the world. Countries where deforestation is most rapid are Burundi (9.0% yearly), Comoros (4.3%), Rwanda (3.9%) and Niger (3.7%). In North Africa, 13% of forest cover was lost between 1972 and 1992 in Nigeria, deforestation of riparian forests and savannas for agricultural purposes represents more than 470,000 hectares per year between 1978 and 1996. For the whole of Africa, 60% of tropical forest cleared between 1990 and 2000 has been converted into permanent agricultural small holdings (FAO, 2007).

In Cameroon the rate of deforestation is 0.9%. In the centre and south regions, the main causes of forest degradation and deforestation are slash and burn agriculture, the direct conversion to industrial plantations (less important since 1989), logging, urbanization, mining and petroleum, construction of hydroelectric dams, poaching, firewood and the transformation of wood into charcoal. In addition, 111 of the 220 logging companies operating in Cameroon in 1996–1997, are engaged in the latter two activities and, produced some 899,245 m^3 of firewood and charcoal out of a total national production for that year of 2,805,930 m^3 (Eba'a, 1998). In Southern Cameroon, 55% of households use firewood, charcoal and sawdust as their main energy sources (Nkamleu, 2000). Such a situation had an impact on the density of animals hunted; following a decreasing gradient (fewer animals were hunted) in a manner inversely proportional to the increase in human population density.

In addition, changes in land use result in rapid degradation of forests and their excessive fragmentation. In the Central Africa sub-region, the proportion of degraded and fragmented forest in Cameroon compared to that of the country's total forest area is high (27.5%). Indeed, the loss of biodiversity continues, mainly because of the combined action of illegal logging, poor farming techniques, bush fires, cutting wood for energy needs, and poaching.

5.3.1 Deforestation in the City of Yaoundé

Yaoundé is located at the heart of the South Cameroonian Plateau. It is built on Precambrian metamorphic rocks that appear profoundly weathered and hardened. Until 1980, biologically diverse forest covered Yaoundé hills and this was an important source of timber and NTFPs. Nowadays, there are only a few small patches and remains threatened by agriculture and construction. Forest and other natural vegetation were almost entirely destroyed for crops, homes, timber, charcoal and traditional medicines. However, Yaoundé forests hills now serve as a place for recreation and sacred rites for indigenous peoples. Degradation is closely related to urbanization.

Uncontrolled urbanization of the city capital is a direct result of population growth. Since 1960, the annual growth rate is recorded to be 9%, representing a doubling of the population every eight years (Franqueville, 1984). According to Timnou (1993), this rate was between 5 to 7% in the 1990s. This strong population growth since independence has naturally been accompanied by a profound change in the urban area of Yaoundé. This was manifested by a rapid spatial extension of the city's periphery, and inner and pericentral city densification. The city gradually spread by fragmentation of formerly agricultural land and has continued to gradually conquer the many hills surrounding its original site.

Achoundong's (1996) findings on the species composition of Yaoundé submontane vegetation are similar to the observations of Aubréville (1932) and Schnell (1971–1976) for the west African mountains. Therefore, the destruction of natural forests for the benefit of family farms, food and cash crops and also housing, have led to a great loss of biodiversity. To these cultural practices, we can add uncontrolled exploitation of forest resources (NTFPs, quarries, timber and firewood). The consequences of such a situation are that all forest reserves (Mount Febe and Messa Mountains, 4,000 ha, Eloundem Mountains, 4,300 ha and Mefou dam, 10,500 ha), have been reduced in size.

LEVEL	Vegetation	Land use, activities	Associated hazards
Top, 970 m asl	Saxicolous groups	Farms (peanuts, cassava, banana, pears) hunting	Erosion
Slope, 820 m asl	Relictual W Afr. montane forest	Outcrop, hunting	Rock fall
Mid-slope, 777 m asl	Afforestation, fruit trees	Construction, mining, cattle rearing	Landslide
Foothill, piedmont, 700 m asl	Crops, agroecosystems	Houses, schools, churches, health centres, shops	Landslide, mudflow
Valley, 600 m asl	River	Houses, schools, churches, health centres, shops	Floods

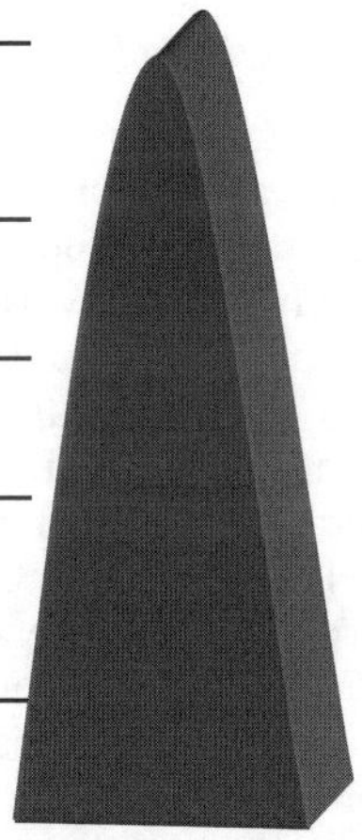

Figure 2. Vertical description of the south-eastern flank of Akok Ndoé (Fékoua, 2011).

So, reserves that covered this "western mountain barrier" have been profoundly altered by human activity (cultivation, grazing, removal of stone quarries), leaving in place a degraded forest (Fékoua, 2011), consisting of saxicolous mosses, groups of rocks, forest recruits and some relictual trees and shrubs (Mandengue, 2009). The inertia of the government, has led local residents to launch a veritable assault on the existing forests, so that they have all but disappeared. Figure 2 shows a stratification of the activities on the slopes of these hills and the associated risks.

Housing is responsible for two negative impacts experienced: the loss of agricultural land and forests, and soil sealing. The main risks caused or aggravated by human activities in the Yaoundé hills are erosion, flood and landslide through mining, excavation, stone removal, sand pits and cultural practices as mentioned by Abéga (2006), Mbeugang Tcheubonsou (2011) and Fékoua (2011).

To conclude, 96% of environmental risks around Yaoundé city are induced by the zoning of Yaoundé hill and to prevent these risks, it is prescribed all buildings located above 800 meters should be removed. Taking into account urban pressure, it is also recommended that the prescribed eviction should be accompanied by reforestation with native species. These measures are the best way to meet the challenges of climate change, global warming as well as flood and landslide in this, the capital city of Cameroon.

5.3.2 Environmental impacts of deforestation through mining and petroleum extraction: coastal region and Adamawa

Cameroon harbors the third largest forest resource in Africa (FAO, 2001; WRI, 2000–2001). The country is well served by surface and subsurface water resources. It also has significant exploitable mineral and oil resources, while research into promising new sites is taking place in the Rio del Rey and the Lake Chad Basin. There are other substantial, practically unexploited, resources such as bauxite (Ngaoundal, Minim-Martap and Fongo Tongo), iron (Mbalam and Kribi), rutile (Akonolinga), tin (Mayo Darlé), nickel and cobalt (Lomié), gold (Betaré Oya) and uranium (Colomine, Poli).

5.3.2.1 Oil exploitation impacts

Oil extraction is involved in the modification of mangrove and coastal forest biodiversity at the genetic, specific, population and ecosystem levels. Additionally, it has contributed to the partial disruption of biodiversity distribution within the coastal region through various activities (surveying, exploration, development, pumping, transportation, loading, unloading, etc.). With 460 km of coastline, the Cameroonian coast is entirely surrounded by operating oil companies, leading to significant effects on the marine environment (Figure 3).

Companies engaged in oil exploration are numerous e.g., COTCO (Cameroon Oil Transportation Company), Total, E&P, SNH (Cameroon National Hydrocarbons Corporation), Perenco, Sonara, Pecten, Euroil, SCDP, CMS Nomeco, and Addax Petroleum Cameroon. They are divided into different areas according to various criteria (production, research, transport, distribution, finance, promotion). Thus, some are more involved in the management of environmental impacts than others, for example, through dug wells.

Oil and gas exploitation in Cameroon indiscriminately affects nearly 30 species of animals and plants. In environmental terms, it increases pollution, for example by

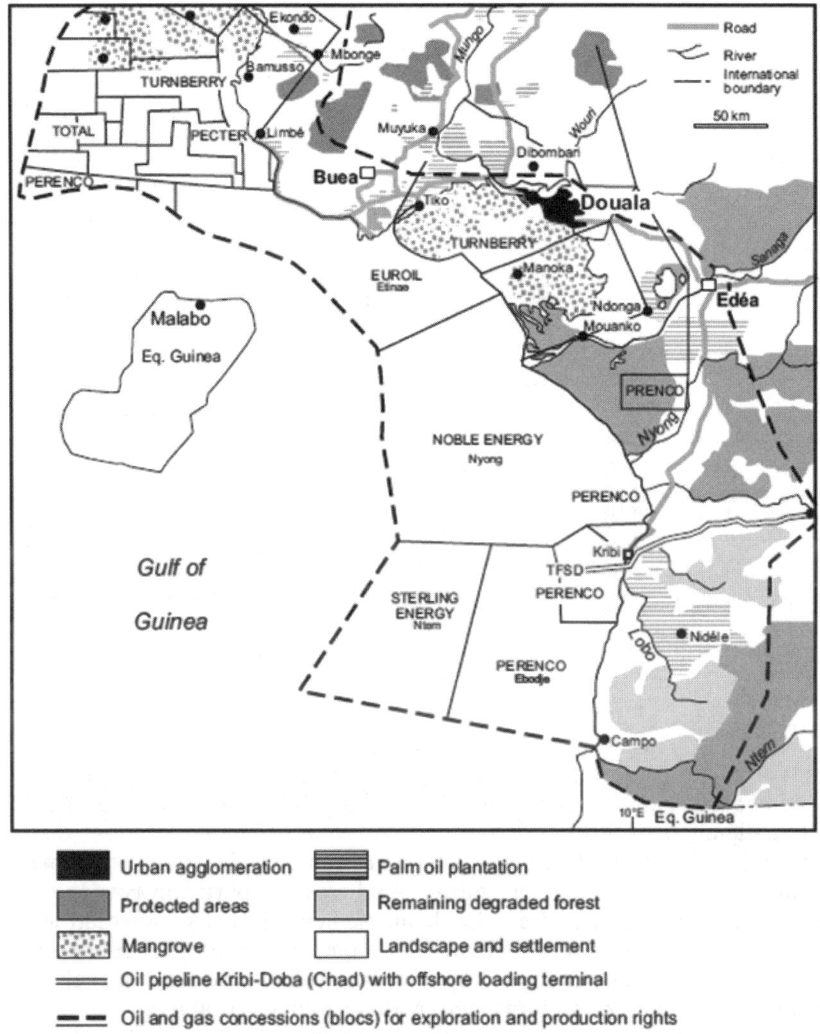

Figure 3. Map of Cameroon's coastal region: besides oil and gas companies (blocs), there are adjacent protected areas, leading to exploitation and exploration mortgage of conservation (after Gold, 2007).

marine oil spills, as was the case in January 15, 2007, near Kribi. In particular, the challenge facing Cameroon with respect to oil and gas exploitation lies in: the under-estimation of the consequences of various projects in coastal areas; compensation for damage or for environmental restoration e.g., in case of leakage or spillage; reintroduction of species; and regeneration of mangroves. Thus, it can be seen that the real and potential environmental impacts of these activities are important, unfortunately, environmental audits and Environmental Impact Assessment (EIA) studies are very casual and not taken seriously. For example, planning for the oil terminal of the Chad-Cameroon pipeline to Kribi, did not take into account the effects of future development of the industrial centre of the city, as the deep water seaport complex facilities on the Grand Batanga, Mboro and Lolabé sites. The 26,000 ha area

Table 1. Some endangered species linked to oil exploration and exploitation activities in the coastal region of Cameroon (after CWCS, 2006; Mbog, 2006; Nanji, 2007).

Operating step	Common name of endangered species	Scientific name
Prospecting	Sole	*Cynoglossus canariensis*
	Sardinella	*Sardinella maderensis*
	Bonga	*Ethmalosa fimbriata*
Planning	Manatee	*Trichechus senegalensis*
	Leatherback turtle	*Dermochelys corialea*
	Mangrove	*Rhizophora racemosa*
Transport	Black Heron	*Egretta ardésiasa*
	Petrel	*Oceanites oceanicus*
	Curtew	*Numenius arquata*

represents 30 km along the sea. Moreover, construction of the Mpolongwe central gas plant is another source of significant negative impacts. Another point worthy of mention, is the negative impact of the production, loading and unloading of crude oil by producers in neighboring countries (Nigeria, Equatorial Guinea) on the ecosystems of the Gulf of Guinea.

As result in the Douala-Edéa Reserve, oil exploration initiated in the 1980s led to the destruction of 228.2 hectares of mangroves, which corresponds to 0.1% of the total of this protected area (Ndjogui, 2009). Destruction consisted in track opening or road widening. This degradation of vegetation cover promoted gradual occupation (24.8 ha) of part of the reserve by local people, who have since significantly destroyed the mangrove forest for the process of smoking fish (Essoungou Kwack, 2009; Ndjogui, 2009; Tchindjang et al., 2010).

Oil activities have contributed to the intensification of various phenomena such as poaching, harvesting, and deforestation. Wild growing, partly invasive species introduced by intensive logging of trees for the implementation of major road works and sea port, construction of various habitats, development of oil sites, caused the destruction of important ecological niches. Apart from drilling mud and tar balls polluting the beaches, some species are threatened by these activities (Table 1).

The main challenge has to do with the regeneration of mangroves and restoration of these areas, and this can't be achieved without a participative approach (Tchindjang *et al.*, 2010).

5.3.2.2 Mining and environment in Cameroon

Mining in Cameroon is largely labor intensive; it consists in the extraction and concentration of minerals using manual methods that are but minimally mechanized. From the environmental point of view, artisanal mining of this type can lead to damage such as: abandonment of non-secure sites, devastated and/or sterilized farmland, degraded forests, polluted rivers prone to erosion, ground water affected by pumping, various forms of anthropogenic pollution, conflict and social change (rush to latest mineral strike site). The challenges are linked to the environmental disturbance caused by this activity and also to the organization of this sector such that it could contribute to poverty alleviation and mining governance. This section focuses on the mining activities around Adamawa (Faro and Deo administrative division).

Faro and Deo administrative division, an area of about 10,475 km², consists of four districts: Tignere, Tignere-Galim, Mayo and Baléo Kontcha, all bordering Nigeria. Vegetation cover includes woodland to shrub. But, in the plains drained by the rivers (Paro, Mayo Beli, Ngoura), vegetation is rather grassland-like with *Pennisetum purpureum* (Sissongo), *Imperata cylindrica, Andropogon* and *Hyparrhenia*. The small stream Mayos and others are bordered by gallery forests having *Daniela oliveri* (Cesalpiniaceae) and *Lophira lanceolata* (Ochnaceae). In flood plains, one can find *Borassus aethiopum*, palmyra palm, which is widely used hereabouts as a building material.

Adamawa plateau is populated by the Fulani, Hausa and Bororo indigenous peoples, of Faro and Deo are Nyem Nyem. The population consists includes semi-breeders and semi-farmers concentrated around rivers (Mayo Sanganaré, Tagouri Mayo, Mayo Dankali), where they find water and arable areas. The population density of the Faro and Deo is about 10.6 inhabitants per km² (BUCREP, 2010).

Geological research on the Adamawa plateau (Eno Belinga, 1966) showed how laterite became mineralized into bauxite or precious stones during the Eocene. This is attributed to the weathering of large granite masses, intersected by intense storkwerks, pegmatite's and veins, which became mineralized into tin, wolframite and corundum. According to Ntep Gweth (2001), Faro and Deo is located in the famous "valuable corridor" running through Cameroon. The mineral potential is diverse and not yet fully known nor exploited. Indeed, gold is present as are precious stones and hot springs rich in chlorine and fluorine. Gold and precious stones are exploited on a small scale. The beginning of gold mining dates back to 1989, while sapphire extraction began in 1998 (Sinnota, 2004).

Gold and sapphire extraction in the Faro and Deo are mined on small scale in many areas: Leggal Goro, Ngoura, Garbaya Bari (Figure 4). There are nearly 300 artisans who have settled here, and at least 100 seasonal workers, follow the discoveries of rich veins (Takoutsi, 2009). This activity has drawn in the surrounding villages, and supports a large population living directly or indirectly from mining related activities. As an indicator of prices here, shoes cost nowadays between 23,000 and 35,000 FCFA (35–55 Euros), according to fluctuations in the international market for gold.

5.3.2.3 Environmental analysis of craft mining in the Faro and Deo

Mining significantly changes the landscape at physical and socio-economic levels. Artisanal mining has direct effects on vegetation. Indeed, the first step before sinking of pits or trenches is the removal of any type of vegetation cover (shrubs or trees, gallery forest or savannas). The trunks, branches and even the leaves are used for various mining purposes (e.g., supporting shafts and galleries; building camp huts and as fuel). Trees that have escaped slaughter have their roots bruised and bared, and, there after, are unable to resist the action of erosion and are prey to the slightest gust of wind.

5.3.2.4 Changes in soils and geomorphology

Soil is the target in mining and it undergoes the most severe pressure. Artisanal mining in the Faro and Deo leads to the degradation and reduction of arable land. The operation is done on river flats and fertile areas. After extraction of the ore, the surface soil mineral horizon is largely lacking and, thus, its fertility is reduced: these actions contribute to the destruction of farmland. The removal of vegetation exposes the soil, depriving it of its triple protection against wind, rain and erosion. It also interrupts the supply of soil organic matter, thereby contributing to rapid degradation of fertility.

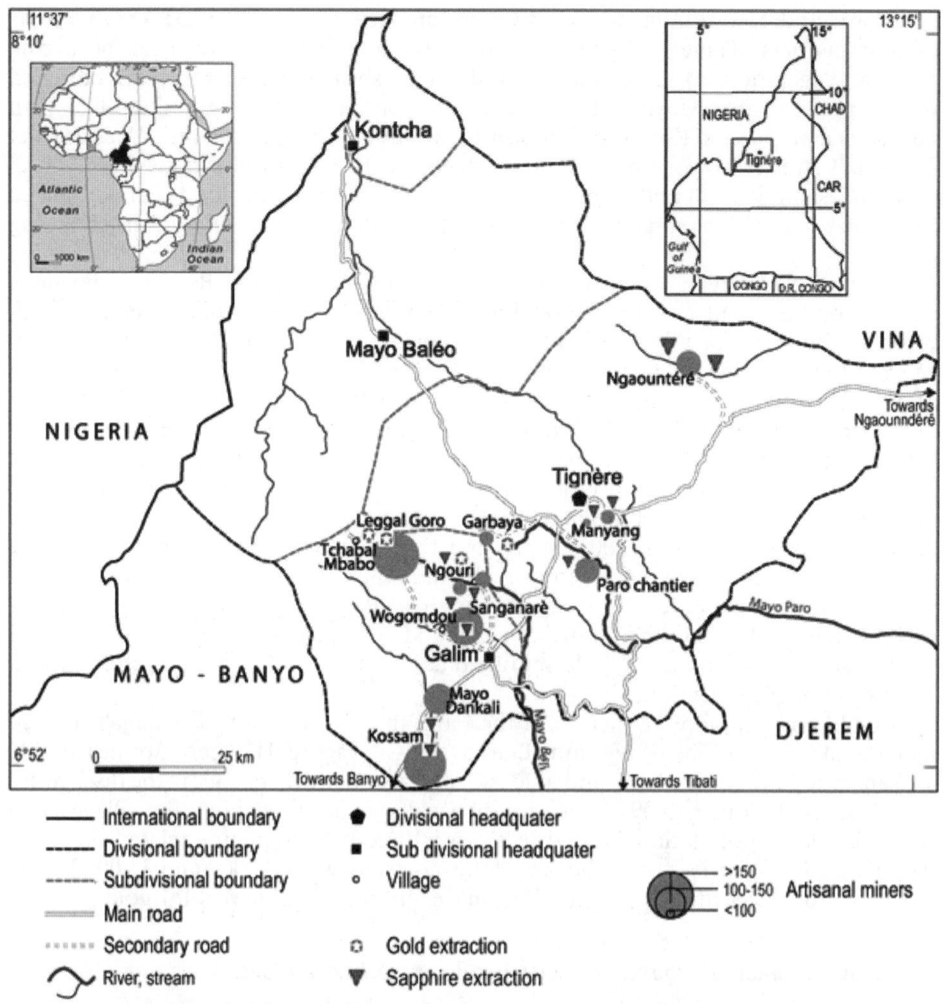

Figure 4. Artisanal mining sites in the Faro and Deo (after Takoutsi, 2009).

The soil microclimate is also changed. The exposure of large, originally covered, areas to insolation causes changes in the rate of evaporation, moisture and soil temperature (Gazel, 1958). Destruction of the remaining humus by runoff, alters soil permeability such that the soil is no longer in phase with the water conditions imposed by the geoecosystem. Consequently, it becomes unstable and changes with respect to its organization and composition. Sapphire extraction causes a large volume of soil to enter the streams (Sinnota, 2004).

For example, in the case of Batié sandpits in Western Cameroon, Tchindjang (2001) and Doumtsop (2011) show that soil degradation depth calculated for sand pit extraction reached 6 to 8 m per year. The volume of material extracted and transported by trucks was around 150 tons per day or 55,000 tons/yr. The volume of suspended losses leading to increased turbidity, measured between 144 to 290 g/l. It is estimated that annually, a total of 15,000 to 20,000 m^3 of sediment are removed.

Photo 1. Artisanal gold mining well at Leggal Goro (depth ca. 80 m) and landslide bank
(Takoutsi, 2009; Sinnota, 2004).

This anthropogenic process accelerates siltation, leading to a sedimentation rate of 2 to 5 cm per day in rivers and reservoirs.

Mining activities cause irreversible destruction of soils, including topsoil. However, reconstitution, that is, the introduction of a thin layer of soil requires not only a 'hyperhumid climate', but also enough slope stability. Under these conditions, weathering is impossible or "mortgaged" in such conditions, because the rate of granitoid basement weathering to produce 1 meter of soil is estimated at 60,000 to 80,000 years under normal conditions, and 10,000 to 15,000 years under accelerated ones (Petit, 1990).

At the geomorphologic scale, artisanal mining leads to rapid change of microrelief. Miners dig wells, trenches, pits, from 2.5 to 80 m deep in the soil, which significantly changes landscape (Photo 1). Subsequently, channels created as a result of concentrated runoff, trigger mass movements, sometimes with loss of life and considerable structural damages along the banks. It also creates slots breakouts, niche separation and slip plane mounds of clayed sand. Gully erosion is most active here. In this way, channels in the topsoil are transformed into ravines, of from 0.5 to 1 m, by regressive erosion. Finally, strips of land undermined at their base, fall into these ravines, which progressively grow into lavakas.

5.3.2.5 *Mining effects on hydrography in the Faro and Deo*

Water in artisanal mining helps in mineral processing, whether for gold or sapphires. It is used for the settling of the ore. After washing, artisanal mining influences the sediment budget of rivers by supplying materials into the bed either directly, by washing the ore, or indirectly, due to erosion, which leads to the pouring of soils and rubble from sinking holes in river banks. All these activities increase siltation in rivers in and around the mined areas so that the rivers take on an abnormally beige to grey color during the day. They also noticeably affect river dynamics and the geometry of the bed. These changes are accompanied by water pollution from particulate matter, clays and other various materials resulting from the activity. Moreover, at the end of the mining operation, miners transform sites and wells into sites for aquaculture. This disruption of the river system has led to increased flooding in rivers having water flow interrupted by wells, as a result of the mass movements from wells, trunks and branches, and stones and sand from the gravel washed and poured downstream. The turbidity of streams increases; and the degradation rate is not much different from that observed in the sandpits sites of Batié (Tchindjang, 2001; Doumtsop, 2011).

Because of mining in the Faro and Deo many economic and social changes have followed and are on going. Outside the mixing of populations (Nigerians, Ghanaians,

Cameroonians, Central African Republicans, Chadians), mining has resulted in significant decreases in agro-pastoral production, which represents the main business of the department (Edimo *et al.*, 2002). The annual reports of ministries in charge of Agriculture and Livestock evoke sapphire exploitation as one of the major causes of the decline in agricultural outputs in the Faro and Deo division. Such a situation poses the problem of unsufficient mining governance.

In oil extraction, the government of Cameroon welds the controlling and, however, this is not the case with mining activities, which are controlled by local populations. In the context of sustainable development, action is needed to optimize production while at the same time ameliorating miner's techniques, such that the result is both profitable and with low to minimal impact on the environment and on the miners themselves.

5.4 CLIMATE CHANGE, DEFORESTATION AND SEDIMENTATION IMPACTS ON WATER AVAILABILITY AT THE EDEA HYDROELECTRIC DAM ON THE SANAGA RIVER

If deforestation and mining contribute to climate change, the construction of large dams involves deforestation and induces similar changes. All these activities form a vicious circle for which the result is increased flooding accompanied by landslides, mudflows, increased sedimentation (hence, reduced water channels), siltation, invasion of river and stream beds by plants.

The hydroelectric dam at Edéa was the first dam built in the equatorial region of Cameroon, in the Sanaga watershed. It was built with a "run off river" architecture, i.e., it has no permanent retaining water reservoir. Water flow during the year is regulated by the retaining dams at Mbakaou, Bamendjing and Mape, but they are located quite distant (400 to 600 km) from the Edéa production unit. Therefore, the 5 to 8 day water transit times provoke the mortgage of electricity production. Moreover, due to the lack of dredging, population growth has led to deforestation and consequent sedimentation, which, in turn, invades the channel and further reduces the amount of water available for production.

The Sanaga basin is occupied by a population, whose activities, outside of water impoundment, also contribute to loss of vegetation cover (deforestation), weakening of the soil. When it rains, large quantities of materials, moving down the slopes towards the stream, are transported into the reservoir. These materials retained by the dam, are deposited and accumulated,thereby threatening the stability of the structure. The following sections assess sedimentation associated with this process, the product of deforestation around the pools and dams. This sedimentation masks the heights of calibrated gauging scales used to measure water levels, which normally form a necessary step in hydrological assessment and monitoring.

5.4.1 The Sanaga River basin: its catchment and the Edéa dam

The Sanaga River's watershed extends from the Sanaga basin (3°45'–7°20'N to 9°45'–15°E) covering a total area of 140,000 km². The Sanaga basin is covered by three climatic zones of varying size, spreading from the Adamawa plateau to the mouth of the Sanaga River (Figure 5).

The tropical Sudanian climate zone (5 to 6-month dry season) ranges from Ngaoundéré to Garoua and Tibati Boulaï, areas within the zone experience 1500 mm

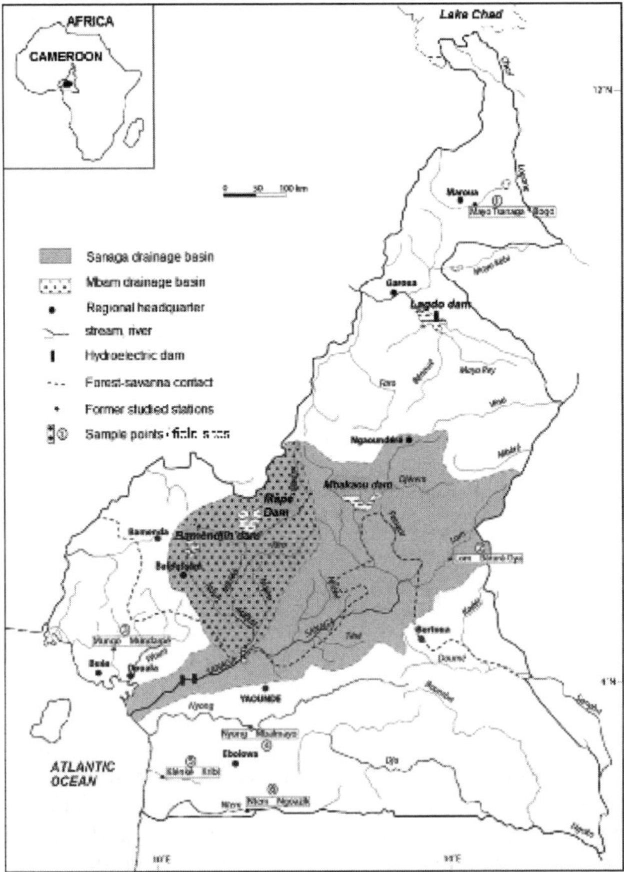

Figure 5. Map of the Sanaga River Basin.

average rainfall per year and average annual temperatures of 25–26° C. The vegetation composed of shrub lands established on lateritic soils indurated or redesigned from igneous intrusions (granites, syenites, anatectic granites). Agricultural activity is centered around cereal crops, legumes and other sources of vegetable protein (e.g., maize, soybeans, peanuts).

The second climate type falling within the Sanaga basin is the equatorial guinean climatic zone (3–4 dry months) having four seasons. It occupies the central part of the basin. It stretches in a N-S direction from Garoua Boulaï to the Republic of Congo border, in the south of Cameroon. In an E-W direction, it extends from the Central African Republic (Bouar) to the massive Bamileke plateau. Rainfall is between 1500 and 2000 mm and average annual temperatures are 23–24° C. From the North to South of this climatic zone, the vegetation passes through all forms of transition, from shrub lands, peri-forestal savannas to dense deciduous forests with Ulmaceae and Sterculiaceae, the last of which are highly degraded due to human activities. The cultivated crops are tubers (cassava, cocoyam, taro), plantain, maize and peanuts. Natural vegetation and crops grow on lateritic soils that have developed on a metamorphic core complex (quartzites, migmatites, charnockites).

The third climate type is an equatorial Cameroonian one (3-month dry season) having two seasons and covers the western portion of the basin. It stretches from the Lower Sanaga towards Kong Wala to the ocean. This area receives 2000–3000 mm of rainfall annually (tropical mountain climate, Suchel, 1972, 1988) and has a mean annual temperature of 2° C. The vegetation in this zone is lowland evergreen and Atlantic forest, which grows on lateritic soils derived from sedimentary and metamorphic rocks. Extensive agriculture is focused on cassava, cocoyam, plantain and oil palm. It is in this zone that Edéa town and dam are located (3°45'N, 10°20'E; 31 m asl).

5.4.2 Climate variability, climate change and water availability

Changes observed in the amount of rainfall within the Sanaga watershed from 1944 to 2008 can be grouped into two periods: the first period (1973–1978) was remarkable for below average precipitation (1,864.3 mm/yr) with some years experiencing particularly low rainfall (1,348.6 mm in 1973–1974, and 1,223 mm in 1976–1977). The second period (1979 to 2008), is characterized by an average rainfall equal to 1585.8 mm/year; this value is below the module. These data suggest that before 1948, the Sanaga River flow at Edéa was natural. From 1948 to 1968, the river's discharge began to be measured. In 1969, the Mbakaou dam, located at 600 km upstream, began functioning and flow rates were regulated. The Mbakaou retention dam on the Djerem River has a capacity of 2.62 billion m³, Bamendjing dam (located at 400 km upstream) on the Nun River, a capacity of 2 billion m³, and the Mapé dam (450 km upstream), a capacity of 3.2 billion m³. The latter two dams were built in 1974 and 1984, respectively to address the problem of drastic decline rates during low flow periods (December, January, February). The development of these three dams, however, failed to produce enough electricity for consumers. The various locks and reserves experience intense evaporation, there is uncontrolled human activity in the watersheds and tributaries of the river (Sigha-Nkamdjou *et al.*, 2005), with increasing silting. Degradation of the vegetation cover exposes the soil to erosion and streams to evaporation, and, hence, of greater relative importance than evapotranspiration. In addition, on the longitudinal profile of the Sanaga, flat areas of the bed receive large deposits of solid materials carried by runoff (Descroix *et al.*, 2005). Moreover, these materials have never been dredged since the Edéa dam was opened in 1974. Therefore, gauging the current gives a less accurate column of water and it effectively prevents an appreciation of low water levels (the measurement calibration scale was clamped to 5.20 meters. IGECO, 1968).

The other reason is that, despite the three control reservoirs (total capacity of 7.82 billion m³), severe and critical low flows were recorded on the river in March 1990 (59 m³/s), 1997 (87 m³/s) and 1998 (73.2 m³/s). The most severely reduced flows were in April 1983 (61 m³/s) and 1998 (13 m³/s) (Bessoh Bell, 2005). These low flows, expected under extreme conditions, resulted from drastically dry seasons and caused a drop in hydroelectric generation, leading to power cuts.

5.4.3 Hydrological parameters of the Sanaga basin

The water balance is equilibrated by the combination of inputs (rainfall, time) and outputs (evaporation, infiltration, runoff during the same period). For inputs, rainfall was studied according to the three climatic zones defined above, and the main measuring stations are shown in Figure 6.

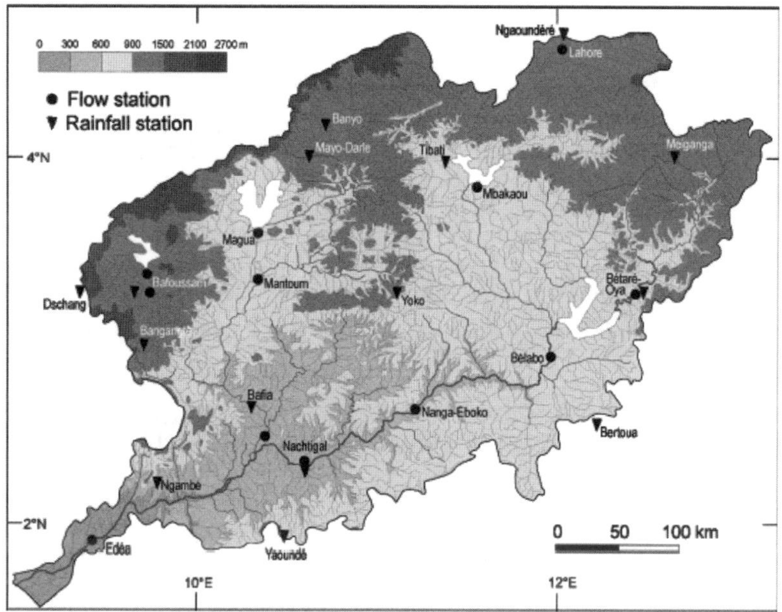

Figure 6. Run-off and rainfall stations within the Sanaga River basin (EDC, 2008).

After analyzing rainfall and calculating various parameters, the following findings emerged:

- There was a decline in annual rainfall between 1944–2008 across the Sanaga River basin. The year 1982, which shows a deficit in the basin and marks a break. In fact, the rainfall values from 1944 to 1982 (1,951.5 mm/yr) are above the mean (1,864.3 mm/yr). The average rainfall (P) in the basin during the study period was obtained, using the Thiessen polygon method given by Equation 1 (Figure 7).

$$P = \frac{S_1 \times P_1}{S} + \frac{S_2 \times P_2}{S} + \ldots + \frac{S_n \times P_n}{S} = \sum_{i=1}^{n} \frac{S_i \times P_i}{S} \tag{1}$$

- A graph of the deviation of rainfall means over the same time frame also shows periods of surplus and deficit, which confirms previous observations (Figure 8).

  From 1944 to 1982, there are 12 decreased rainfall years as opposed to 25 surplus years. Conversely, from 1982 to 2008, there were 22 decreased rainfall years as opposed to four years of surplus. The 1944–1982 period appears to be the wettest, with an average of 65.78% as compared to 15.38% rainfall for the 1982–2008 average period, in which there was no deficit. It is clear that a drought existed from 1982 to 2008, with a rainfall reduction of 57.81%.
- The observed natural flows show the same trends as for rainfall, marked by a disruption year, i.e., 1982 (Figure 7). The hydrograph (Figure 9) was obtained from multiple surveys of gauging and calibrations performed from 1944 to 2008.

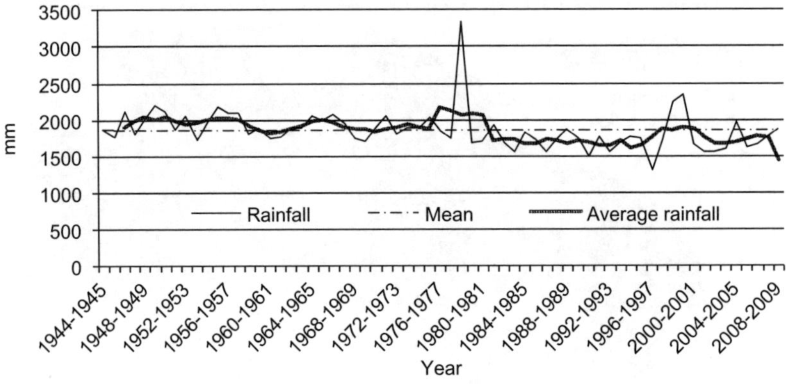

Figure 7. Mean rainfall for the Sanaga Basin (Data from Douala Meteorological Services; CRH-Centre de Recherches hydrologiques à Yaoundé, hydrological data; AES SONEL data, 2005).

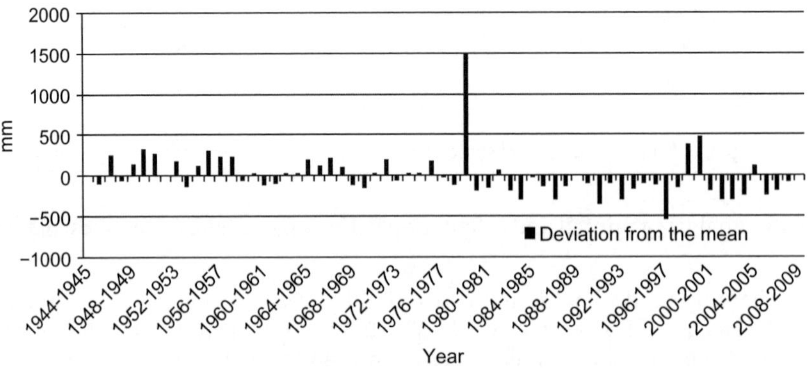

Figure 8. Deviation from rainfall means within the Sanaga Basin (Data from Douala Meteorological Services; CRH-Centre de Recherches hydrologiques à Yaoundé, hydrological data; AES SONEL data, 2005).

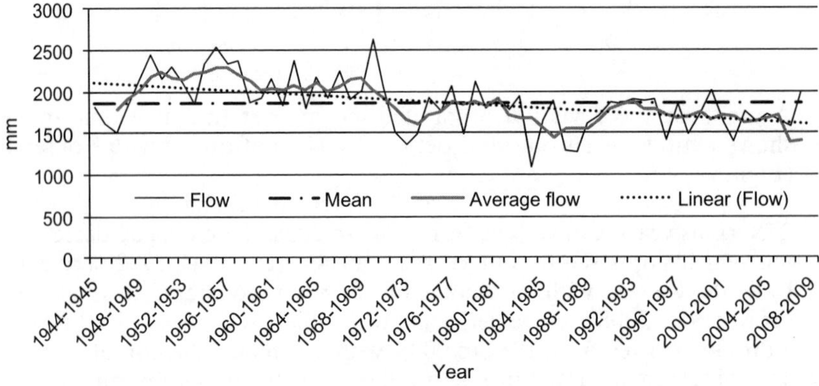

Figure 9. Evolution of natural flows of the Sanaga River at Edéa (Data from Douala Meteorological Services; CRH-Centre de Recherches hydrologiques à Yaoundé, hydrological data; AES SONEL data, 2005).

The annual flow rate during the observation period was 1,864.3 m³/s. It represents a specific flow of 14.2l/s/km². The maximum annual flow rate was 2,625.04 m³/s in 1969–1970 and the minimum was 1,098.42 m³/s in 1983–1984. The graph above shows that from 1944 to 2008, annual natural flows have gradually decreased from an annual average of 2,025.5 m³/s in 1944–1982 to 1,739.5 m³/s in 1982–2008. The absolute difference is 286 m³/s. The deviation from the mean shows the same trends as for the curves of rainfall.

- One can notice that the average amount of runoff during the first period (1944–1982) was 479.1 mm against 407.5 mm for the 1982–2008 period. For the whole study period (64 years), the Sanaga watershed is characterized by decreasing amounts of water with a relative value of 13.1%.
- The flow deficit confirms previous observations and shows a large break in 1982 (Figure 10), while the flow rate varies in an inverse manner (Figure 11). This situation can be explained by the fact that the hydrological features as shown by Liénou *et al.* (2008) are that of a tropical regime, characteristic of a river that has its origin in the tropics and has its outlet in the equatorial zone.
- Evapotranspiration, calculated according to Thornthwaite (Equation 2) show an average of 1,204 mm, which is a below average rainfall value (1,859.7 mm).

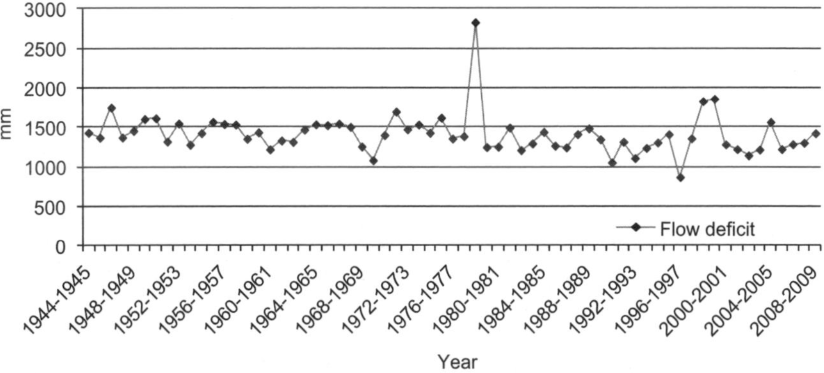

Figure 10. Flow deficit of the Sanaga river at Edéa (Data recorded at Douala Meteorological Services; CRH-Centre de Recherches hydrologiques à Yaoundé, hydrological data; AES SONEL data, 2005).

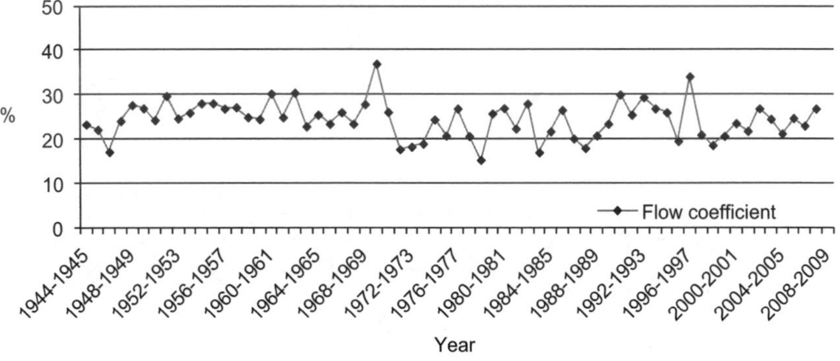

Figure 11. Runoff coefficient of the Sanaga river at Edéa (Data from Douala Meteorological Services; CRH-Centre de Recherches hydrologiques à Yaoundé, hydrological data; AES SONEL data, 2005).

It represents 64.74% of the precipitation collected in the Sanaga River basin. The evapotranspiration curve resembles the flow deficit graph and also shows a peak in 1982 (Figure 12).

$$ETP = 1,6 \left[\frac{10T}{I}\right]^{\alpha} \times F \ (\lambda) \tag{2}$$

In terms of hydrological parameters, it appears that all rainfall graphs and related flow graphs show decreasing trends in the river basin, indicating high vulnerability to climate change and climatic deregulation. Analysis of the relationships between rainfall, flow and evapotranspiration raises the problem of water availability in the Edéa dam. The interannual relationship between rainfall and flows is calculated by means of Equation 3.

$$Cr = \frac{\sum_{i=1}^{n} \left(X_i - \overline{X}\right) \left(Y_i - \overline{Y}\right)}{\sqrt{\sum_{i=1}^{n} \left(X_i - \overline{X}\right)^2 \left(Y_i - \overline{Y}\right)}^2} \tag{3}$$

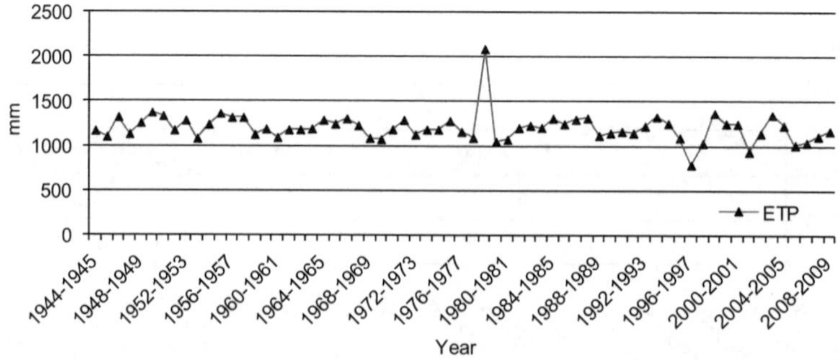

Figure 12. Interannual trend of potential evapotranspiration within the Sanaga basin (Data from Douala Meteorological Services; CRH-Centre de Recherches hydrologiques à Yaoundé, hydrological data; AES SONEL data, 2005).

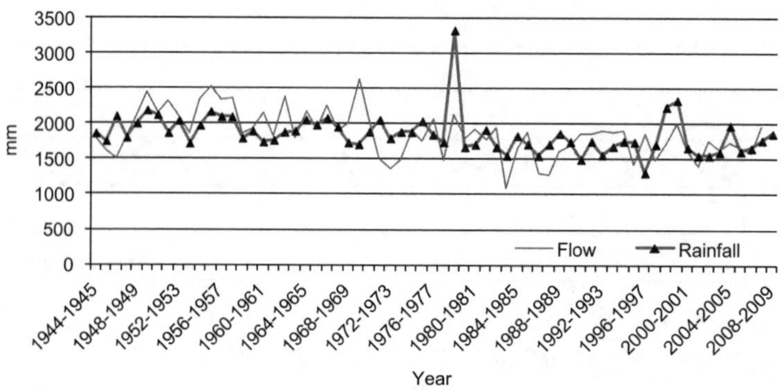

Figure 13. Interannual relationship between rainfall and flows (Data from Douala Meteorological Services; CRH-Centre de Recherches hydrologiques à Yaoundé, hydrological data; AES SONEL data, 2005).

Figure 13 is a graph representing two parameters, rainfall and flows. Nevertheless, the correlation coefficient (0.1535) is weak. It means that these two parameters are not interrelated, and there is no causal relationship between them. Overall, the average rainfall curve is above that of flows, even though there exists some exceptional overlaps observed in 1946–1947, 1972–1973, 1978–1979, 1983–1984, 1987–1988. This phenomenon results from the lack of flow related to the spatiotemporal distribution of precipitation within the Sanaga basin.

The assessment of the seasonal relationship (Figure 14), presents a bimodal rainfall curve (May to September), and two notches in June and December; conversely, the flow curve is unimodal (October). The average monthly rainfall curve depicts an equatorial climate, while the flow one reflects a tropical system.

5.4.4 Relationship between evapotranspiration and water availability in the Sanaga basin

The availability of water (R) is the difference between recorded rainfall (P) and potential evapotranspiration (ETP) according to the relation R = P–ETP. The average amount of water available in the Sanaga basin during the studied period (1944–2008), is 651.7 mm. The minimum is 253.19 mm in 2003–2004 and the maximum is 1,244.8 mm in 1978–1979. Though these amounts of available water are relatively low, they are seemingly enough to meet consumer needs within the Sanaga basin. However, if we

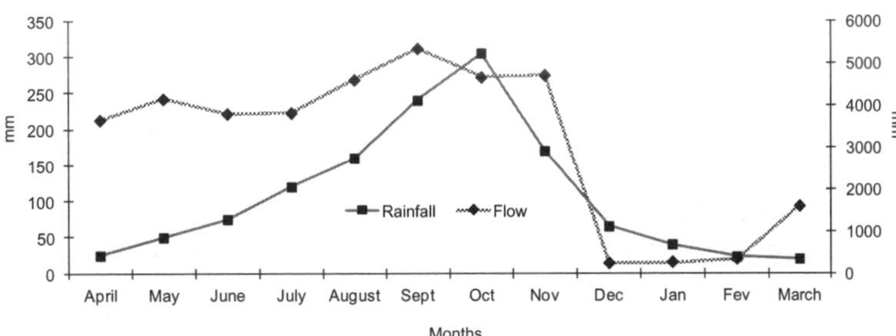

Figure 14. Seasonal relationship between rainfall and flows (Data from Douala Meteorological Services; CRH-Centre de Recherches hydrologiques à Yaoundé, hydrological data; AES SONEL data, 2005).

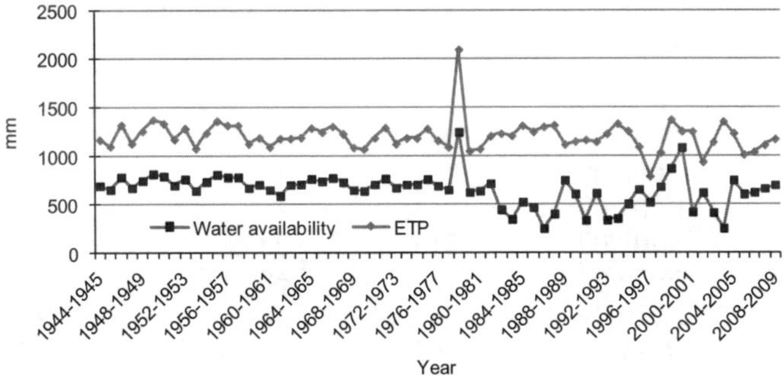

Figure 15. Relationship between potential evapotranspiration and water availability (Data from Douala Meteorological Services; CRH-Centre de Recherches hydrologiques à Yaoundé, hydrological data; AES SONEL data, 2005).

look at the relationship evapotranspiration/water availability, we noticed that from 1944 to 1982, water availability was greater than evapotranspiration and the reverse from 1982 to 2008 (Figure 15).

Therefore, evapotranspiration–even more important than rainfall since 1982– does not preclude the availability of water and one must seek the causes of water shortage in electricity provision elsewhere.

5.4.5 Sedimentation/silting rates

The position of the upper and middle Sanaga River basin in mountainous zones promotes great runoff (slope effect) and increases erosion. This has been assessed by the analysis of suspended solids (TSS), turbidity and sedimentation.

Water samples were collected with 1.5 liter bottles from the riverbed at Edéa between April 29, 2009 and March 28, 2010. Laboratory analysis consisted in the determination of suspended solids (TSS) and solid flow rates, to calculate specific degradation (Table 2).

From these data, the total amount of materials passing through Edéa during the study period was estimated at $80{,}725 \times 3{,}600 \times 24 \times 365.25 \times 64 = 163{,}039{,}184.64$ tons giving an approximate volume of about 125,414.7 hm³ MY since the construction of the dam.

80,725	= average annual sediment load
	= 3,600 unit time per second (60×60)
24	= one day is 24 hours
One year	= 365.25 (365 jours1/4)
64	= number of years of study (1944–2008)

Data on average turbidity of the water at Edéa from 1944 to 2009? is shown in Figure 16.

The quantities of materials carried by runoff over the year represent the specific degradation of the Sanaga basin. Also, from the turbidity data we estimated the layer

Table 2. Suspended solids, sediment discharge and specific degradation of the Sanaga River at Edéa (Laboratory analysis, 2010).

Samples	Collection date	TSS (mg/l)	Solid flows kg/s	Specific degradation (t/km²)
SED 1	29/04/09	33,6	15,78	0,311
SED 2	25/05/09	54,2	43,97	0,895
SED 3	18/06/09	11,8	14,18	0,279
SED 4	17/07/09	81,6	150,45	3,064
SED 5	28/08/09	81,6	223,85	4,56
SED 6	26/09/09	16	68,71	1,354
SED 7	24/10/09	92,8	489	9,96
SED 8	26/11/09	102,8	320,62	6,32
SED 9	26/12/09	17,4	21,7	0,442
SED 10	23/01/10	7,6	5,06	0,103
SED 11	22/02/10	6,4	2,51	0,047
SED 12	28/03/10	13,6	4,37	0,089

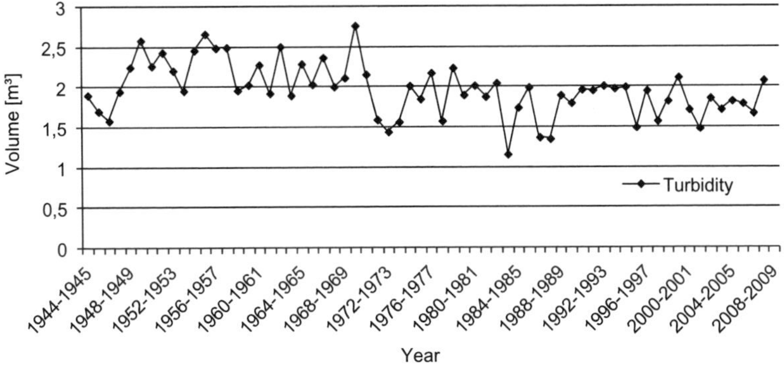

Figure 16. Average yearly turbidity of Sanaga River at Edéa (Data from Douala Meteorological Services; CRH-Centre de Recherches hydrologiques à Yaoundé, hydrological data; AES SONEL data, 2005).

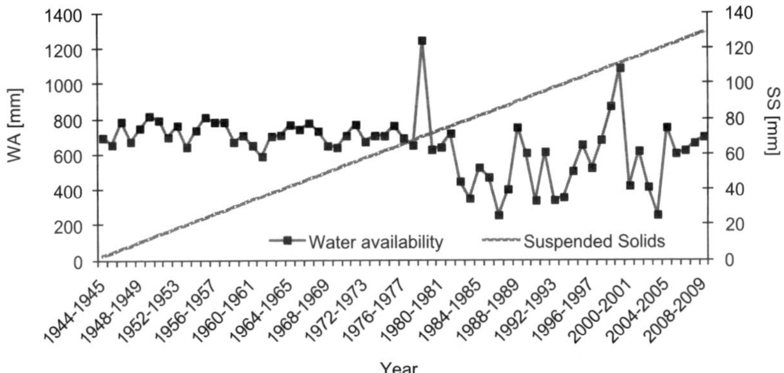

Figure 17. Graph of suspended solids and water availability from 1944 to 2009? (Data from Douala Meteorological services; CRH-Centre de Recherches hydrologiques à Yaoundé, hydrological data; AES SONEL data, 2005).

(thickness) of land that is carried away each year by runoff and its cumulative action. Finally, we determined and correlated the thickness of soil degraded or removed in relationship with water availability (Figure 17).

5.4.5.1 Causes of sedimentation/silting

Data on suspended solids, turbidity, sediment, transport and specific degradation, give an idea of sedimentation in the Sanaga River basin. Sedimentation is due to three causes identified by complementary physical, chemical and biological analyses.

Physically, the migrations of materials are mainly provided by runoff, which transports material down slopes. Deposits frequently encountered in this basin consist of sand and clay-sand. Table 3 below presents the combination of physico-chemical properties of collected samples.

The relatively low pH in turn, confirms the nature of the drained land dominated by acid rock (granites and metamorphic rocks). The low pH of February is related to low water at this time of the year, which leads to an increase in the concentration of acidic components (HCO_3^-). From February to May, pH increases, then it decreases

Table 3. Physico-chemical properties of water from the Sanaga at Edéa
(Laboratory analysis, 2010).

Samples	Temp. (°C)	pH	Conductivity (µS/cm)	Alkaline (µeq/l)
		Test parameters		
SED 1	26	6,76	39,9	269
SED 2	26	6,89	40,5	266
SED 3	26	6,64	38,3	266
SED 4	26	6,54	32,8	225
SED 5	26	6,36	42,1	231
SED 6	26	6,52	32	
SED 7	26	6,46	32,9	217
SED 8	26	6,47	41,5	
SED 9	26	6,57	38,7	214
SED 10	26	6,37	34,6	
SED 11	26	5,89	35,1	236
SED 12	26	6,93	36,7	

Table 4. Major ions tested (April 2009–March 2010) (Laboratory analysis, 2010).

Tests					Samples							
IONS (mg/l)	SED 1 Apr.	SED 2 May	SED 3 Jun.	SED 4 Jul.	SED 5 Aug.	SED 6 Sep.	SED 7 Oct.	SED 8 Nov.	SED 9 Dec.	SED 10 Jan.	SED 11 Feb.	SED 12 Mar.
Na^+	2,10		2,08		2,78		1,88		2,35		1,75	
NH_4^+	0		0		0		0		0		0	
K^+	1,64		1,62		1,90		1,40		1,42		1,27	
Mg^{++}	1,20		1,23		0,87		0,83		1,16		1,16	
Ca^{++}	2,00		1,94		1,75		1,52		1,69		1,72	
F^-	0,06		0,07		0,08		0,06		0,06		0,07	
Cl^-	0,60		0,41		1,38		0,44		0,85		0,30	
NO_3^-	2,88		1,67		0,94		1,66		1,25		0,30	
PO_4^{3-}	0		0		0		0		0		0	
SO_4^{2-}	0,98		0,83		1,04		0,54		0,26		0,21	
HCO_3^-	16,41	16,23	16,23	13,73	14,09		13,24		13,05		14,40	
Ca^+/An^-	0,347 /		0,357 /		0,416 /		0,353 /		0,428 /		0,386 /	

from June to August, following the increase in flows during the short rainy season.
It decreases slightly in August to October related to human activities carried out
within the basin, with a lag time of one month. The conductivity varies from 32 to
42.1 S/cm. This suggests that the Sanaga River is not charged and still carries good
quality water.

Chemically, sedimentation depends on the nature of the elements present (Fe, Al,
Na, etc.). The relationship of cations to anions, in order of magnitude, indicates faster
sedimentation in December and August (Table 4).

Biologically, plant roots concentrate solutions by pumping infiltrated water, causing the flocculation of colloids and deposition of dissolved substances. By this means acids can be biodegraded. This can lead to the release and deposition of metals, a process by which the physico-chemical properties and quality of water may be more or less modified.

5.4.5.2 Discussion

Due to the high population on the west side of the Sanaga River, most human activities are concentrated here, and have an influence on river flow. Activities are practiced on a variety of soils, which thereafter become exposed to erosion. Soil types are reflect variations in the petrographic nature of the rocks and the diversity of plant species that cover the basin as a whole. In the foregoing paragraphs, it has been established that runoff carries large amounts of materials that accumulate at the dams and reservoirs reducing their depth and capacity, and lead to reservoir filling and siltation. The height of these deposits reduced the water column available because it affects the costs of calibration. The average specific degradation was estimated at 27,424 tons per km²/yr. The accumulation of material in the dam over the years has caused siltation and filling of the tank transforming the dam into a perched dam.

However, the decrease in rainfall caused a decline in production of electricity against a growing demand of a growing population. Decrease in runoff from the Sanaga continued since 1973. This reduction in runoff preceded the decrease in rainfall and can mainly be explained by sedimentation and silting. Temperature increase of 0.5° C within the basin due to global warming, could be offset by abundant vegetation and drainage density.

However, the lack of trees along the river banks and near the dam (caused by logging or deforestation) contributes to water turbidity, sedimentation, and modification of the landscape. Sedimentation contributes to both the rising of water during the wet season and lowering during the dry season. The rise of the water causes bank erosion and destruction of certain mostly riparian plant species. Increasing sedimentation can influence several hazards including landslides and floods and to prevent this, the following steps must be taken:

- Reforest (using native species) the entire watershed of the river.
- If reforestation is to be effective, then dredging of reservoirs must be performed every ten years.
- An integrated approach to the management of soil and water that takes into account all the components (physical, biological, socioeconomic, human) of the watershed, and is based on changing rainfall must be adopted, and different hydrometric stations installed.
- Realize an environmental audit of ten dams constructed in order to control and mitigate some impacts.
- Prevent hydroclimatic risks (floods, landslides), health risks, societal, technological, by adopting a campaign of environmental awareness for the population and teaching appropriate action.
- Developing an environmental management system based on international standards as ISO 14001.

Consequently, the multiplication of thermal electricity units and hydroelectric dams (Mekin, Lom Pangar, Natchigal; Mem Vele; added to the existing dams Edéa, Songloulou, Lagdo for production, Mbakaou, Bamendjing and Mape for regulation) in Cameroon for energizing the country is not a solution to the problem. Because, the more there are, the greater sedimentation will increase, and with it other corollaries like landslide, erosion, and floods. So, the country must manage the existing dam by dredging. Dams are going to exacerbate climate change because of increasing deforestation and evaporation, whose cumulative effects have brought rainfall and flow depletion and favored global warming resulting in increasing local temperatures.

5.5 NATURAL HAZARDS LINKED TO DEFORESTATION AND CLIMATE CHANGE

5.5.1 Typology of natural hazards in Cameroon

Cameroon is exposed to five principal hazards, depending upon climatic regime change and local geomorphological configuration. Specifically: earthquakes, volcanism, floods, landslides, drought and insect infestations. With respect to the quality of randomness, earthquakes are the greatest hazard, however, with respect to vulnerability, volcanism is more important. 'By vulnerability, we mean "the characteristics of a person or group and their situation that influence their capacity to anticipate, cope with, resist and recover from impact of a natural hazard' (Wisner *et al.*, 2004:11). Figure 18 shows a compilation of available information on hazards observed in Cameroon since the 19th century.

In this context, volcanism appears to be more deadly than the earthquake in Cameroon, while paradoxically, worldwide, earthquakes are most deadly. Finally, for all events recorded in Cameroon, volcanism has lead to a total of 1883 deaths as opposed to none for earthquake, 119 for landslides and 115 for floods (Figure 19).

5.5.2 Landslide and its socio-environmental consequences

5.5.2.1 Historical context of landslide in Cameroon

Literature on landslide is abundant and the many studies allow us to assess its likely consequences for Cameroon (Hewitt, 1997; Caine, 1980; Dikau *et al.*, 1996; Varnes, 1978, Cruden and Varnes, 1996). Landslide is a generic term designating a geomorphological process of mass displacement of a volume of material along a slope. The mode of displacement of the material can be a fall, a topple, a slip or slide,

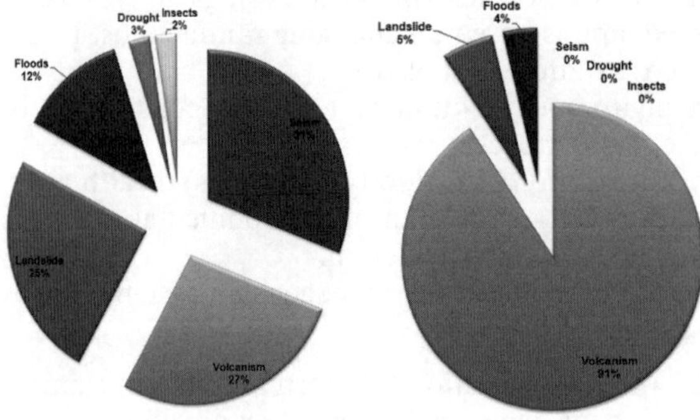

Figure 18. Percentage of types of hazards (left) and distribution of victims (right) in Cameroon
(Tchindjang, 2012).

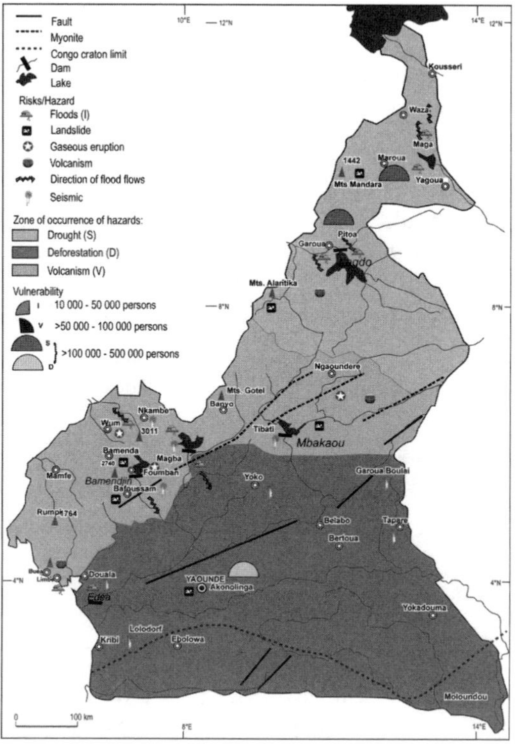

Figure 19. Main hazards recognized in Cameroon (Tchindjang, 2012).

an expansion or spread or a discharge or flow. Landslides are differentiated based on the type and volume of material moved as well as the mode of transport. Landslides, originally associated with mountainous regions can also occur in lowlands. They are usually caused by a combination of geological factors: water-soaked regolith, discontinuity, and schistosity fault contrast in relation to permeability and rigidity of the material (earthquake), geomorphological affects (tectonics and volcanism, subsidence, erosion and groundwater extraction, freezing and thawing, etc.) and anthropogenic affects (deforestation, agricultural activities, mining, urbanization, excavation slopes).

In Cameroon, growing concern and interest in mass movements dates from the 1950s, this was the time that it was marked in the collective memory in the western highlands, and it was around the 1990s, that it became a topic among policy makers. In fact, in the highlands, expansion of the economy that accompanied expansion of cash crops, led to deforestation, reduction of land available for food crops and bare, unstable slopes. Therefore, once inaccessible hill peaks and mountains, began to be cultivated. Moreover, the descent of the Bororos, pastoral tribes from the northern regions, in search of pastures and their installation in the heart of the western highlands, has lead to overgrazing and degradation of watersheds (e.g., mountains of Oku, Jakiri, Mont Bangou, Sabga). Unfortunately, these movements are poorly recorded, not only because of the remoteness of the lands in relation to research centers, but also because of the limited means and interest of researchers. For this reason, the Centre for Research on the Epidemiology of Disasters (CRED; some events recorded in Cameroon only since 1980) database lists only a single landslide in the Waban (2003), which killed 22 people. We used for our research a variety of sources (Ayonghe *et al.*, 2002; Lambi, 2001; Tsou Martial, 2007), and fieldwork to compile an array of these changes observed in Cameroon.

The deadliest events recorded in Table 5 are the landslides of 2001 in Limbe (24 dead) and those of 2003 in Magha (22 dead). Although the Magha event is the deadliest of all (22 victims for a single shift complex). Of the 23 events, 12 occurred in September, 5 in August, 3 in July, 2 in June and 1 in October. It is clear from this table that investigations should make use of historical and local memory to identify additional landslides and integrate them into the database. It should be noted that we

Table 5. Some mass movements recorded in Cameroon since 1954 (Data compiled from multiple CRED, 2007; Ayonghe *et al.*, 2002; Lambi, 2001; Tsou, 2007, field observations).

Date	Location	Damage type	Rock type	Probable cause
September, 1954	Nchingang, Lebialem SW	Floods, houses abandoned & bridges destroyed,	Volcanic rocks (trachytes acid tuffs)	Rainfall, human activities
September 1956	Beine, Lebialem SW	3 dead; bridges, houses and plantations destroyed, floods.	Volcanic rocks (trachytes acid tuffs)	Rainfall, human activities
September 1957	Fomenji, Lebialem SW	12 dead, bridges and farmlands destroyed	Volcanic rocks (trachytes acid tuffs)	Rainfall, human activities
August 1973	Fonengé, Lebialem SW	1 dead and displacement of 200–300 inhabitants.	Volcanic rocks (trachytes acid tuffs)	Rainfall, human activities
June 1988	Bamboko, Melong mt Manengouba	8 dead, houses & farmlands destroyed	Pyroclasts	Rainfall, human activities
July 1991	Limbe, Mt Cameroun SW	1 dead, 1 house destroyed	Pyroclasts of a volcanic cone	Rainfall, grading on slope of the cone
September 1991	Pinyin, Santa, Bamenda	3 landslides recorded, 12 dead, 2 houses destroyed	Rhyolites & weathered pyroclasts	Rainfall & agriculture on slopes
September 1992	Fomenji, Abi, Fonengé (SW)	12 dead; bridges, houses & farmlands destroyed.	Volcanic rocks (trachytes acid tuffs)	Rainfall, human activities
September 1994	Fomenji, fotang, Fonengé	6 bridges & crops destroyed.	Volcanic rocks (trachytes acid tuffs)	Rainfall, human activities
September 1995	Bafaka, Ndian, Mt Roumpi	57 slides, 3 dead, 1 house, farmland & tracts of forest destroyed	Sandstone, trachytes, basalts, tuffs & pyroclasts	Seismic activity (magnitude VII), rainfall
September 1997	Sho, Belo (NW)	2 dead; 1 house & much farmland destroyed, road traffic severed	Pyroclasts in a subsident sector	Rainfall, perched water table, human activities

(Continued)

Table 5. (*Continued*).

September 1997	Gouata, Dschang (Mt Bamboutos)	1 dead, farmland destroyed	Granitic regolith waterlogged	Rainfall
September 1997	Batié	Farmland destroyed	Granitic regolith waterlogged	Rainfalll and sandpits
July 1998	Bingo, Belo	5 dead, 3 houses destroyed	Weathered rhyolites	Rainfall
August 1998	Bamumba, wabane Lebialem	5 injured, 11 houses, bridges & farmland destroyed by ca. 11 landslides	Granitoïds regolith waterlogged	Rainfall and possibly slight tremor
August 1998	Abi, Ako, Atsuela, Babong	1 injured; damage		
September 1998	Anjin, Belo	2 dead, 1 house & much farmland destroyed	Pyroclasts waterlogged	Rainfall, human activities on slope
September 2000	Rom Nwah	20 slides with 6 dead, 17 injured, 7 houses destroyed	Weathered granite and granitoïds	Seismic activity (magnitude IV) and rainfall
June 2001	Limbe	43 slides with 24 dead, 2,800 homeless, 120 houses destroyed	Basalt, pyroclasts and tuffs waterlogged	Rainfall and possible tremor
July 20, 2003	Magha'a	22 dead; 50 cattle killed; houses, farmland & tracks destroyed	Volcanic rocks (trachytes acid tuffs)	Rainfall, human activities
October 2008	Moume, Bafang	1 dead, 12 houses destroyed &100 homeless	Migmatites regolith waterlogged	Rainfall
August 2009	Bamenda	Main road severed	Volcanic rock	Subsidence and rainfall
August 2010	Bamenda	5 dead, 40 homeless, 5 houses destroyed	Trachytes	Rainfall, slope failure & human activities

have selected only the data actually recorded and available. It is, therefore, difficult to quantify the number of landslides in Cameroon. Sometimes the only witnesses to events are the people involved and who later may inform scientists or authorities. For example, there is recorded for Fogwe Nji (1999), on the Ring Road, 75 slides (from June to August 1991 to 1996) for sites ranging from 1240 to 1900 m and slopes measuring between 15° and 27°.

Another challenge has to do with the quantification of phenomena and localizing tools. This led us to plot the trends (Figure 20) of the landslide hazard in Cameroon as well as exposed populations. The Cameroon volcanic line and the mountainous backbone appear to be the preferred site of mass movements. Roughly speaking, the wet highlands of Cameroon are the most likely sites for this hazard. If in the past, such events were relatively limited, today, factors such as high population densities on these fertile lands, intensification of agricultural activities coupled with the unbridled exploitation of land, major construction of roads, abandonment of cropping patterns and ancestral stabilizers such as anti-erosion fences and terraces, aggravate and desta-bilize the structure of these environments, leading them to experience recurring move-ments. Moreover, even if some events were recorded in history, their magnitude and frequency remains unknown even today.

As for the Cameroonian highlands, reference data begin in 1954. From 1954 to 2007, there were more than 21 deadly landslides and almost 119 deaths (Figure 20). The evolution curves and linear trend show that we can expect an increase in the number of mass movement and victims in the future. The environmental consequences of landslides are recurrent flooding and increase of erosion and deforestation on the slopes.

5.5.3 Increase of flood hazards

In view of recent climate change, the country is susceptible to floods, which have become increasingly recurrent over the past thirty years. Geological conditions are favorable to the occurrence of this hazard, but, in fact, it is a recent phenomenon, induced by and reflecting deforestation and climate change. Indeed, until the last 30 years in Cameroon, historical research did not take in account natural hazards faced by groups or individuals. Even though research (by this we mean a specific approach to questions, critical thinking and analysis of data) allows a better understanding of the reactions of societies to threats and disasters they have suffered, the disaster is defined by the extent of damage to persons and property. There is not necessarily any correlation between the size of a hazard and the extent of damage. Cameroon, due to several natural and human factors, is a country experiencing flooding everywhere.

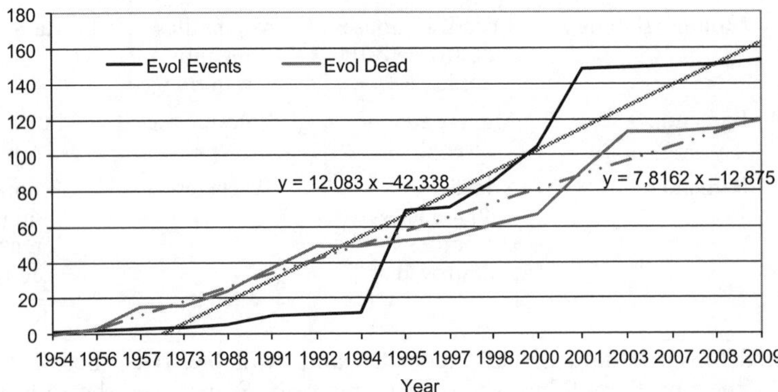

Figure 20. Evolution of mass movement recorded in Cameroon since 1954 (Data compiled from multiple events CRED, 2007; Ayonghe *et al.*, 2002; Lambi, 2001; Tsou, 2007, field observations; Tchindjang, 2012).

Flood risks are ubiquitous in Cameroon. But the most vulnerable sectors are urban quarters generally located in marshy zones, which paradoxically represent the most populated areas. It should be noted that heavy storms are the triggering factors while climate change and the lack of forest cover appear to be underlying and aggravating. This relationship also holds true for the breaking of a hydroelectric or hydro-agricultural dam.

Flood risk is due to a combination of two factors: hazard and vulnerability (habitat, population densities, infrastructures, equipment, psychological factors, land use, land cover). Thus, the consequences of this hazard will be negligible in uninhabited, non-urban areas. So, flooding hazards are, generally speaking, more often perceived in urban areas than in rural ones. A case in point is that of Nkolbisson (a district of Yaoundé) in April 2008. A flood occurred related to heavy rains that hit the mountainous part of the city, drained by the Mefou River. The causes are related to the unplanned occupation of prohibited areas for construction, including mountainous areas (850–1200 m asl) devoted to green space. Floods will doubtless increase in the near future due to unplanned urbanization; the main challenge is to secure and drain urban areas (Figure 21).

Excess water causes the soil to become saturated, especially in the slums and marshy areas, where the water table is already quite close to the surface. Moreover, even if floods are common in some Cameroonian cities like Yaoundé, Douala, Maroua, Garoua, most of the time, people have developed techniques to reduce, as much as possible, damage and loss of life by drowning. This said, they still remain victims of poor sanitation conditions, which become especially deadly after flooding because of contamination of water and food. With its high susceptibility to flooding, Cameroon has recorded losses of lives and suffered other damage (Table 6).

It is also clear from this table that each region of Cameroon has experienced at least one flood, affecting one or more families. Flooding has sometimes caused temporary or permanent displacement of populations (Yaoundé-Nkolbisson 2008, 400 persons were displaced). In addition, it is important to mention the responsibility of risk incurred by the populations themselves. In Douala, most of the time, potentially affected and homeless people do not withdraw, except in case of extreme danger. If they do leave, they come back right after the event, and a final departure from the site is often the result of forced eviction by the government. The most important and

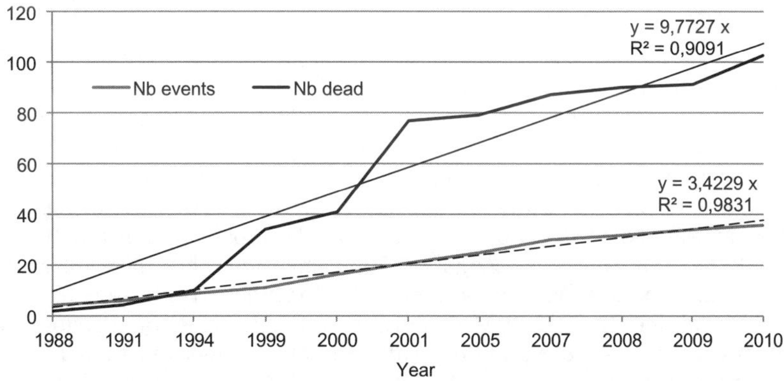

Figure 21. Trend of the major floods recorded in Cameroon since 1988 (CRED, 2007; fieldwork records; Tchindjang, 2012).

Table 6. Some cases of flooding between 1988 and 2010 (CRED, 2007; Mbeugang Tcheubonsou, 2011).

Dates	Location	Damage
1988	Douala	2 dead, 1,000 affected.
1991	Maroua	2 dead, 200 homeless
1994	Maroua Mokolo, Logone & Chari	6 dead
1999	Wouro-Haoussa, Demsa	24 dead, 1,000 homeless
October 1999	Lagdo	Many human victims, 650 animals dead, 1,144 ha of crops destroyed
2000	Douala, Bonaberi, Bépanda, Nylon districts	3 dead, 500 affected
3 August 2000	Douala	40 homeless
19 August 2000	Maroua	4 dead, many homeless, bridges & houses destroyed
8 February 2000	Yaoundé (Mfoundi & Central Post office)	Many shops flooded
2001	Limbé, Sud West	30 dead, 1,500 homeless
2–3 August 2001	Douala	4 dead
9 August 2001	Bamenda	
19 August 2001	Maroua	2 dead
29 September 2001	Bertoua	
2004	Douala	Provoked cholera epidemic
2005	Douala	2 dead
2007	Mokolo, Sarki-Fada, Tongo, Tourou, Dza-Wandaï, Idanang, Mendeze, Nassarawo, Domayo, Djimeta, Bonguelé, Guimengayak and Zimangayak	8 dead, 17 injured & traumatized, 1,220 homeless, total of 1,440 affected.
6 April 2008	Yaoundé (Nkolbisson)	2 deads, much damaged equipment
October 2009	Yaoundé (Nsiméyong II)	1 dead
July 2010	Pouss (Far Nord)	More than 12 dead, much damaged equipment

dramatic event for the far north, occurred in July 26, 2010. Floods killed 12 people and made 500 homeless, who were displaced sought shelter in a school class room. The event was due to a violent storm that destroyed houses in Pouss. Apart from the sad loss of life, there was collateral damage to rice and other crops. The reason that this flood was so deadly may be related to the fact that in this region, a vast alluvial plain watered by the Logone, is among the most populous in the country. In Yaoundé Ekozoa watershed, Zogning (2005) mentions that there were 11 deaths for the 1980–2003 period. According to his fieldwork, these deaths were distributed as follows: 5 Mfoundi (streets), 5 at Mokolo Elobi and 1 in Messa (as a result of drowning). All these examples prove that floods are present in Cameroon, even if they do not seem to have reached critical levels. Authorities and the government should not wait to reach this critical level before reacting, prevention is better, and it is clear that in the future there will be increased environmental risk of flooding.

5.6 CHALLENGES LINKED TO CLIMATE CHANGE, DEFORESTATION AND ENVIRONMENTAL RISKS

5.6.1 Towards afforestation and reforestation: The case of Kaélé

In northern cities of Cameroon, such as Garoua and Maroua, people use mainly firewood for household purposes. Ntsama Atangana (2009) estimated that the wood volume used was 152.64 t/yr. The anarchical way of collecting firewood creates great pressure on this resource. In those cities, more than 94% of households use firewood, 90% charcoal and 64% wood. Wood consumption is more than 12 times the amount of gas used by households. The individual wood requirement per inhabitant is 2–3 kg/day. This situation has brought the government to renew the Green Sahel Operation.

The Green Sahel Operation, initiated and launched in 1960, had only limited success, especially in 1975, which resulted in an interruption. The program as prescribed in the Forest and Environment Sector Program (FESP) in 2002, was relaunched in 2008 in an effort to mitigate the effects of global warming and reduce desert encroachment. These actions are aligned with the ratification by Cameroon of the Conventions on Climate Change (UNFCCC), the implementation of resolutions of the United Nations Convention to Combat Desertification (UNCCD, 2003) ratified by Cameroon 27 May 1997, and the Convention on Biological Diversity (CBD). It should be noted that those three conventions create a synergy around reforestation, agroforestry and afforestation (Ribot 2004, desertification and forests). This revival also helps in coping with climate change.

The Green Sahel Operation was relaunched in 2008 in northern Cameroon after the mounting by Cameroon of a National Action Program to Combat Desertification (PAN/LCD) launched in 2006. It is administered by the Government of Cameroon (owner, developer) and MINEP (Ministry of Environment and Nature Protection), which is the prime contractor, and is funded by the Public Fund Investment Budget. This operation aims to prevent and reduce degradation in arid, semi-arid and dry lands, and to restore degraded soils. In addition, it also aims to restore and improve the fertility of degraded and marginal lands, strengthen the vegetation cover in the Sahel, deter firewood cutting coupled with the distribution of improved stoves, and finally, sensibilization of populations respecting the fight against desertification. Tree planting began in March 2008, in an area of about 1,500 ha in Lera (Mayo Kani Division), and 1,500 ha in Maltam (Mada) in the Logone and Chari, 1,000 ha at Goussor in 2009, 3,500 ha in 2010 in several villages (Bipaing, Berkede in the Mayo-Kani, Adjir-Mora in the Mayo-Sava, Djiril Bogo in the Diamaré, of Katikilé-Darack and Djénéné Goulfey in the Logone and Chari and Ouro-Daban-Yagoua in Mayo-Danay). After two years, a budget of relining voted for each site, allowed the use of 60,000 plants for relining sites of Lera, Maltam (Mada) and Kalfou (Mayo Danay).

5.6.1.1 Inventory of afforested sites at Kaélé

Sustainable management of ecosystems, in general in the far north, is subject to strong attacks. If the southern part of Cameroon has a rich and diverse plant community, which is potentially immune to desertification, the northern part, marked by a long dry season and rainfall concentrated in 3 to 6 months is more exposed to the impacts of climate variations. This relatively extreme climate, when associated with overgrazing, pressure on wood resources for fuel wood and poor agricultural practices, has led to intensified land degradation in this region and increased erosion such that the land has gradually become infertile. The gleysols and vertisols found there consist of swelling clays characterized by large shrinkage cracks; they are waterproof and cause flooding in the rainy season. However, this Cameroonian region has a potentially rich fauna and other natural resources (sand, limestone) not found in sufficient quantities in the south.

Kaélé, our main observation site, is located in the Mayo Kani division. It provides more than half of the wood and charcoal consumed in Maroua city. It receives an average total rainfall of 800 mm/yr. For this environmentally sensitive area, the causes of drought are both natural and anthropogenic. Natural factors include: wind erosion, rain, water erosion leading to land degradation and loss of productive potential.

Anthropogenic causes are linked with the extension of farmlands, poor farming practices, bush fires, logging and charcoal production, and overgrazing. It is an environmentally sensitive area due to over exploitation of the soil for agriculture and pastoralism (Boutrais, 1995). Agricultural and pastoral overloading leads the soils to suffer the effects of erosion and to gradually decrease in productive capacity/fertility, with a trend toward sterilization of the soil. When irreversible stages of this process are reached, one can observe large areas of barren soils, locally called "hardé", which are actually hardened lands, unsuitable for agricultural activities. Finally, added to this, in the region, the production of firewood and charcoal is the most important form of exploitation of woody vegetation. It is in this context that the Green Sahel Operation was begun in order to afforest degraded lands.

5.6.1.2 Methodology for afforestation and observation

Recourse to afforestation techniques also requires the control of water. The essential technique is to plant native species that once existed on the choosen sites we visited. The first site was Berkede, having an area of 500 ha where 16,000 trees were planted in 2010. The second site, was in the village of Goussor, which has an area of 1,000 ha where 160,000 seedlings were planted during the 2009 rainy season. The third site was located in Lera, having an area of 1,500 ha where 240,000 plants were planted in March 2008. Before planting, technicians will dig 50 cm of subsoil to extract and destroy the remains of hardened soil. Then, with the help of a geophysicist, they dug a borehole at each site (using a power hand pump) of more than 80 m into the granite bedrock to obtain water for irrigation of seedlings during their growth phase.

Prior to this, an awareness campaign was launched for local populations to sensitize them and ensure their effective participation in the implementation and monitoring of these operations. The participation rate reached more than 80%. Moreover, to ensure the effective engagement of local populations and to avoid further pressure on wood resources the administration distributed to stakeholders improved stoves (Table 7).

The same participatory methods were used for the establishment of post-planting monitoring activities. For this operation, involved local populations received 50,000 FCFA (76 Euros) monthly per individual. Two hundred young people had been recruited to monitor the trees. In addition, 800 people were requisitioned for watering at a rate of 30,000 FCFA (46 Euros) per month for 6 months.

5.6.1.3 Results and discussion

At Lera, we noticed the return of wildlife and a revival of economic activities related to NTFPs. The remaining problems to solve were bush fires, vandalism, deferred grazing, nocturnal pasture and the lack of signage (species) at different sites. We must, therefore, stress that afforestation is an important contribution to human adaptation to climate change. All the studied sites should be complemented with conservation

Table 7. Afforestation site of Kaélé (Fieldwork, July 2010–2011).

Year and location	Wooded area in ha	Number of seedlings	Relining in 2010	Dissemination and promotion of improved stoves	Price in FCFA, Euros & US dollars
March 2008 Lera 10°09'19" & 14°35'25,9"	1,500	240,000	60,000	21300	800,000,000 1,219,592€ 1,600,000 $
2009 Gousor 10°09'12,2" & 14°32'25,8"	3,000	160,000		13076	800,000,000 1,219,592€ 1,600,000 $
2010 Berkédé 10°03'47,3" & 14°24'23,8"	3,500	80,000		15384	1,020,000,000 1,554,980 € 2,040,000 $
Total	8,000	480,000	60,000	49760	2,620,000,000 3,994,164€ 5240,000 $

measures in order to protect the plants for sustainable use as a community forest by the municipality. Among the diverse approaches in the battle against desertification, we have choosen to highlight four.

- Protection, restoration and/or water conservation of the soil. The techniques deployed in this framework are to facilitate the retention of water by soil in order to reduce the erosive effects of runoff. In the context of low slope, it consisted of facilitating the infiltration of water into the plot by the practice of contour line culture, the installation of stop bands, metaling and others.
- The fight against wind erosion: the techniques in this domain aim to "fix" the ground, so to prevent soil mobilization by wind and then to reduce its effects. In such a context, recovery of the herbaceous and shrubby vegetation and also their enclosure (against men and animals) is critical, and that is what we observed at the Mayo Kani. They can also implement hedges plant used as windbreaks by natural recovery of a shrub or tree (afforestation is central).
- Fighting against bush fires and reducing the burden of firewood. It aims at minimizing the effects of fire by the practice of early firing. It appears important to innovate by the implementation of technologies that can help in the promotion of timber during the carbonization or combustion processes. Also, it is important to reduce the demand for wood by replacing fossil energy by solar sources or improved stoves.
- Afforestation/reforestation constitutes a significant and optimal solution to climate change. It appears to be the most effective response for containing or incorporating the three previous processes analyzed. This response is the most manageable and more efficient through the selection, improvement of the species, fertilization and irrigation. The failure of the first Green Sahel Operation in 1970–1980 was related to an emphasis on planting exotic species (including eucalyptus), which destroyed and affected the original landscape. Nowadays, the Green Sahel framework plays an important role in the micro-climate and in wind break. It is best to advise participating farmers who receive plants for agroforestry and forest management in the agricultural and pastoral areas.

Moreover, integration of plants into pasture and farming processes gives an advantage to the farmer, who benefits from timber, fruit, fodder and even fertilizing his farmland.

5.6.2 Forest governance and reforestation in the City of Yaoundé: Issue of forest governance and administration

Admittedly, the previous afforestation was carried out in dry conditions whereas southern Cameroonian wetlands or sub-humid farmlands suffer from intense deforestation for reasons identified earlier. The operation was conducted in northern Cameroon while the city capital,Yaoundé, is exposed to threats and hazards associated with deforestation (landslides, floods). It is urgently recommended that the Yaoundé hills are reforestated in order to rejuvenate and restore the landscape. Indeed, the results of the reforestation of the Gbazabangui hills in Central Africa Republic (Tagboka Yangana, 2009), show that in one decade (1997–2007), 'on a total highly degraded areas of 70 hectares exposed to high risk, 42 hectares have been rehabilitated' Biodiversity of the hill has been restored with the return of fungi, butterflies and caterpillars, duikers, guenons, mangabeys, boa constrictors and cane rats. Rock falling in the eastern front, southeast and southwest that caused accidents have ceased. Runoff volume decreased, resulting in the decrease of flooding in areas such as Benzvi and Gbakondja. Annual public expenditures for unblocking drainage channels have been reduced by over half. They rose from 565 million in FCFA/year (861,337 Euros) in 1997 to 252 million of FCFA/year (384,172 Euros) in 2007. In addition, the financing of micro- and multiple campaigns have helped in reducing food shortages in neighboring quarters and human pressures on the resources of the Gbazabangui reserve.

If such a project can be applied to the city, it will certainly restore the natural vegetation and biodiversity, which will encourage the development of ecotourism in this town through the restoration of mountain ecosystems of Yaoundé hills. The results would consist in the return of wildlife, reduced flooding, rock fall and landslides.

Table 8. Main local species that can be planted on Yaoundé hills (CUY/DST/SPJ Yaoundé, 2001, 2009).

	Scientific name	Trade name	Common local name
Main species	*Terminalia ivorensis*	Framiré	Lidia
	Terminalia superba	Fraké	Akom
	Canarium schweinfurthi	Aïélé	Abel/Otui
	Anthrocarium klaineana	Onzambili	Angongui
	Detarium macrocarpum	Mambodé	Amouk
	Afzelia pachyloba	Pachyloba	Mbanga afum
	Chlorofora excelsa	Iroko	Abang
	Inga edulis	Longeon	
	Vitex ciliata	Vitex	Evoula
Secondary species	*Garcinia kola*	Bitter cola	Oyié
	Cola acuminata	Kola	Abeu
	Tricocipha acuminata	Mvut	Mvut
	Cola pachycarpa	Kola des singes	Ekom
	Mirianthus arborea	Ananas des singes	Engokom

In addition, Law No. 94/01 of 20 January 1994 based on forestry and wildlife, in Article 33, prescribed a reforestation rate of 800 m² per 1,000 inhabitants. This means that for Yaoundé city, which has approximately 1.8 million inhabitants, the forest area should be around 1,440,000 m² or about 14,400 ha. The former forest reserves of Yaoundé, instead of being reforested, were destroyed under the eye of the authorities who did nothing to stop or limit the disaster. Yaoundé no longer has half of forest that it had at the time of the adoption of the 1994 Act. So, in the case of Yaoundé, the capital, reforestation is the best solution to protect the town, to adapt to climate change, to restore the soil, and to stabilize slopes in order to reduce mass movement and protect human life. To this end, local plants have been identified by the municipality (Table 8).

5.6.3 Forest governance and REDD

Cameroon's forests, which cover 19.6 million hectares or 41.3% of the country and represent 11% of the Congo Basin forests, represent an opportunity for Cameroon to use through the mechanism for Reducing Emissions from Deforestation and forest Degradation (REDD). Karsenty and Ongolo, (2011) stress that the originality of REDD lay on the fact it is an incentive since the essence of the proposal is to reward states for what they achieve in the fight against deforestation. This mechanism is founded on the hypothesis that developing countries 'pay' an opportunity cost to conserve their forests and would otherwise prefer other choices and to convert their woodlands to other uses. The basic idea is, therefore, to pay rents to these countries to compensate for the anticipated foregone revenues. This opportunity is in line with the national policy of sustainable forest management enshrined in the 1994 Forestry Law and the Environment of 1996 and in terms of convergence of the Commission des Forêts d'Afrique Centrale (COMIFAC).

Thus, the international community by promoting the role of forests and trees in the stabilization of global Green House Gases (GHG) in the atmosphere, provides Cameroon an opportunity to modernize agriculture and livestock development and ecotourism. Moreover, the adoption of a REDD strategy with specific action plans that reduce pressure on natural formations would improve the protection of biodiversity, air quality, water, health as well as fight against erosion, illegal logging and desertification.

Nevertheless, many analysts and NGOs worry that REDD may be nothing more than an instrument to protect the lucrative stocks of carbon that forests represent to the exclusion of the local people. These fears are rooted in the turbulent history of the creation of protected areas in several developing countries and the resulting forced exclusion of local populations. But we think and as suggested by Ngombala Samba's (2011) work, that payments for environmental services (PES) (through rapid rural appraisal), appear to be essential tools in the fight against deforestation as well as against rural poverty and exclusion.

5.6.4 A complex mining governance

Mining in Cameroon is controlled by the State, and the Ministry of Mines and Technological Development (MINIMIDT), represents the management institution. As such, it relies on the agenda of the Framework for Support and Promotion of Handicrafts Mining (CAPAM). At a legal level, Law No. 2001/001 of 16 April 2001 known as the Mining Code was revised and completed by Law No. 2010/011 of July 29, 2010. One can add to this Decree No. 07/2003 of 25 064/PM on the establishment, organization and operation of CAPAM. These laws recognize small mines (mechanized company) and artisanal miners (Table 9) grouped into Common Initiative Groups (CIG), but it does not recognize indigenous or local populations operation of sites. In an effort to right this, the CAPAM has tried to implement six pilot sites, namely Betare-Oya, Ngoura site (Colomina), Kambel site (Batouri), Bindiba site (Garoua Boulai), Mobilong site (Yokadouma) and Ketté site (Beke). Taking in consideration the results of Takoutsi's (2009) and Nono's (2011) research in Adamawa and Betaré Oya, the following recommendations are made.

Many ore deposits (gold, sapphire) are mined by artisans working by hand and production rarely reaches one-third of potential reserves. In some cases, the deep part of the deposit is irretrievably lost as a result of an uncontrolled work surface. All these mortgage the sustainability of this activity; nevertheless, there are some socio-economic benefits for local people and stakeholders, even those in more remote areas. After our analysis of the role and responsibility of the state in mining, it becomes evident that the institutions in charge of mining activities do not have sufficient nor appropriate operational resources. These institutions are unable to control the activity and collection of state taxes (ad valorem tax on gold is 5%, that of the sapphire and diamond is 8%). Therefore, the role devoted to artisanal mining in the sustainable development of the regions involved will essentially come from the strategy and organization of this activity economically, environmentally and socially (health, education, food availability). More, emphasis should be given to social issues and the environmental consequences of mining before moving gradually towards the economic importance of the activity. All these require planning and awareness of the local people.

Well organized and planned artisanal mining can be a powerful way to fight poverty. This said, with the advent of industrialization in the mining sector in Cameroon, an important question that arises concerns the place of the traditional mining industry in this new context? It may be that better organization of artisanal miners' produces greater more long-lasting benefits to community development than poorly planned and implemented industrialization? Hence the issue of mining governance is summarized in the Table 9.

Despite these mining laws, there are still aspects of artisanal mining of gold, sapphire and other precious stones that do not obey World Bank directives. This is due to the lack of environmental assessment studies before the launching of the activity (OP 4.01 World Bank Policy). Indeed, operating companies and artisanal miners have never had any environmental compliance certificates. So, an incompatibility exists between the framework of Law No. 96/12 of August 5, 1996 and the mining Code (2001 and 2010). The lack of environmental impact assessment measures readily explains failures to restore mining sites after exploitation. It also explains why there is child labor in mines, something that leads to educational wastage, later health problems, and other social and medical evils. While we are on the topic, there are other failures to satisfy World Bank operational policies, such as OP 4.04 Natural Habitats, OP 4.09 on Integrated Pest Management, OP 4.11 on Cultural Heritage; OP 4.20 on Indigenous Groups, OP 4.30 on Involuntary Settlement, OP 4.36 on Forests and OP 4.60 on Projects in areas under dispute or conflicts.

Hence, in Adamawa (eastern Cameroon), the flowering of mining companies has not led to the development of the concerned villages and regions. Apart from C&K mining (a Korean company) in Betare Oya, which tries to fulfil its responsibilies *vis a vis* local people and existing laws, other companies e.g., Caminco, Locamat,

Table 9. Mining Governance in Cameroon (after Takoutsi, 2009 and Nono, 2010; fieldwork Adamawa region).

Actors & types of operators	Cameroonian state	MINIMIDT	CAPAM	Law No 2001/001	Law No 2010/011
Clandestine Collectors	Difficult to collect mining taxes; growing numbers of illegal collectors; gold being channelled into informal networks	Difficult to monitor and identify illegal collectors	Battle with illegal collectors who offer competitive prices	Nothing in this Act takes into account this type of collector, they are outlaws	Nothing in this Act takes into account this type of collector, they are outlaws
Artisanal miners	State and artisanal miners do not have enough mineral wealth	Ministry is not strict with artisanal miners; does not force them to respect legal regulations	CAPAM tries to group them into GICAMINE in view of the channel or in the formal circuit	All artisans must have a permit to mine, which most do not	Breakthough law of 2010. It clearly defines the rights and duties of artisanal miners in this activity
Mechanized gold mining company	State adopts the policy of mass recovery of gold and encourages the establishment of mining companies	Ministry issued several permits to mechanized artisanal mining companies	It encourages collaborative work between mining companies and federations GICAMINE	No special provisions for mechanized artisanal mining of gold	Law of 2010 breaks new ground by recognizing artisanal mechanized mining under the label of small mine
Companies operating industrial gold mines	State promotes industrialization of this sector as the industrial exploitation of gold is not yet effective in Cameroon	Operating permits are issued and the discovery of mineable ore is expected	This structure is invested in research partnerships in the context of industrial use of gold	Includes provisions for industrial gold operations	Law of 2010 amends and supplements certain provisions of 2001 Mining Code: encourages industrial exploitation of gold (and other precious metals?)
Local population & municipality	Local people through local councils and municipalities need more multi-faceted support from State for socio-economic development financed by resources and benefits from gold exploitation	Ministry prefers that development activities are undertaken by local people	Tries to work for the well being of local people, although much remains to be done!	Law emphasizes local residents development concerns and sets out provisions for protection of the environment	Law promotes sustainable development

AK mining, Imperial-Mining, and Somec have not respected the prescribed specifications prior to their installation in 2008. Though sustainable development is not a panacea because it takes in account poverty alleviation.

5.7 GENERAL CONCLUSION

Cameroon is characterized by highly contrasting physical features including 402 km of coastline and mountain ranges punctuated by peaks over 3,000 m asl. Disturbances related to climate change appear increasingly in Africa and particularly in Cameroon, which is, in a sense, Africa in miniature (Afrique en miniature). These events result in a high recurrence of extreme weather such as droughts and floods and consequent landslide. Analysis of hydrological and metrological data showed that there are rainfall decreases generally all over the country with decreasing number of rainy days.

This has a significant impact on the activities of the rural populations. It appears that the impacts of climate change are reflected in the reduction of agricultural production, which could eventually be accompanied by food shortage, possible scenarios include: scarcity of pasture leading to a proliferation of conflicts between herders and farmers all in search of natural resources, an increase in outbreaks associated with hot weather and water, increase in migration of sub-regional populations in search of better conditions for survival. The scarcity of water resources due to lower rainfall and deforestation also disrupt energy production in Lagdo dam on the Benue, and Edéa dam on the Sanaga River. These dams, as shown for Edéa, are used below their potential because of worsening climatic conditions and inappropriate management.

According to transient Global Circulation Models, the average temperature in Cameroon is predicted to increase as a result of global warming. Agricultural production in Cameroon is characterized by low levels of input (e.g., quality seeds, fertilizers, pesticides and herbicides) due to farmers low purchasing power and equally low levels of government subsidies (Molua and Utomakili, 1998). Therefore, when considering projected climate change, one may reasonably ask whether Cameroonian farmers can continue farming in the same way that they have done for generations. This paper has tried to explore the challenges of climate change in Cameroon. Using a retrospective approach in the palaeo-geomorphological context, this paper has focused on prospects for environmental protection and conservation. This preventive viewpoint was used to analyze some risks then, used as a framework for landscape protection. It raises some questions, too. Given that climate change will influence many aspects of the ecology of species as they attempt to adapt or move with the changing climate. The unanswered and difficult questions are how many species in Africa are likely to face extinction as a result of the speed of habitat change and what differences in vulnerability to climate change might we expect between large-bodied and small-bodied species, over and above the effects of generation time? In terms of agricultural activities, slash and burn agriculture, bush fires, forest fires and subsistence farming not only facilitate predation since predators easily spot and attack the nests of birds, but also lead to the fragmentation of the habitat of these birds, some of which are endangered species like the mount Cameroon francolin or Francolinus camerunensis (*www.birdlife.org, 2011*).

In general, climate change coupled or not with deforestation jeopardizes food security (e.g., changes in fruit maturation times), security of land tenure with conflicts that are clogging the courts and provoke the migration or displacement of populations (human and animal species, including birds). The past million years represent a period of profound global climate shifts well expressed in the African climate. Long-term fluctuations have been primarily paced by orbital variations of the Earth and were closely linked to the onset and amplification of the high-latitude glacial cycles. At all timescales, climate changes in the African tropics are related to widespread, if not global, disruptions in the Earth's climate system.

Moreover, the deforestation crisis has brought major changes in forest ecosystems, nonetheless, modern methods of agriculture can reduce pressure on resources. However, this requires the development of alternative activities for conservation such as ecotourism. In the context of the fight against poverty, ecotourism seems to be a minor sideline. But it also raises the question of the costs of systemic economic opportunities linked to the forest uses, including payment for environmental services of forest-related activities. If such a vision were realized, reforestation activities would be done not only by forestry-related ministerial institutions (Ministry of Forests and Wildlife, Ministry of Environment, Protection of the Nature and Sustainable Development), but also by all institutions whose activities contribute more or less to deforestation (Ministry of Energy and Water Resources, Ministry of Mines and Technological Development, Ministry of Agriculture and Rural Development).

In terms of increasing natural hazards, whose occurrence is inherent in all human activities as a result of climate change, much remains to be done to build a risk management system in Cameroon. Four elements can help to explain the current poor understanding of risks and threats in Cameroon.

- Lack of instruments for measuring and recording (non-existent database), which limits our ability to assess risk.
- Poverty increases vulnerability and exposure to hazards: the population has not adapted to hazardous conditions because of the close correlation between vulnerability, population density and economic resources.
- Awareness campaigns educating people about risks are lacking, this both exposes the population to greater vulnerability, and at the same time it, prevents potential adaptation to climate change.
- People, as a result of ignorance, indifference, distrust etc., exacerbate risk through urbanization and occupation of the river valleys.

Howsoever, we are convinced that in an urban milieu like Yaoundé, and as underlined by Smith (1996), housing over development and modifications of hill slopes for buildings can lead to slope failure after 15–30 years. The cost of reconstruction to provide stability will be so high that it will not be affordable either for the people or the State, hence, the sole realistic option for rehabilitation of the milieu is reforestation or afforestation, coupled to protection of steep slopes by rigorous law-enforced zoning.

REFERENCES

Abéga, R. 2006, *Profil environnemental des sites non constructibles à Yaoundé: cas de Messa-carrière*, (Mém. DESS, Université de Dschang, FASA, CRESA, Yaoundé, juin 2006).

Achoundong, G. 1996, Les forêts sommitales au Cameroun: végétation et flore des collines de Yaoundé. *Bois et forêts des tropiques,* **247**, pp. 37–52.

Almeida, F.F.M. and Black, R. 1967, *Comparaison structurale entre le nord-est du Brésil et l'ouest Africain*. Proceedings of the Symposium on Continental Drift in the southern hemisphere, Montevideo.

Aubréville, A. 1932, La forêt de la Côte-d'Ivoire. *Bulletin du comité d'études historiques et scientifiques de l'AOF,* **15**(2–3), pp. 205–249.

Ayonghe, S.N., Suh, C.E., Ntasin, E.B., Samalang, P. and Fantong, W. 2002, Hydrologically, seismically and tectonically triggered landslides along the Cameroon Volcanic Line, Cameroon. *Africa Geosciences Review,* **9**(4), pp. 325–335.

Bessoh Bell, S. 2005, *L'influence des fluctuations hydroclimatiques sur les productions de l'énergie hydroélectrique et de l'aluminium*: cas du barrage d'Edéa, (Mémoire maîtrise, Université de Yaoundé).

Birdlife, 2011, Mount Cameroon Francolin (*Francolinus camerunensis*). (www.birdlife.org/datazone/speciesfactsheet.php?id=185)

Bonnefille, R. 1983, Evidence for a cooler and drier climate in Ethiopian Uplands towards 2.5 Myr ago, *Nature,* **303**, pp. 487–491.

Boukéké, D.B. 1994, *Structures crustales d'Afrique centrale déduites des anomalies gravimétriques et magnétiques: le domaine précambrien de la République Centrafricaine et du Sud Cameroun.* (Thèse de géophysique, Paris, Université Paris XI Orsay: Editions ORSTOM, TDM 126).

Boulangé, B. 1984, Les formations bauxitiques latéritiques de côte d'Ivoire. Les faciès, leur transformation, leur distribution et l'évolution des modelés. *Travaux & Documents de l'ORSTOM*, Paris, **175**, 1–362.

Boutrais, J. 1995, *Hautes terres d'élevage au Cameroun.* Thèse d'Etat, **3**, (Paris: ORSTOM Editions, Coll. Etudes et Thèses).

Bricquet, J.P., Mahé, G., Bamba, F. and Olivry, J.C. 1996, Changements climatiques récents et modification du régime hydrologique du fleuve Niger à Koulikoro (Mali). *Publikation AISH,* **238**, pp. 157–166.

Bricquet, J.P., Bamba, F., Mahe, G., Toure, M. and Olivry, J.C. 1997, Evolution récente des ressources en eau de l'Afrique Atlantique. *Revue des Sciences de l'Eau,* **3**, pp. 321–337.

BUCREP, 2010, *Recensement général de la population et de l'habitat 2005. Rapport de présentation des résultats définitifs* (Yaoundé, mars 2010).

Burke, K.C. and Gunnell, Y. 2008, The African erosion surface: a continental scale synthesis of geomorphology, tectonics and environmental change over the past 180 million years. *The Geological Society of America* (Boulder, Colorado USA), pp. 1–66.

Burke, K. and Whiteman, A.J. 1972, Uplift, rifting and the breakup of Africa. In *Proceedings from the NATO Conference on Continental Drift,* **2,** London, edited by Tarling, D.H. (Newcastle: Academic Press), pp. 735–745.

Caine, N. 1980, The rainfall intensity-duration control of shallow landslides and debris flows. *Geografiska Annal,* **62A**, pp. 23–27.

CIFOR, 2009, *La forêt et au-delà.* CIFOR, Bogor, Indonésie, (http://www.cifor.cgiar.org/fileadmin/templatesnew/res/documents/brochures/CIFORBrochure2009-F.pdf [29 février 2010]).

Cornacchia, M. and Dars, R. 1983, Un trait structural majeur du continent africain. Les linéaments centrafricains du Cameroun au Golfe d'Aden. *Bulletin de la SociétéGéologique de France,* **25**, pp. 101–109.

CRED, 2007, *Data version on Cameroon*: v03.07 Created on: Oct-9-2007. Source: "EM-DAT: The OFDA/CRED International Disaster Database. (Université Catholique de Louvain—Brussels—Belgium, www.em-dat.net).

Cruden, D.M. and Varnes, D.J. 1996, Landslides types and processes. In: *Landslides investigation and mitigation,* edited by Turner, A.K. and Schuster R.L., Transport Research Board, National Research Council, National Academy Press Special Report **247**, pp. 36–75.

CUY, 2001, *Schéma Directeur d'Aménagement Urbain (SDAU) de Yaoundé.* (unpublished)

CUY, 2009, *Plan de Développement Local (PDL) de Yaoundé, Yaoundé* (unpublished).

Déruelle, B., Moreau, C. and Nkonguin Nsifa, E. 1983, La dernière éruption du mont Cameroun dans son contexte structural. *Revue de Géographie du Cameroun,* **11**(2), pp. 39–46.

Déruelle, B., Ngounouno, I. and Demaiffe, D. 2007, The 'Cameroon Hot Line' (CHL): A unique example of active alkaline intraplate structure in both oceanic and continental lithospheres. *Comptes Rendus Geoscience,* **339**(9), pp. 589–600.

Descroix, L., Gautier, E., Besnier, A.L., Okechukwu Amogu, Viramontes, D. and Gonzalez Barrios J.L. 2005, Sediments budget as evidence of land-use changes. In *Sediments Budgets,* **2**, edited by Horowitz, A. and Walling, D.E. (IAHS Press), pp. 262–263.

Dewolf, Y. and Bourrie, G. 2008, *Les formations superficielles. Genèse—Typologie— Classification—Paysages et environnements—Ressources et risques.* (Paris: Ellipses, Edition Marketing).

Dikau, R., Brunsden, D., Schrott, L. and Ibsen, M.L. 1996, *Landslide recognition. Identification, movements and causes,* (Chichester, England: Wiley).

Doumtsop, M. 2011, *Evolution des sablières de l'Ouest Cameroun et leurs conséquences environnementales et socio économiques: cas de Batié de 1985 à 2010.* (Mémoire de Master Université de Yaoundé I. département de Géographie, Yaoundé, février 2011).

Dumont, J.F. 1986, Identification par télédétection de l'accident de la Sanaga (Cameroun). Sa position dans le contexte des grands accidents d'Afrique Centrale et de la limite nord du craton Congolais. *Géodynamique* **1**(1), pp. 13–19.

Dumont, J.F. 1987, Etude structurale des bordures Nord et Sud du plateau de l'Adamaoua: influence du contexte atlantique. *Géodynamique,* **2**(1), pp. 55–68.

Eba'a Atyi, R. 1998, *Cameroon's logging industry: structure, economic importance and effects of devaluation.* (CIFOR, Bogor, Indonesia: CIFOR Occasional paper, **14**).

EDC, 2008, *Projet Lom Pangar. Rapport d'activité annuelle, (jan-déc.2008).*

Edimo, A., Ndogmo, J.P. and Eyike, E. 2002, *L'exploitation artisanale anarchique et illégale du saphir dans l'Adamaoua,* (Rapport d'enquête technique MINMEE/DMG).

Eisenberg, J. 2009, *Morphogenese der Flusseinzugsgebiete von Nyong und Ntem in Süd Kamerun unter Berüchsichtigung neotektonischer Vorgänge,* (PhD-thesis, Goethe-University Frankfurt).

Eno Belinga, S.M. 1966, *Contribution à l'étude géologique, minéralogique et géochimique des formations bauxitiques de l'Adamaoua (Cameroun),* (Thèse 3ème cycle, Université, Paris).

Essoungou Kwack, J.N. 2009, *Evaluation de l'impact de l'approche participative dans la restauration des écosystèmes de mangrove de la réserve de faune de Douala,* (Mémoire de Master II Professionnel, Université de Dschang, CRESA FORET BOIS Yaoundé, mars 2009).

FAO–Food and Agriculture Organization of the United Nations, 2001, *State of the World forests,* (http://www.fao.org/docrep/003/y0900e/y0900e00.htm).

FAO–Food and Agriculture Organization of the United Nations, 2007, *Situation des forêts du monde 2007,* (Rome).

FAO–Food and Agriculture Organization of the United Nations, 2008, *Forestry,* (www.fao.org/forestry).

Fékoua, D. 2011, *Anthropisation et risques environnementaux sur les collines de Yaoundé.* (Mémoire de Master Professionnel, Université de Dschang, CRESA FORET BOIS Yaoundé, août 2011).

Franqueville, A. 1984, *Yaoundé, construire une capitale,* (Paris: Editions de l'ORSTOM).

Gasse, F. 2000, Hydrological changes in the African tropics since the Last Glacial Maximum, *Quaternary Science Reviews,* **19**(1–5), pp. 189–211.

Gasse, F. 2002, Diatom-inferred salinity and carbonate oxygen isotopes in Holocene waterbodies of the western Sahara and Sahel (Africa), *Quaternary Science Reviews,* **21**(7), pp. 737–767.

Gasse, F. 2006, Climate and hydrological changes in tropical Africa during the past million years. *Comptes Rendus Palevol,* **5**(1–2), pp. 35–43.

Gasse, F. and Roberts, N. 2004, Late Quaternary hydrologic changes in the arid and semi-arid belt of northern Africa. Implications for past atmospheric circulation. In: *The Hadley Circulation: Present, Past and Future,* edited by Diaz, H.F. and Bradley, R.S. (Dordrecht, The Netherlands: Springer), pp. 313–345.

Gasse, F. and Van Campo, E. 1994, Abrupt post-glacial climate events in West Asia and North Africa monsoon domains. *Earth and Planetary Science Letters,* **126**(4), pp. 435–456.

Gazel, J. 1958, *Carte géologique du Cameroun.* (Bulletin **2**, Dir. Mines et Géol. Cameroun).

Giresse, P., Maley. J. and Brenac, P. 1994, Late Quaternary palaeoenvironments in the Lake Barombi Mbo (West Cameroon) deduced from pollen and carbon isotopes of organic matter. *Palaeogeography, palaeolimatology, palaeoecology,* **107**, pp. 65–78

Giresse, P., Mvoubou, M., Maley, J. and Ngomanda, A. 2008, Late –Holocene equatorial environments inferred from deposition processes, carbon isotopes of organic matter, and pollen in three shallow lakes of Gabon, west-central Africa. *Journal of paleolimnology,* **41**(2), pp. 369–392.

Grandin, G. 1976, *Aplanissements cuirasses et enrichissements des gisements de manganese de quelques regions d'Afrique de l'Ouest.* (Paris: ORSTOM).

Hewitt, K. 1997, *Regions of risk: a geographical introduction to disasters.* (London: Longman).

Hulme, M., Doherty, R., Ngara, T., New, M. and Lister, D. 2001, African climate change: 1900–2100. *Climate Research,* **17**(2), pp. 154–168.

IGECO Electricité de France, 1968, *Etude de la crue exceptionnelle en vue du calage du plancher de la salle des machines à Edéa,* (Paris: ORSTOM).

IPCC, Climate Change, 2001a, The scientific basis. Contribution of Working Group I. In *The Third Assessment Report of the Intergovernmental Panel on Climate Change,* edited by Houghton, J.T., Ding, Y., Griggs, D.J., Noguer, M., Van der Linden, P.J., Dai, X., Maskell, K. and Johnson, C.A. (Cambridge: Cambridge University Press).

IPCC, Climate Change 2001b, Impact, adaptation and vulnerability. Contribution of Working Group II. In *The Third Assessment Report of the Intergovernmental Panel on Climate Change,* edited by McCarthy, J.J., Canziani, O.F., Leary, N.A., Dokken, D.J. and White, K.S., (Cambridge: Cambridge University Press).

Kadomura, H. 1984, *Natural man induced environmental changes in Tropical Africa. Case Studies in Cameroon and Kenya; A preliminary report of the tropical African geomorphology and late Quaternary palaeoenvironments research projects (1982–1983),* (Sapporo-Hokkaido).

Kadomura, H. 1995, Palaeoecological and palaeohydrological changes in the humid tropics during the last 20.000 years, with references to Equatorial Africa. In *Global Continental Palaeohydrology,* edited by Gregory, K.J., Starkel, L. and Baker, V.R., (London: John Wiley and sons Ltd.), pp. 177–202.

Kadomura, H. and Hori, N. 1990, Environmental implications of slope deposits in humid tropical Africa: Evidence from southern Cameroon and western Kenya. *Geographical reports* (Tokyo: Metropolitan University), **25**, pp. 213–236.

Kadomura, H. and Kiyonaga, J. 1994, Origin of Grassfields landscape in the West Cameroon Highlands. *Savannization processes in Tropical Africa* (Tokyo: Metropolitan University, Department of geography), **2**, pp. 47–85.

Karsenty, A. and Ongolo, S. 2011, Can "fragile states" decide to reduce their deforestation? The inappropriate use of the theory of incentives with respect to the REED mechanism. *Forest Policy and Economics*, in Press, available on-line 16 June 2011.

Lambi, C.M. 2001, The impacts of human activities on land degradation in some highland regions of Cameroon: implications for development. In: *Environmental Issues, Problems and Prospects*; edited by Lambi, C.M. and Eze Bassey, E. (Bamenda: Unique Printers), pp. 45–66.

Lézine, A.-M. and Cazet, J.P. 2005, High-resolution pollen record from core KW31, Gulf of Guinea, documents the history of lowland forests of West equatorial Africa since 40,000 yr ago. *Quaternary Research,* **66,** pp. 432–443.

Liénou, G. 2007, *Impacts de la variabilité climatique sur les ressources en eau et le transport des matières en suspension de quelques bassins versants représentatifs des unités climatiques au Cameroun,* (Thèse Doctorat, Université de Yaoundé I, Cameroun).

Liénou, G., Mahé, G., Olivry, J.C., Naah, E., Servat, E., Sigha-Nkamdjou, L., Sighomnou, D., Ndam Ngoupayou, J., Ekodeck, G.E. and Paturel, J.E. 2005, Régimes des flux des matières solides en suspension au Cameroun: revue et synthèse à l'échelle des principaux écosystèmes; diversité climatique et actions anthropiques. *Hydrological Sciences Journal,* **50**(1), pp. 111–124.

Liénou, G., Mahé, G., Paturel, J.E., Servat, E., Sighomnou, D., Ekodeck, G.E., Dezetter, A. and Dieulin, C. 2008, Evolution des régimes hydrologiques en région équatoriale camerounaise: un impact de la variabilité climatique en Afrique équatoriale? *Hydrological Sciences Journal,* **53**(4), pp. 789–801.

Mahé, G., Lerique, J. and Olivry, J.C. 1990, L'Ogooué au Gabon. Reconstitution des débits manquants et mise en évidence de variations climatiques à l'équateur. *Hydrologie Continentale,* **5**(2), pp. 105–124.

Mahé, G. and Olivry, J.C. 1991, Changements climatiques et variations des écoulements en Afrique occidentale et centrale, du mensuel à l'interannuel. In: Hydrology for the water management of large river basins, edited by Van de Ven, F.H.M., Gutknecht, D., Loucks, D.P. and Salewicz, K.A. (Congrès AISH Vienne, Autriche, 13–15 août 1991), Publication AISH, 201, pp. 163–172.

Maley, J. 2001, The impact of arid phases on the African Rain Forest through Geological History. In *African Rain Forest Ecology and Conservation*, edited by Weber, W., White, L., Vedder A. and Naughton-Treves, L., (London: Yale University Press) pp. 68–87.

Maley, J. and Brenac, P. 1987, Analyses polliniques préliminaires du quaternaire récent de l'ouest Cameroun: mise en évidence de refuges forestiers et discussion des problèmes paléo climatiques. *Mémoires et Travaux de l'Ecole Practique des Hautes Etudes,* Institut de. Montpellier, **17**, pp. 129–142.

Maley, J. and Brenac, P. 1998, Vegetation dynamics, palaeoenvironments and climatic changes in the forests of West Cameroon during the last 28,000 years BP. *Review of Palaeobotany and Palynology,* **99**, pp. 157–187.

Mandengue, S. 2009, *La micro foresterie en milieux urbain et périurbain camerounais: enjeux et perspective (le cas de la ville de Yaoundé),* (Mémoire Master Professionnel en environnement. CUD, FLESH, Département de Géographie).

Mbeugang Tcheubonsou, E.M. 2011, *Evaluation et la gestion des risques naturels dans le secteur NO de Yaoundé: approches méthodologiques,* (Mémoire de Master. Université de Yaoundé I. Département de Géographie. Yaoundé, août 2011).

Mbog, D.M. 2006, *Rapport d'étude du projet de gestion participative et conservation de la diversité biologique des mangroves*. Projet/TCP/CMR/FAO-A-.

Mittermeier, R.A., Myers, N., Thomsen, J.B., da Fonseca, G.A.B. and Olivieri, S. 1998, Biodiversity hotspots and major tropical wilderness areas: approaches to setting conservation priorities. *Conservation Biology*, **12**, pp. 516–520.

Molua, E.L. 2003, *Global climate change and Cameroon's Agriculture: evaluating the economic impacts*, (PhD dissertation, Institute of Agricultural Economics, Georg-August University Göttingen, Germany).

Molua, E.L. 2006, Climate trends in Cameroon: implications for agricultural management.*Climate Research,* **30**, pp. 255–262.

Molua, E.L. and Utomakili, J.B. 1998, An Analysis of Resource-Use Efficiency in Banana Production in the South West Province of Cameroon, *International Journal of Tropical Agriculture,* **16**, pp. 113–118.

Molua, E.L. and Lambi, C.M. 2006, The economic impact of climate change on agriculture in Cameroon, (CEEPA Discussion paper No. 17, Centre for Environmental Economics and Policy in Africa, University of Pretoria).

Moreau, C., Regnoult, J.M., Déruelle, B. and Robineau, B. 1987, A new tectonic model for the Cameroon Line, Central Africa. *Tectonophysics*, **139**, pp. 317–334.

Nanji, R.O. 2007, *Assessment at the fisheries resources of fishermen living around the Sanaga Estuary (Douala-Edéa Wildlife reserve)*, (DESS Dissertation).

Ndjogui, E.T. 2009, *Identification des indicateurs d'aménagement des aires protégées du Cameroun: Le cas de la Réserve de Faune de Douala Edéa*, (Mémoire de DEA de Géographie, Option: Gestion Des ressources naturelles, Yaoundé, mai 2009).

Ngako, V. 1999, *Les déformations continentales panafricaines en Afrique Centrale. Résultat d'un poinçonnement de type himalayen*, (Thèse Doctorat d'état, Université de Yaoundé I).

Ngako, V., Jégouzo, P. and Nzenti, J.P. 1991, Le cisaillement centre camerounais. Rôle structural et géodynamique dans l'orogénèse panafricaine. *Comptes Rendus de l'Académie des Sciences,* Paris, **313**, pp. 457–463.

Ngombala Samba, Z. 2011, *Evaluation des capacités pour la mise en œuvre de paiement des services environnementaux en Centrafrique: cas des services offerts par les forêts*, (Mémoire de Master Professionnel, Université de Dschang, CRESA FORET BOIS Yaoundé, avril 2011).

Nicholson, S.E. and Grist, J.P. 2001, A conceptual model for understanding rainfall variability in theWest African Sahel on interannual and interdecadal timescales. *International Journal of Climatology,* **21**(14), pp. 1733–1757.

Nkamleu, G.B.N. 2000, *Les enjeux économiques de la gestion forestière au Cameroun*, (Thèse de Doctorat d'état. Université de Yaoundé II).

Nono, C.A. 2011, *Impacts de l'exploitation artisanale de l'or sur l'environnement et le développement socio économique à Bétaré Oya/Est Cameroun*, (Mémoire de Master Professionnel, Université de Dschang, CRESA FORET BOIS Yaoundé, août 2011).

Ntep Gweth, P. 2001, *Ressources minérales du Cameroun avec carte thématique*, pp. 195–197.

Ntsama Atangana, J. 2009, *La problématique du bois de feu dans la ville de Garoua au nord Cameroun*, (Mémoire de Master II Professionnel, Université de Dschang, CRESA FORET BOIS Yaoundé, mars 2009).

Olivry, J.C. 1987, Les conséquences durables de la sécheresse actuelle sur l'écoulement du fleuve Sénégal et l'hypersalinisation de la basse Casamance. In: The Influence of Climate Change and Climate Variability on the Hydrology Regime and Water Resources, edited by Solomon, S.L., Beran, M. and Hogg, W. (Proceedings of the Vancouver Symposium, August 1987), *Publication IAHS*, **168**, pp. 501–512.

Olivry, J.C., Bricquet, J.P. and Mahé, G. 1993, Vers un appauvrissement durable des ressources en eau de l'Afrique humide? In: Hydrology in Warm Humid Regions, edited by Gladwell, J.S. (Proceedings of the Yokohama Symposium, July 1993), *Publication IAHS,* **216**, pp. 67–78.

Olivry, J.C., Bricquet, J.P., Bamba, F. and Diarra, M. 1995, Le régime hydrologique du Niger supérieur et le déficit des deux demières décennies. In: *Actes du colloque PEG1-GBF* (Paris: Coll. Colloques et Séminaires, ORSTOM, novembre 1993), pp. 251–266.

Petit, M. 1990, *Géographie physique tropicale. Approches aux études physiques du milieu.* Morphogenèse-Paysages, (Paris: Karthala-ACCT), p. 351.

Poudjom Djomani, Y.H., Diament, M. and Albouy Y. 1992, Mechanical behaviour of the lithosphere beneath the Adamawa Uplift (Cameroon, West Africa) based on gravity data. *Journal of African Earth Science,* **15**(l), pp. 81–90.

Ribot, J. 2004, *Waiting for Democracy The Politics of Choice in Natural Resource Decentralization,* (Washington DC: World Resources Institute, WRI Report).

Sangen, M. 2009, *Physiogeographische Untersuchungen zur pleistozänen und holozänen Umweltgeschichte an Alluvionen des Ntem-Binnedeltas und alluvialer Sedimente der Flüsse Boumba, Ngoko, Nyong und Sanaga in Süd-Kamerun,* (Unpublished PhD-thesis, GoetheUniversity Frankfurt).

Schnell, R. 1971–1976, *Introduction à la phytogéographie des pays tropicaux,* (Paris: Gauthier-Villars), **2**. "Les milieux, les groupements végétaux", **3**, p. 451. "*La flore et la végétation de l'Afrique tropicale*", p. 459.

Servat, E., Paturel, J.-E., Lubes-Niel, H., Kouame, B., Masson, J.M., Travaglio, M. and Marieu, B. 1999, De différents aspects de la variabilité de la pluviométrie en Afrique de l'ouest et centrale non sahélienne. *Revue des sciences de l'eau,* **12**(2), pp. 363–387.

Sigha-Nkamdjou, L., Sighomnou, D., Liénou, G., Ndam, J.R., Bello, M., Kamgang, R., Ekodeck, G.E., Ouafo, M.R., Mahe, G., Paturel, J.E. and Servat, E. 2005, Impact des modifications climatiques et anthropiques sur les flux de matières de quelques bassins fluviaux du Cameroun. *AISH Publication,* **292**, pp. 291–300.

Sighomnou, D. 2004, *Analyse et redéfinition des régimes climatiques et hydrologiques du Cameroun: perspectives d'évolution des ressources en eau,* (Université Yaoundé I, Thèse doctorat d'Etat).

Sinnota, M. 2004, L'impact phyto-géomorphologique de l'exploitation du saphir dans l'Adamaoua—cas de Paro Lawel, (Mémoire de Maîtrise, Université de Ngaoundéré).

Sircoulon, J. 1976, Les données hydropluviométrique de la sècheresse récente en Afrique intertropicale. Comparaison avec les sécheresses 1913 et 1940. *Cahiers ORSTOM, série Hydrologie,* **13**(2), pp. 75–174.

Smith, K. 1996, *Environmental hazards. Assessing risk and reducing disaster,* (London, New York: Routledge).

Suchel, J.B. 1972, *La répartition des pluies et les régimes pluviométriques au Cameroun; contribution à l'étude des climats de l'Afrique tropicale,* (CEGET-Université Fédérale du Cameroun), pp. 1–287.

Suchel, J.B. 1988, *Les climats du Cameroun.* (Université de St. Etienne), p. 1188.

Tagboka Yangana, B. 2009, *Impact des actions de régénération des zones dégradées de la Réserve Spéciale de Forêt de Gbazabangui (RCA),* (Mémoire de Master II Professionnel, Université de Dschang, CRESA FORET BOIS Yaoundé, mars 2009), p. 52.

Takoutsi, R.C. 2009, *Analyse environnementale de l'artisanat dans le FARO DEO. Perspectives pour le développement,* (Mémoire de Master II Professionnel, Université de Dschang, CRESA FORET BOIS Yaoundé, mars 2009), p. 101.

Tchindjang, M. 1996, *Le Bamiléké central et ses bordures: Morphologie Régionale et dynamique des versants.* Etude géomorphologique. (Thèse de Doctorat Publiée 3, Université de Paris 7), p. 867.

Tchindjang, M. 2001, L'industrie minière d'exploitation des sables dans les altérites du granite panafricain de Batié. In *Readings in Geography*, pp. 83–103.

Tchindjang, M. 2012, *Paradoxes et risques dans les hautes terres camerounaises, multifonctionnalités naturelles et sous valorisation humaine.* (HDR, **3**, Université, Paris VII, Diderot), p. 265.

Tchindjang, M., Ndjogui, E.T. and Ngambi, J.R. 2010, Essai de construction participative des indicateurs d'interactions dans la Réserve de Faune de Douala-Edéa au Cameroun. *International Journal of Advanced Studies and Research in Africa*, **1**(3), pp. 202–227.

Tchoua, F. 1974, *Contribution à l'étude géologique et pétrologique de quelques volcans de la Ligne du Cameroun*, (Thèse d'Etat, Université de Clermont-Ferrand), p. 337.

Timnou, J.P. 1993, Migration, urbanisation et développement au Cameroun. *Les cahiers de l'IFORD*, **4**.

Tsou Ndzitouo, M. 2007, *Impacts environnementaux et risques induits par les activités agropastorales à Magha'a (Sud Ouest Cameroun)*, (Mémoire de Maîtrise 2007, Université de Yaoundé I), p. 101.

Tyson, P.D., Lee-Thorp, J., Holmgren, K. and Thackeray, J.F. 2002, Changing gradients of climate change in southern Africa during the past millennium: implications for population movements. *Climate Change,* **52**, 129–135.

UNCCD, 2003, La convention des Nations Unies sur la lutte contre la désertification et sa dimension politique. http://www.unccd.int/parliament/data/bginfo/PDUNCCD(fre).pdf

Varnes, D.J. 1978, Slope movement types and processes. In: *Special Report 176: Landslides: Analysis and Control,* edited by Schuster, R.L. and Krizek, R.J., (Washington D.C.: Transportation and Road Research Board, National Academy of Science), pp. 11–33.

Verschuren, D., Laird, K.R. and Cumming, B.F. 2000, Rainfall and drought in equatorial East Africa during the past 1100 years. *Nature,* **403**, pp. 410–414.

WCS, 2006, *Report of activities 1998–2006*, (Mouanko: Cameroon Wildlife Society).

Wisner, B., Blaikie, P., Cannon, T. and Davis, I. 2004, *At risk. Natural hazards, people's vulnerability and disasters*, (London and New York: Routledge), p. 470.

WRI, 2000–2001, *People and ecosystems: The fraying web of life.* (World Bank), p. 276. http://www.wri.org/publication/world-resources-2000-2001-people-and-ecosystems-fraying-web-life

Zogning, O. 2005, *Risques d'inondation à Yaoundé: le cas de la zone de confluence du Mfoundi au centre-ville et des quartiers péricentraux du bassin versant de l'Ekozoa*, (Mémoire de Maîtrise, Université de Yaoundé I); p. 169.

Regional Index

Subject Index